Edward Szczerbicki and Ngoc Thanh Nguyen (Eds.)

Smart Information and Knowledge Management

T0143058

Studies in Computational Intelligence, Volume 260

Editor-in-Chief
Prof. Janusz Kacprzyk
Systems Research Institute
Polish Academy of Sciences
ul. Newelska 6
01-447 Warsaw
Poland
E-mail: kacprzyk@ibspan.waw.pl

Further volumes of this series can be found on our homepage:
springer.com

Vol. 238. Li Niu, Jie Lu, and Guangquan Zhang
Cognition-Driven Decision Support for Business Intelligence,
2009
ISBN 978-3-642-03207-3

Vol. 239. Zong Woo Geem (Ed.)
*Harmony Search Algorithms for Structural Design
Optimization,* 2009
ISBN 978-3-642-03449-7

Vol. 240. Dimitri Plemenos and Georgios Miaoulis (Eds.)
Intelligent Computer Graphics 2009, 2009
ISBN 978-3-642-03451-0

Vol. 241. János Fodor and Janusz Kacprzyk (Eds.)
Aspects of Soft Computing, Intelligent Robotics and Control,
2009
ISBN 978-3-642-03632-3

Vol. 242. Carlos Artemio Coello Coello,
Satchidananda Dehuri, and Susmita Ghosh (Eds.)
*Swarm Intelligence for Multi-objective Problems in Data
Mining,* 2009
ISBN 978-3-642-03624-8

Vol. 243. Imre J. Rudas, János Fodor, and
Janusz Kacprzyk (Eds.)
Towards Intelligent Engineering and Information Technology,
2009
ISBN 978-3-642-03736-8

Vol. 244. Ngoc Thanh Nguyen, Rados law Piotr Katarzyniak,
and Adam Janiak (Eds.)
New Challenges in Computational Collective Intelligence, 2009
ISBN 978-3-642-03957-7

Vol. 245. Oleg Okun and Giorgio Valentini (Eds.)
*Applications of Supervised and Unsupervised Ensemble
Methods,* 2009
ISBN 978-3-642-03998-0

Vol. 246. Thanasis Daradoumis, Santi Caballé,
Joan Manuel Marquès, and Fatos Xhafa (Eds.)
Intelligent Collaborative e-Learning Systems and Applications,
2009
ISBN 978-3-642-04000-9

Vol. 247. Monica Bianchini, Marco Maggini, Franco Scarselli,
and Lakhmi C. Jain (Eds.)
Innovations in Neural Information Paradigms and Applications,
2009
ISBN 978-3-642-04002-3

Vol. 248. Chee Peng Lim, Lakhmi C. Jain, and
Satchidananda Dehuri (Eds.)
Innovations in Swarm Intelligence, 2009
ISBN 978-3-642-04224-9

Vol. 249. Wesam Ashour Barbakh, Ying Wu, and Colin Fyfe
*Non-Standard Parameter Adaptation for Exploratory Data
Analysis,* 2009
ISBN 978-3-642-04004-7

Vol. 250. Raymond Chiong and Sandeep Dhakal (Eds.)
*Natural Intelligence for Scheduling, Planning and Packing
Problems,* 2009
ISBN 978-3-642-04038-2

Vol. 251. Zbigniew W. Ras and William Ribarsky (Eds.)
Advances in Information and Intelligent Systems, 2009
ISBN 978-3-642-04140-2

Vol. 252. Ngoc Thanh Nguyen and Edward Szczerbicki (Eds.)
Intelligent Systems for Knowledge Management, 2009
ISBN 978-3-642-04169-3

Vol. 253. Roger Lee and Naohiro Ishii (Eds.)
*Software Engineering Research, Management and Applications
2009,* 2009
ISBN 978-3-642-05440-2

Vol. 254. Kyandoghere Kyamakya, Wolfgang A. Halang,
Herwig Unger, Jean Chamberlain Chedjou,
Nikolai F. Rulkov, and Zhong Li (Eds.)
Recent Advances in Nonlinear Dynamics and Synchronization,
2009
ISBN 978-3-642-04226-3

Vol. 255. Catarina Silva and Bernardete Ribeiro
Inductive Inference for Large Scale Text Classification, 2009
ISBN 978-3-642-04532-5

Vol. 256. Patricia Melin, Janusz Kacprzyk, and
Witold Pedrycz (Eds.)
*Bio-inspired Hybrid Intelligent Systems for Image Analysis and
Pattern Recognition,* 2009
ISBN 978-3-642-04515-8

Vol. 257. Oscar Castillo, Witold Pedrycz, and
Janusz Kacprzyk (Eds.)
*Evolutionary Design of Intelligent Systems in Modeling,
Simulation and Control,* 2009
ISBN 978-3-642-04513-4

Vol. 258. Leonardo Franco, David A. Elizondo, and
José M. Jerez (Eds.)
Constructive Neural Networks, 2009
ISBN 978-3-642-04511-0

Vol. 259. Kasthurirangan Gopalakrishnan, Halil Ceylan, and
Nii O. Attoh-Okine (Eds.)
*Intelligent and Soft Computing in Infrastructure Systems
Engineering,* 2009
ISBN 978-3-642-04585-1

Vol. 260. Edward Szczerbicki and Ngoc Thanh Nguyen (Eds.)
Smart Information and Knowledge Management, 2009
ISBN 978-3-642-04583-7

Edward Szczerbicki and Ngoc Thanh Nguyen (Eds.)

Smart Information and Knowledge Management

Advances, Challenges, and Critical Issues

 Springer

Prof. Edward Szczerbicki
Faculty of Management and Economics
Gdansk University of Technology,
Str. Narutowicza 11/12
80-233 Gdansk,
Poland
E-mail: Edward.Szczerbicki@zie.pg.gda.pl

Prof. Ngoc Thanh Nguyen
Wroclaw University of Technology
Institute of Informatics
Str. Wyb. Wyspianskiego 27
50-370 Wroclaw
Poland
E-mail: Ngoc-Thanh.Nguyen@pwr.wroc.pl

ISBN 978-3-642-26188-6   e-ISBN 978-3-642-04584-4

DOI 10.1007/978-3-642-04584-4

Studies in Computational Intelligence   ISSN 1860-949X

Typeset & Cover Design: Scientific Publishing Services Pvt. Ltd., Chennai, India.

Printed in acid-free paper

9 8 7 6 5 4 3 2 1

springer.com

Preface

This is our second book related to intelligent information and knowledge management which is based on the concept of developing a platform to share ideas. The contributors to this book, the authors of carefully selected and peer revived Chapters, are academics, educators, and practitioners who are researching and implementing in real life exciting developments associated with our never ending quest to vanquish challenges of our knowledge-based society.

The opening Chapter of the book titled *Immuno-Inspired Knowledge Management for Ad Hoc Wireless Networks* is authored by Martin Drozda, Sven Schaust, and Helena Szczerbicka. The Authors present the cutting edge research in the new, still emerging area of biologically inspired immune systems development. This research has already showed promising results in the field of smart misbehavior detection and classification of data in general. In this Chapter the Authors perform a very comprehensive overview of the recent developments in the area of biologically inspired classification approaches to detect possible threats and misbehavior, especially in the area of ad hoc networks. They also discuss exciting challenges in translating functionality of biological immune system to technical systems. The Chapter that follows is titled *Immune Decomposition and Decomposability Analysis of Complex Design Problems with a Graph Theoretic Complexity Measure* and is authored by Mahmoud Efatmaneshnik, Carl Reidsema, Jacek Marczyk, and Asghar Tabatabaei Balaei. The authors introduce a new approach to decomposition challenge of large, complex problems in general, and design problems in particular. A measure of decomposition quality is introduced and its application in problem classification is highlighted. The measure proposed by the Authors is complexity based (real complexity) and can be employed for both disjoint and overlap decompositions. After discussing the advantages of their approach and illustrating it with examples, the Authors conclude with distinct future directions of their research. The next Chapter is titled *Towards a Formal Model of Knowledge Sharing in Complex Systems* and is authored by Nadim Obeid and Asma Moubaiddin. The Authors develop and discuss a novel multi-agent system that assists in the process of knowledge sharing between various groups of workers and decision-makers. In the proposed platform each agent is a knowledge broker and controller responsible for specialized knowledge needs. The Authors use Partial Information State approach for agents' knowledge representation and present a multi-agent based model of argumentation and dialogue for knowledge sharing. The approach is illustrated with examples and the Chapter concludes with clear vision of further research towards application of the proposed platform to support strategic and tactic reasoning for rational agents. The Chapter that follows is titled *Influence of the Working Strategy on A-Team Performance* and is authored by Dariusz Barbucha, Ireneusz Czarnowski, Piotr Jędrzejowicz, Ewa Ratajczak-Ropel, and Iza Wierzbowska. The Authors first introduce and discuss the issues related to A-Team (defined as a problem solving system of autonomous agents and common memory) performance during the process of agents' cooperation in an attempt to improve a solution to a given problem. Then the Chapter focuses on investigation of influence of different strategies on A-Team performance. To implement various strategies the A-Team platform called JABAT is used by the Authors. Different working strategies are evaluated by computational experiment using several benchmark data sets. The Authors show that designing effective working strategy can considerably improve the performance of an A-Team system and formulate some rules useful in A-Team development and implementation. The next Chapter is titled *Incremental Declarative Process Mining* and is authored by Massimiliano Cattafi, Evelina Lamma, Fabrizio Riguzzi, and Sergio Storari. In this Chapter the Authors introduce and discuss the current state of the art in research developments and directions related to the area of Business Processes Management (BPM) that studies how to describe, analyze, preserve and improve organisational processes. They focus on subfield of BPM called Process Mining (PM) which aims at inferring a model of the processes from past experience. The Authors introduce and define a novel activity as part of PM – Incremental Process Mining. To solve this problem, they modify the process mining that actually performs theory revision. Through the illustrative experimental results the Authors show that incremental revision of an existing theory or a model can be superior over learning or developing a new one from scratch. Conclusions of the Chapter define new research directions leading to even deeper understanding of the difference between learning from scratch and revision. The following Chapter titled *A Survey on Recommender Systems for News Data* is authored by Hugo L. Borges and Ana C. Lorena. The Chapter introduces to the reader the research area called Recommender Systems (RS) which focuses on the structure of evaluations. The Authors present and discuss very rich and interesting origins of RS that are linked to cognitive science, approximation theory, information retrieval, prediction theories and are also related to management science and to the process of modelling options made by consumers in marketing. Next they introduce formal definition of RS and show that it is one of the major tools that helps to reduce information overload by acting as a filter and by customizing content for the final user. The Authors illustrate their arguments with some interesting news recommendation related cases. The next Chapter is titled *Negotiation Strategies with Incomplete Information and Social and Cognitive System for Intelligent Human-Agent Interaction* and authored by Amine Chohra, Arash Bahrammirzaee, and Kurosh Madani. The Authors divide their Chapter into two parts. The first part aims to develop negotiation strategies for autonomous agents with incomplete information, where negotiation behaviors are suggested to be used in combination. The suggested combination of behaviors allows agents to improve the negotiation process in terms of agent utilities, number of rounds, and percentage of agreements that were reached. The second part of this Chapter aims to develop a SOcial and COgnitive SYStem (SOCOSYS) for learning negotiation strategies from human-agent or agent-agent interactions. Both parts of the Chapter suggest interesting avenues of future research toward development of fully intelligent negotiation system that should combine in an efficient way behaviours developed in this Chapter in the first and second part with fuzzy and prediction behaviours. The following Chapter is titled *Intelligent Knowledge-Based Model for IT Support Organization Evolution* and is authored by Jakub Chabik, Cezary Orłowski, and Tomasz Sitek. The Authors develop, from conceptualization to implementation, a smart knowledge-based model for predicting the state of the IT support organization. As such organizations face the complex problem of predicting and managing their transformation process, the focus of the Authors of this Chapter is to develop support tools that can add the above tasks. The tools the Authors propose are based on fuzzy modeling mechanisms and reasoning using uncertain and incomplete knowledge and information. After developing the tools, the Authors illustrate their application through the experimental case study related to the banking sector. The next Chapter addresses one of the fascinating future challenges of our digital society – digital preservation. It is titled *Modeling Context for Digital Preservation* and authored by Holger Brocks, Alfred Kranstedt, Gerald Jäschke and Matthias Hemmje. The Authors, after introducing the notion of context in digital preservation, identify relevant components of context and the interrelationship between the different aspects of context, and propose a generic model to represent the required context information. They chose the Open Archival Information System (OAIS) as the conceptual framework for introducing context to digital preservation. The formalism the Authors propose in a very comprehensive way relates to ontologies and business process models. The Chapter presents exemplary illustrations of the proposed formalism. The two cases presented are the scenarios in the domain of document production, archival, access and reuse within digital archives and libraries. The scenarios show that the context of archiving process adds eminent information with respect to preservation. This context information should be modeled and preserved together with the content data as an essential prerequisite for future access, understanding and reuse of data. The Chapter that follows is titled *UML2SQL—a Tool for Model-Driven Development of Data Access Layer* and authored by Leszek Siwik, Krzysztof Lewandowski, Adam Woś, Rafał Dreżewski, and Marek Kisiel-Dorohinicki. The Authors of this Chapter address a number of issues related to a tool that is used for model driven development of data access layer. The tool is called   UML2SQL which is an open source application and includes an object query language allowing for behavior modeling based on Unified Modelling Language (UML) activity diagrams, and thus effectively linking structural and behavioral aspects of the system development. The Authors present and discuss in a comprehensive way the UML2SQL architecture, its processes and schemes which make this tool a distinct one in the data access domain. The Chapter includes also an illustrative example of UML2SQL implementation. The next Chapter is titled *Fuzzy Motivations in Behavior Based Agents* and authored by Tomas V. Arredondo. The Author presents a fuzzy logic based approach for providing biologically inspired motivations to be used by agents in evolutionary behaviour learning. In the presented approach, fuzzy logic provides a fitness measure used in the generation of agents with complex behaviours which respond to user expectations of previously specified motivations. The developed technique is shown as a simple but very powerful method for agents to acquire diverse behaviours as well as providing an intuitive user interface framework. The approach developed is supported by illustrative examples related to navigation, route planning, and robotics. The following Chapter is titled *Designing optimal operational-point trajectories using an intelligent sub-strategy agent-based approach* and authored by Zdzislaw Kowalczuk and Krzysztof E. Olinski. The Chapter introduces and discusses an intelligent sub-strategy agent-based approach to optimization, in which the search for the optimal solution is performed simultaneously by a group of agents. In this novel approach presented by the

Authors, compilation of the partial solutions delivered by the agents, which results in shifting the operational point from the initial to the designed terminal point, forms the final solution being sought. The Chapter includes illustrative examples and discusses further research steps that would consider sensitivity of this approach to the specification of elementary control strategies and to the applied arrangement of the decision points in the state space. Another direction suggested by the Authors of further developments in this area relates to more complex agent behaviours and interacting mechanisms at higher level of intelligence. The Chapter that comes next is titled *An Ontology-based System for Knowledge Management and Learning in Neuropediatric Physiotherapy* and authored by Luciana V. Castilho and Heitor S. Lopes. The Authors begin with very comprehensive literature review in the area of knowledge management and ontologies setting the research needs and background for their own proposition. Next, they propose a novel methodology for modelling and developing an ontology-based system for knowledge management in the domain of Neuropediatric Physiotherapy together with its application aiming at supporting learning in this domain. The related knowledge acquisition process involved knowledge capture from domain experts and compilation of information that could be gathered from reference textbooks. The knowledge that was captured was represented as ontology. Knowledge base was developed allowing for its reuse for substantial enhancement of educational and learning processes in the area of Physiotherapy. The last Chapter of this selection is titled *Mining Causal Relationships in Multidimensional Time Series* and authored by Yasser Mohammad and Toyoaki Nishida. The Authors develop and introduce a novel approach to mine multidimensional time-series data for causal relationships. The idea of analyzing meaningful events in the time series rather than by analyzing the time series numerical values directly is the main novelty feature of the proposed system. The Authors include mechanisms supporting discovery of causal relations based on automatically discovered recurring patterns in the input time series. The mechanisms are integrated variety of data mining techniques. The proposed approach is evaluated using both synthetic and real world data showing its superiority over standard procedures that are currently used. Comprehensive conclusion of the Chapter discusses future directions of this promising research.

   The very briefly introduced Chapters of this book represent a sample of an effort to provide guidelines to develop tools for intelligent processing of knowledge and information that is available to decision makers acting in information rich environments of our knowledge based society. The guide does not presume to give ultimate answers but it poses models, approaches, and case studies to explore, explain and address the complexities and challenges of modern knowledge administration issues.

Edward Szczerbicki
Gdansk University of Technology, Gdansk, Poland

Ngoc Thanh Nguyen
Wroclaw University of Technology, Wroclaw, Poland

Table of Contents

Immuno-inspired Knowledge Management for Ad Hoc Wireless Networks

Martin Drozda, Sven Schaust, and Helena Szczerbicka

Simulation and Modeling Group
Dept. of Computer Science
Leibniz University of Hannover
Welfengarten 1, 30167 Hannover, Germany
{drozda,svs,hsz}@sim.uni-hannover.de

Abstract. During the last years approaches inspired by biological immune systems showed promising results in the field of misbehavior detection and classification of data in general. In this chapter we give a comprehensive overview on the recent developments in the area of biologically inspired classification approaches of possible threats and misbehavior, especially in the area of ad hoc networks. We discuss numerous immuno related approaches, such as negative selection, B-cell cloning, Dendritic cell algorithm or Danger signals. We review present approaches and address their applicability to ad hoc networks. We further discuss challenges in translating functionality of the biological immune system to technical systems.

1   Introduction

The Biological immune system (BIS) protects its host against extraneous agents that could cause harm. The BIS can be found in living organisms ranging from plants to vertebrates. Even though the BIS complexity in these various life forms can be very different, the common goal is to sustain survival in often unforeseeable conditions. The BIS can be succinctly described as a protection system that is based both on prior experience possibly accumulated over an evolutionary time span and short-term adaptation that occurs during the lifetime of the host. Whereas for example plants only rely on the former type of protection, the innate immunity, more complex life forms additionally employ the latter type, the adaptive immunity. It is important to note that the existence of the adaptive immune system, its internal mechanisms and interplay with the innate immunity is also a result of evolutionary priming.

Due to the efficiency in protecting its host, the BIS has become an inspiration for designing protection systems. Besides this area, it has become an inspiration in areas such as machine learning, optimization, scheduling etc. The technical counterpart of the BIS, Artificial Immune Systems (AISs) have become an independent research direction within the field of computational intelligence.

Computer and network security is an area which gained an increased interest, manly due to the omnipresence of the Internet. It is the role of secure protocols to

E. Szczerbicki & N.T. Nguyen (Eds.): Smart Infor. & Knowledge Management, SCI 260, pp. 1–26.
springerlink.com   © Springer-Verlag Berlin Heidelberg 2010

guarantee data integrity and user authentication. Unfortunately, flaws in secure protocols are continuously being found and exploited as the Internet experience shows. The history of security of home and small mobile computing platforms points out that such attacks can disrupt or even completely interrupt the normal operations of networks [1]. There are several thousand families of viruses and worms recorded. Some of these families consist of several thousands or even tens of thousand viruses that were created by virus kits, a specialized software aimed at automated virus creation.

Ad hoc wireless networks do not rely on any centralized or fixed infrastructure. Instead each participating wireless device acts as a router for forwarding data packets. The advantage of such an architecture is the ease of deployment. This advantage is however at the cost of a more complex maintenance. Additionally, in many scenarios it is expected that the participating wireless devices will be resource restricted due to their reliance on battery power. This further implicates energy aware hardware with a lesser computational throughput.

The above limitations establish the basic motivation for designing autonomous detection and response systems that aim at offering an additional line of defense to the employed secure protocols. Such systems should provide several layers of functionality including the following [2]:

(i) distributed self-learning and self-tuning with the aspiration to minimize the need for human intervention and maintenance,
(ii) active response with focus on attenuation and possibly elimination of negative effects of misbehavior on the network.

In the following we will introduce the basic concepts of the Biological immune system, related computational paradigms and applications to the security of ad hoc wireless networks.

2   The Biological Immune System

The Biological immune system (BIS) of vertebrates is a remarkable defense system, able to quickly recognize the presence of foreign microorganisms and thereafter eliminate these *pathogens*. Pathogens are common microorganisms such as viruses, bacteria, fungi or parasites. To protect the human body against diseases the immune system must, according to Janeway [3], fulfill four main tasks: *Immunological recognition*, *Immune effector functions*, *Immune regulation* and *Immune memory*. These tasks are fulfilled by the following two parts of the BIS:

- the *innate immune system*, which is able to recognize the presence of a pathogen or tissue injury, and is able to signal this to the *adaptive* immune system.
- the *adaptive immune system*, which can develop an immunological memory after exposure to a pathogen. Such an immunological memory serves as a basis for a stronger immune response, should a re-exposure happen.

2.1   Innate Immune System

The innate immune system offers a rapid response on exposure to an infectious microorganism. The reaction is based on the availability of *phagocytes* such as *neutrophils, macrophages, mast cells* and *dendritic cells* which can be found in almost all tissues and in the blood circulation. As most infections occur in the tissues, it is the neutrophils and macrophages which act as the "second" line of defense, right after the skin which prevents pathogens from entering the body. Once a pathogen has entered the body, it is surrounded and enclosed by the present neutrophils and macrophages. The process of enclosure causes the macrophages to secrete *cytokines* and *chemokines*. These two classes of proteins are able to attract even more neutrophils and similar cells in order to contain the area of infection. This can eventually lead to a more specific immune reaction. This process is called *inflammation*.

Macrophages are cells which reside in most of the body's tissue and are therefore one of the first cells that make contact with foreign microorganisms. Immediately after a pathogen has entered the body by passing its initial barriers, such as the skin, macrophages are able to recognize the cell and start an immune reaction. This reaction is typically the enclosure of the foreign cell and the elimination with toxic forms of oxygen. At the same time the macrophages mobilize more cells of the innate and adaptive immune systems by secreting cytokines and other stimulating proteins. This process attracts several other innate immune cells such as *natural killer cells* or *dendritic cells*. Once these cells get into the infected area and therefore become exposed to the pathogen, the adaptive immune system gets stimulated by the dendritic cells reaction. Macrophages also act as *antigen presenting cells* (APC) stimulating the adaptive immune system, displaying cell fragments of the pathogen on their surface. *Antigen* is that part of a pathogen that can be easily processed by an APC. An example of antigen are cell coats, capsules or cell walls.

Dendritic Cells are also part of the innate immune system and also act as APCs stimulating the adaptive immune system. They exist in small quantities in the tissues and in the blood. Dendritic cells (DCs) [4] induce immunological tolerance, and activate and regulate the adaptive system. DCs exist in several stages. In the precursor stage, they are able to secrete inflammatory or anti-viral *cytokines*. Immature DCs can capture and analyze a foreign cell. Upon contact with a pathogen, they are activated and differentiate into either a specialized APC or a mature DC. Secretion of inflammatory cytokines helps expose presence of harmful pathogens in the body. Once DCs are activated they migrate into *lymphatic tissue* areas where they interact with the specific cells of the adaptive immune system.

2.2   Inflammatory Signals

As mentioned before several inflammatory signals are used to activate and attract cells from the immune system. Typically these signals are based upon cytokines and chemokines. Chemokines are a group of about 50 different proteins with different receptor capabilities. Their role is for example to start the production of toxic forms of oxygen in order to eliminate pathogens and to attract more immune cells. Cytokines are also responsible for the inflammation in the infected area and for attracting and activating more immune cells. The process of inflammation is necessary to contain the infected area and to stimulate specific actions against the detected microorganisms.

2.3   Adaptive Immune System

Besides the innate immune reactions, vertebrates have developed a more specific system able to combat microorganisms which get past the first and second line of defense. Part of this system are *lymphocytes* which recirculate in the blood and the lymphatic system, hence their name. These types of cells have specific receptors on their surface, thus recognizing only specific pathogens.

The lymphatic system is a network of tissues and vessels. This network transports extracellular fluids through the body. Part of the system are lymphoid organs and lymph nodes, which are areas of tissue convergence. Thymus and bone marrow are the central lymphoid organs, playing a major role in the production of T- and B-lymphocytes (T-cells and B-cells). Both cells migrate from these areas into the bloodstream which transports them towards the lymph nodes, where they leave the bloodstream and enter the lymph node. Lymphocytes recirculate between the blood and these organs until they are exposed to their specific APCs. APCs from the innate system travel to the lymph nodes and induce the activation of the pathogen specific lymphocytes. After the activation, lymphocytes undergo a period of proliferation and mutation, leaving afterwards the lymph node towards the infectious area.

B-cells are able to recognize pathogens without the extra help of an APC. Once a B-cell binds to an antigen, it will proliferate and differentiate into a *plasma cell*. These cells are able to produce specific antibodies, which will mark an infected cell as a pathogen. This significantly streamlines pathogen elimination. Antibodies are therefore found in the fluid component of blood and in extracellular fluids. Once they recognize a pathogen they bind it, effectively preventing any pathogen replication. In case of bacteria the binding is not sufficient as they still can replicate. It is therefore necessary that the antibodies attract phagocytes to eliminate the pathogen.

T-cells are able to recognize pathogen fragments produced by APCs. When a positive recognition occurs, a signaling process, that can mobilize other players of the BIS, is performed. A T-cells gets activated by the first encounter with antigen, and thereafter proliferates and differentiates into one of its different functional types. These functional types are *cytoxic (killer) T-cells*, activating *helper*

T-cells and controlling *regulatory T-cells*. While the first type kills infected cells and intercellular pathogens, helper T-cells offer essential stimulation signals that activate B-cells to differentiate and produce antibodies. The regulatory T-cell suppresses the activity of lymphocytes and helps control the immune response.

A schematic overview of the immune response mechanism is shown in Fig. 1. This figure depicts different evolution pathways of T-cells and B-cells as well as the interdependence of these two types of lymphocytes. Notice that a pathogen detection and elimination is often done in an orchestrated way in which various immune cells cooperate. Such a cooperation has a tremendous impact on the robustness and reliability of the BIS. In fact, a pathogen elimination by either the innate or adaptive immune system often requires some form of mutual stimulation or *co-stimulation*. For example, a pathogen does not get eliminated by a player of the adaptive immune system as long as a co-stimulation is not received from the innate immune system. Such a co-stimulation can indicate that the given pathogen causes harm to the host and thus must be eliminated.

The guiding role of the innate immune system was investigated by Charles Janeway in late 1980' [3]. It led to the formulation of the infectious non-self model. This theory states that only infectious foreign cells will get eliminated by the BIS. *Non-self cells* are cells that are foreign to the body, whereas *self cells* are a body's own cells such as immune cells, blood cells etc. Based on the observations of Janeway, Polly Matzinger formulated the Danger theory [5]. Her theory recognizes the role of danger signals emitted by cells undergoing injury, stress or death. The Danger theory challenged the predominant "self vs non-self" model

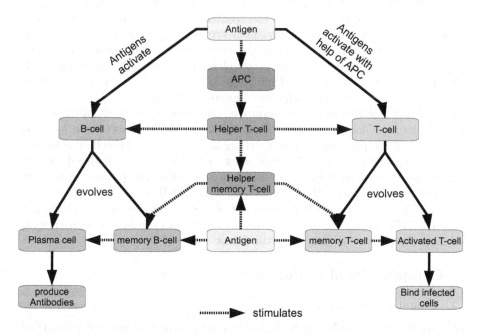

Fig. 1. Schematic of the immune responses [7].

due to F.M. Burnet from the 1950' [6]. The "self vs non-self" model claimed that all non-self cells get eliminated by the BIS.

2.4   Memory of the Adaptive Immune System

In order to provide a more efficient immunological response after a re-exposure to a pathogen, both B-cells and T-cells can mature and become *memory B-cells* and *memory T-cells*. In contrast to normal B- and T-cells, memory cells live much longer, thus allowing the host to keep its immunological memory over a long period.

2.5   Selection Principles of the Immune System

Pathogen recognition is based on pattern matching and key-lock mechanisms. *Positive* and *negative* selection principles are used for priming T-cells and B-cells.

T-cells mature in the thymus in two stages called (i) *positive* and (ii) *negative selection.*

(i) Immature T-cells in the thymus get first tested on self cell reactivity. T-cells that do not react with self cells at all are subject to removal.
(ii) T-cells that survived the first step are tested on whether they can react with self cells too strongly. If this is true, the result is again removal.

Remaining T-cells are released into the body. Such T-cells are mature but remain *naive* until their activation. In order these two stages to be achievable, the thymus is protected by a blood-thymic barrier that is able to keep this organ pathogen-free. As a result, mature T-cells are reactive with cells that could not be present in the thymus, i.e. with non-self cells. They are also weakly self reactive but they are unable to get activated by a self cell. Unfortunately, the repertoire of self cells in thymus does not have to be always complete. This can lead to reactivity with self cells, i.e. to *autoimmune* reactions (false positives). The weak self reactivity of T-cells makes interaction with other immuno-cells possible.

Creating of new B-cells with an improved pathogen recognition ability is done through cloning. Cloning of B-cells allows for some "error". This error is believed to be inversely proportional to the matching ability of a given B-cell. The weaker is the ability, the greater is the error allowance. The process of B-cell cloning with an error allowance is called *somatic hypermutation*. The purpose of somatic hypermutation is to diversify the matching ability of B-cells.

Both T-cells and B-cells that were not able to recognize any pathogen get removed. This happens to most B-cells and T-cells and can be understood as an instance of positive selection (only the fittest stay).

3   Computational Paradigms

3.1   Negative Selection

Negative selection is a learning mechanism applied in training and priming of T-cells in the thymus. In the computational approach to negative selection due to D'haeseleer et al. [8], a complement to an n-dimensional vector set is constructed. This is done by producing random vectors and testing them against vectors in the original vector set. If a random vector does not match anything in the original set, according to some matching rule, it becomes a member of the complement (detector) set. These vectors are then used to identify anomalies (faults/misbehavior). The negative selection algorithm has thus two phases:

(i) Learning (detector generation) phase.
(ii) Detection phase.

Elberfeld and Textor [9] recently showed that negative selection can be done in polynomial time, if the n-dimensional vectors are represented as bit vectors, and the matching and testing is done using the *r-contiguous bits matching rule* [8].

Definition 1. *Two bit strings of equal length match under the r-contiguous matching rule if there exists a substring of length r at position p in each of them and these substrings are identical.*

Although negative selection based on bit vectors was shown to be computable in polynomial time, whether an efficient algorithm will ever be proposed for other types of negative selection is not clear. For example, there is no efficient approach for negative selection based on real-valued vectors due to Ji and Dasgupta [10].

3.2   B-Cell Cloning Algorithm

The B-Cell Algorithm (BCA) was introduced by Kelsey and Timmis [11] as a new immune system inspired approach to solve optimization problems using population generation with a special form of mutation. Each B-cell is a n-dimensional bit vector. Each B-cell within the BCA is evaluated by an objective function $g(x)$. Once a B-cell has been evaluated, it is cloned, placing its clones into a *clonal pool*. Mutation and adaption is done only within this pool. According to [11] the size of the population and the clonal pool are typically identical. Diversity is maintained by selecting one clone at random, randomly changing each element of the vector with a predefined probability. The algorithm then uses a mutation operator called *contiguous somatic hypermutation*. This approach is based on randomly selecting a start point of a contiguous region together with a random length. This area of the bit vector is then subject to mutation. This is in contrast to selecting multiple random entries over the complete vector, as performed by other mutation approaches. Finally each clone is evaluated with respect to $g(x)$, replacing its parent when having a better fitness value.

3.3   Danger Theory and Dendritic Cell Algorithm

The computational counterpart to the Danger theory of Matzinger was formulated by Aickelin et al. [12]. Within the Danger theory the following types of signals were proposed:

(i) PAMP (pathogen associated molecular pattern) signal: connected with pathogen (misbehavior) presence.
(ii) Danger signal: connected with a sudden change in behavior.
(iii) Safe signal: connected with the expected system's behavior.
(iv) Amplifying signal: this signal can amplify the effects of the previous three signals.

The amplifying signal is inspired by the function of cytokines in activating and attracting immune cells to the area of infection.

Motivated by the role of dendritic cells in processing the signals produced within the BIS, Greensmith et al. [13] proposed the *Dendritic cell algorithm.* By computing the output signal concentration, it models the various maturity stages of dendritic cells, taking into consideration several different signals and their weights:

$$C_{[csm,semi,mat]} = \frac{(W_P C_P) + (W_S C_S) + (W_D C_D)(1 + IC)}{W_P + W_S + W_D} \tag{1}$$

where

C_P = PAMP signal, W_P = PAMP signal weight
C_S = safe signal, W_S = safe signal weight
C_D = danger signal, W_D = danger signal weight
IC = amplifying (inflammatory cytokine) signal
C_{csm}, C_{semi}, C_{mat} is signal concentration corresponding to the three different maturity stages of dendritic cells, where csm = co-stimulatory, $semi$ = semi-mature, mat = mature.

The Dendritic cell algorithm is thus based on a signal processing function with signals and their weights corresponding to features (network measurements) and their relevance. This algorithm proceeds as follows. A pool of dendritic cells is created. A small subset of these dendritic cells is presented with an antigen (feature vector). The output of Eq. 1 gets computed. Based on predefined fuzzy thresholds, the dendritic cells either default to the co-stimulatory stage or will migrate to semi-mature or mature stage. If an antigen drives a significant number of dendritic cells into the mature stage, it will be marked as malignant, i.e. representing a misbehavior. Mapping features to the various types of signals and suggesting maturation thresholds is problem specific.

4   The Problem of Misbehavior Detection in Ad Hoc Networks

In this section a brief introduction to ad hoc networks and related protocols is given. We also review several misbehavior types and give an assessment of their applicability to ad hoc networks.

Definition 2. *Let $N = (n(t), e(t))$ be a net where $n(t), e(t)$ are the set of nodes and edges at time t, respectively. Nodes correspond to wireless devices that wish to communicate with each other. An edge between two nodes A and B is said to exist when A is within the radio transmission range of B and vice versa.*

Besides general ad hoc networks we also will take into consideration a subclass which is known as *sensor networks*.

Definition 3. *A sensor network is a static ad hoc network deployed with the goal to monitor environmental or physical conditions such as humidity, temperature, motion or noise. Within a sensor network, nodes are assumed to have energy and computational constraints.*

Each node in an ad hoc wireless network has the ability to observe a variety of protocol states and events. Based on these states and events, performance measures or *features* can be computed.

4.1   Protocols and Assumptions

We now review several protocols, mechanisms, assumptions and definitions.

AODV [14] is an on-demand routing protocol that starts a route search only when a route to a destination is needed. This is done by flooding the network with RREQ (= Route Request) control packets. The destination node or an intermediate node that knows a route to the destination will reply with a RREP (= Route Reply) control packet. This RREP follows the route back to the source node and updates routing tables at each node that it traverses. A RERR (= Route Error) packet is sent to the connection originator when a node finds out that the next node on the forwarding path is not replying.

At the MAC (Medium access control) layer, the medium reservation is often contention based. In order to transmit a data packet, the IEEE 802.11 MAC protocol [15] uses carrier sensing with an RTS-CTS-DATA-ACK handshake (RTS = Ready to send, CTS = Clear to send, ACK = Acknowledgment). Should the medium not be available or the handshake fails, an exponential back-off algorithm is used. This algorithm computes the size of a *contention window*. From this window a random value is drawn and this determines when the data packet transmission will be again attempted. This is additionally combined with a mechanism that makes it easier for neighboring nodes to estimate transmission durations. This is done by exchange of duration values and their subsequent storing in a data structure known as Network allocation vector (NAV).

In *promiscuous mode*, a node listens to the on-going traffic among other nodes in the neighborhood and collects information from the overheard packets. Promiscuous mode is energy inefficient because it prevents the wireless interface from entering sleep mode, forcing it into either idle or receive mode. There is also extra overhead caused by analyzing all overheard packets. According to [16], power consumption in idle and receive modes is about 12-20 higher than in sleep mode. Promiscuous mode requires that omnidirectional antennas are used.

Depending on the capabilities of the network nodes, the *Transmission control protocol* (TCP) [17] can be also part of the protocol stack used. Its task is to provide a reliable end-to-end connection control. TCP provides mechanisms for opening and closing connections among nodes. The reception of each data packet gets acknowledged by the destination node. TCP uses a *congestion window* which determines the number of data packets that can be sent to a destination without an acknowledgment. The congestion window size gets increased, if data packets get timely acknowledged. In the presence of congestion, the congestion window size gets decreased. The congestion window is thus an adaptive mechanism for controlling the number of data packets that are allowed to traverse a network at the same time without an acknowledgment.

4.2   Misbehavior

None of the protocols just discussed can prevent the occurrence of misuse or attack. Unfortunately, there is a large number of misbehavior types towards networks and communication in general. We will now present some misbehavior examples which either aim at disrupting data traffic or at gaining some undeserved advantage.

The most simplistic misbehavior type is the *packet forwarding misbehavior*, where a node is either dropping, delaying, duplicating or jamming a received data packet. One possibility for enforcing correct data packet forwarding is data transmission overhearing by neighboring nodes in promiscuous mode [18]. This possibility is, as we already pointed out, energy inefficient.

A *selfish* node might wish to gain an exclusive access to the wireless medium. This behavior can be achieved through manipulation of the network allocation vector as defined in IEEE 802.11 class of protocols or through manipulation of the contention window. At the transport layer, manipulation of the *congestion window* of a TCP connection can have similar effects. Another possibility for executing selfishness is to ignore incoming communication requests from other nodes (RTS). A motivation for such a behavior can be to save battery power and thus to gain a longer lifespan [19].

In *overloading* an attacker injects data packets that he knows are invalid (e.g. with an incorrect correction code). A node being a target of this misbehavior will get into a busy-trashing mode [20].

Several other attacks target routing tables and route caches in order to create network inconsistencies. This can be done by injecting bogus RREP packets or by advertising unreachable nodes as the next hop [21,22,23,24]. Another example of routing manipulation is the *gratuitous detour* attack [25]. The goal of this attack is to make routes over the attacking node to appear longer. Since many routing protocols would choose the shortest path between a connection source and destination, this node can avoid forwarding data packets and thus save battery power.

Impersonification or IP spoofing is another prominent class of attacks. An instance of this attack is manipulation of the data packet origin [21]. Instead of hiding behind the address of an existing node, an attacker might also create several fictitious nodes. This is called the *Sybil attack* [26]. Fictitious nodes can increase the hop count from the source to the destination node. Again, since many routing protocols favor the shortest route for data packet forwarding, the attacking node might be able to avoid data packet forwarding. This attack can additionally lead to damaging network inconsistencies.

In a *black* or *gray hole* attack [25], the attacking node would reply to any received RREQ packet with a RREP packet. This increases the probability that this node would lie on an active forwarding route. This distorts the distances in the network and allows the attacking node to drop data packets. In a black hole attack all data packets get dropped, whereas in a gray hole attack data packets would get dropped according to a dropping probability or in an adaptive way.

Rushing attacks were introduced in [27]. This type of misbehavior can be used together with routing protocols that utilize the RREQ-RREP handshake. It is prescribed that only the first RREQ packet is broadcast by a given node. This means, a node that manages to broadcast a RREQ packet as the first one, will most likely be included in a forwarding route. In this attack RREQ packet broadcasting gets done with different priorities, i.e. some RREQ packets get delayed whereas others get immediately sent or rushed.

Attacks can not only be performed by a single node but also by several nodes in collusion. *Wormholes* [28] are an example of this type of attacks. In the wormhole attack, two nodes are linked with a private high-speed data connection. This can potentially distort the topology of the network and attackers may be able to create a virtual cut vertex. This attack can also be combined with black or grey holes, as these first attract the traffic, allowing the wormhole to transfer the traffic via the private link.

At the transport layer, a well-known denial-of-service (DOS) attack is possible. In the *TCP SYN flooding* attack, an attacker attempts to open a large number of connections, strictly larger than what the destination can serve. The connection buffer of the host overflows. As a consequence, it is not possible to serve any legitimate connections [29,30]. Another attack is possible through the manipulation of the TCP acknowledgment mechanism: *ACK division, DupACK spoofing, and optimistic ACKing*. This can have an impact on the size of the congestion window and thus impact the data throughput of a given network. *JellyFish attacks* were introduced in [31]. They also target the congestion control of TCP-like protocols. These attacks obey all the rules of TCP. Three kinds of JellyFish (JF) attacks were discussed in [31]: JF reorder attack, JF periodic dropping attack, and JF delay variance attack.

Different kinds of attacks have been presented in this section to motivate the necessity of mechanisms which try to detect and mitigate misbehavior in (ad-hoc) networks.

4.3 Applying AIS to Misbehavior Detection

In this section several AIS based approaches for misbehavior detection and network security in general will be presented.

Applying Self/Non-self Discrimination to Network Security: The pioneering work in adapting the BIS to networking has been done by Stephanie Forrest and her group at the University of New Mexico. In one of the first BIS inspired works, Hofmeyr and Forrest [32] described an AIS able to detect anomalies in a wired TCP/IP network. Co-stimulation was in their setup done by a human operator who was given 24 hours to confirm a detected attack.

Applying AIS to mobile Ad Hoc Networks: Sarafijanović and Le Boudec [33,34] introduced an AIS for misbehavior detection in mobile ad hoc wireless networks. They used four different features based on the network layer of the OSI protocol stack. They were able to achieve a detection rate of about 55%; they only considered simple packet dropping with different rates as misbehavior. A co-stimulation in the form of a danger signal emitted by a connection source was used to inform nodes on the forwarding path about perceived data packet loss. Such signal could then be correlated with local detection results and suppress false positives.

The AIS architecture of Sarafijanović and Le Boudec is shown in Fig. 2. The main modules are Virtual thymus module, Danger signal module, Clustering module and Clonal selection module. The Virtual thymus module implements the negative selection process. Its purpose is to prime misbehavior detectors. The Danger signal module emits a danger signal, if a connection source node receives no data packet acknowledgment as prescribed by the TCP protocol. The Clustering module is basically a statistical analysis module. It implements a data clustering algorithm that is used to determine whether an observed

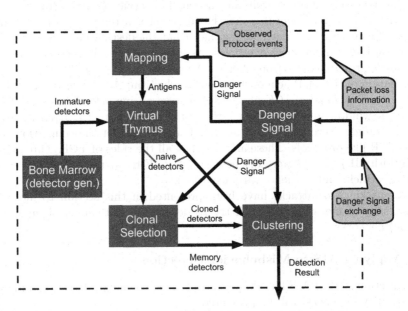

Fig. 2. AIS as proposed by Sarafijanović and Le Boudec.

behavior can be classified as misbehavior. The Clonal selection module attempts to improve the quality of the computed detectors. Other parts of their AIS allow for (received) danger signal processing, feature computation and other necessary information flow.

Applying AIS to Wireless Sensor Networks: An AIS for sensor networks was proposed by Drozda et al. in [35,36]. The implemented misbehavior was packet dropping; the detection rate was about 70%. The negative selection process was used to compute misbehavior detectors. They observed that only a small fraction of detectors was ever used in the misbehavior detection. This is due to the limiting nature of communication protocols that severely restrict the possible state space. The notion of complementary features was introduced. A feature is said to be *complementary*, if it depends on another feature in a significant way. An example of complementary features are features that

(i) measure wireless medium contention and
(ii) actual data packet dropping.

If sufficiently many data packets get dropped, it will result in a decrease in wireless medium contention. Clearly such two features are expected to be correlated. This means, an attacker that does data packet dropping must also manipulate the level of medium contention in order to stay unnoticed. This makes the detection process more robust and suggests that using highly correlated features can be of advantage. This principle was investigated in [37] but could not be demonstrated under a data traffic model with a low data packet injection rate.

Schaust et al. [38] investigated the influence of various data traffic models and injection rates on the performance of the AIS introduced in [35,36]. This AIS is available as AIS-Lib [2], an add-on module to the Jist/SWANS network simulator [39].

Applying AIS to Hybrid Sensor Networks: Bokareva et al. [40] proposed a self-healing hybrid sensor network architecture using several mechanisms from the adaptive immune system. The architecture consists of simple sensing nodes, advanced monitoring nodes, a base station (data collection point) and high-performance databases (computers). The latter two entities are inspired by the role of the thymus and lymph nodes in the BIS. The goal was to achieve a system that incorporates fault tolerance and adaptability to pathogens. Therefore sensor nodes are watched by monitoring nodes. Monitoring nodes have the capability to communicate directly with the base station. The task of the base station is to collect the data from the network and then to coordinate possible reactions to a discovered misbehavior.

Applying AIS to Secure a Routing Protocol: Mazahr and Farooq [41] proposed an AIS based approach to secure a mobile ad hoc network routing protocol called *BeeAdHoc*. The system makes use of the negative selection mechanism together with a learning phase and a detection phase as proposed by D'haeseleer et al. [8]. Three types of antigens were modeled: *scout* antigen and two types of *forager* antigens. The purpose of these antigens is to estimate abnormalities within their BeeAdHoc protocol. All antigens are represented as bit strings consisting of four features with the same length. Detectors are bit strings of the same size as the antigens. The matching between antigens and detectors is performed using a matching function with either Euclidean, Hamming or Manhattan distance as a measure. A mobile ad hoc network consisting of 49 nodes was simulated and evaluated.

The classification techniques proposed in [32,33,36,41] are based on negative selection. An approach based solely on the Danger theory avoiding the inefficiency of the negative selection was proposed by Kim et al. in [42]. Several types of danger signals, each having a different function are employed in order to detect routing manipulation in sensor wireless networks. The authors did not undertake any performance analysis of their approach.

Drozda et al. [37] investigated the suitability of a feature evaluation process based on a wrapper approach [43] to produce a suitable set of features for misbehavior detection. A co-stimulation inspired architecture with the aim to decrease the false positives rate while stimulating energy efficiency is proposed and evaluated. This architecture is further discussed in Section 5.

We would like to note that even though the BIS seems to offer a good inspiration for improving misbehavior detection in ad hoc and sensor networks, approaches based on machine learning and similar methods received much more attention; see [44,45,46] and the references therein.

4.4   Other Application Areas

In this section some approaches which do not aim at network security are presented. This demonstrates the general applicability of the AIS approach towards different kinds of problems.

Applying Innate Immunity to Specknets: Davoudani et al. [47] investigated the applicability of the innate immune system mechanisms to miniature device networks (*Specknets*). Their goals was to compute the state of a monitored environment. Specknets consist of tiny devices (specks), which are able to sense and process data, and to communicate over a short range. The functionality of dendritic cells was mapped onto *scouting messages* (SMs). Specks were classified into *tissue specks* and *integration specks*. Tissue specks contain sensors for monitoring the environment, whereas integration specks collect information from SMs, analyze it and determine an appropriate response. SMs originate at integration specks, traverse the network, collect data from tissue specks, and eventually return to its start point.

Applying Adaptive Immunity to Data Gathering in WSN: Atakan and Akan [48] proposed a method to determine a minimum number of nodes necessary to transmit data to a sink or base station in a wireless sensor network. The approach is based on the adaptive immune system and a B-cell stimulation model. The resulting method is called *Distributed Node and Rate Selection*

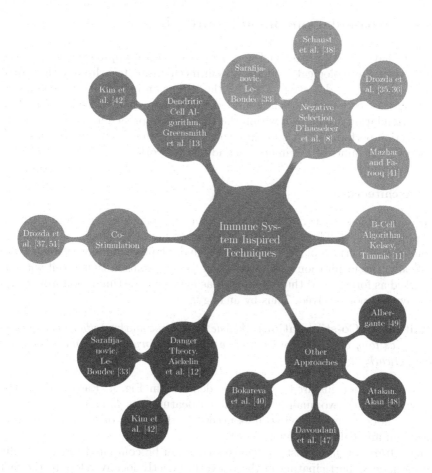

Fig. 3. Publication mindmap of Immune system inspired approaches.

(DNRS). DNRS selects the most appropriate sensor node for transmission and, at the same time, regulates its reporting frequencies. B-cells are mapped to sensor nodes and antibodies to sensor data. The antibody density is mapped to the reporting frequency rate, while pathogens are mapped to event sources. Natural cell extinction is mapped to packet loss.

Applying AIS to Wireless Discussion Forums: Albergante [49] described a protocol for moderating Wireless Discussion Forums (WDF). This protocol is based on AIS principles. Participants of the WDF are mapped to cells, whereas the AIS has the role of detecting infected cells, which in the case of a WDF is a participant which ignores the rules of a forum. A negative and a positive feedback mechanism is used to model the stimulation of immune system cells. An artificial lymphatic system was introduced to control the lifespan of immune cells.

A publications mindmap of the approaches discussed in this section is shown in Fig. 3.

5   A Co-stimulation Based Approach

So far we have briefly introduced and discussed several approaches which take advantage of the biological immune system mechanisms to achieve either some form of misbehavior detection or to use the BIS as an inspiration for communication. In this section we present a more detailed case study which is inspired by the co-stimulation mechanism within the BIS. This approach intends to increase the detection performance, taking into consideration the specific limitations such as energy constraints of the underlying ad hoc network.

5.1   Architecture

To achieve a robust detection performance, a co-stimulation inspired mechanism is considered in [37]. This mechanism attempts to mimic the ability of communication between various players of the innate and adaptive immune system. More specifically, the inspiration was based upon the observation that a cell will not be marked as foreign and thus eliminated as long as a co-stimulation from other BIS players is not received. Thus by analogy:

Definition 4 (Co-stimulation). *A node should not mark another node as misbehaving as long as it does not receive a co-stimulatory signal from other parts of the network.*

The co-stimulation inspired approach is depicted in Fig. 4. Each node in the network computes two qualitatively different feature sets f_0 and f_1. In order to classify a node as misbehaving, it is necessary that both of these feature sets indicate an anomalous behavior.

The feature set f_0 consists of features that can be computed strictly locally by each node s_i participating in the network. Strictly locally refers to the fact that such a computation does not require cooperation from any other node, i.e. only traffic measurements done by node s_i are necessary. The negative side is the necessity to operate in promiscuous mode. An example of a feature that belongs to the f_0 feature set is the watchdog feature described in [18].

Let $S, s_1, ..., s_i, s_{i+1}, s_{i+2}, ..., D$ be the path from S to D determined by a routing protocol, where S is the source node, D is the destination node. The *watchdog* feature is said to be the ratio of data packets sent from s_i to s_{i+1} and then subsequently forwarded to s_{i+2}. Whether a given data packet is further forwarded to s_{i+2} is determined by s_i in promiscuous node by overhearing the traffic between s_{i+1} and s_{i+2}.

To the f_1 feature set belong features that need to be computed at several nodes and then subsequently jointly evaluated. An example is a feature that measures the data packet interarrival delay at a node. It is clear that if the node s_{i+1} drops a data packet, this will be reflected in an increased interarrival delay at the node s_{i+2}. This measurement has to be compared with the corresponding measurement at the node s_i in order to be able to conclude whether the increase in interarrival delay was significant. Besides energy efficiency, the feature sets f_0

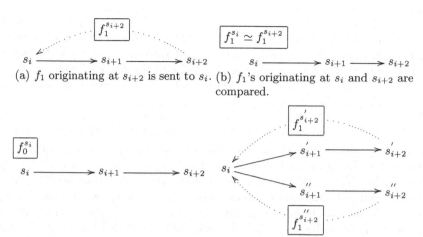

(a) f_1 originating at s_{i+2} is sent to s_i. (b) f_1's originating at s_i and s_{i+2} are compared.

(c) *Co-stimulation:* if (b) indicates that s_{i+1} is misbehaving, f_0 will be computed and evaluated.

(d) If s_i forwards data packets for several connections, multiple f_1's are computed and sent.

Fig. 4. Immuno-inspired misbehavior detection approach.

and f_1 have a very different expressive power. For example to show that s_{i+1} is misbehaving, either f_0 computed at s_i is necessary or f_1 computed at both s_i and s_{i+2} is necessary.

Definition 5. *Strictly local (or point) detection is done by a single node s_i without any cooperation with other nodes. The node s_i thus only computes features based on the properties of (i) its own data traffic, or (ii) neighboring nodes' data traffic observed in promiscuous mode.*

Definition 6. *Local detection is done co-operatively by several nodes through exchange of data traffic measurements. Such data traffic measurements need to be jointly evaluated by one of the participating nodes.*

Drozda et al. [37] investigated 24 different point and local features with respect to their suitability for misbehavior detection, protocol requirements and energy efficiency.

A co-stimulation is introduced in the following way. Each node in the network computes the feature set f_1. This feature set is then proliferated upstream (towards the connection source); see Fig. 4(a). Each f_1 travels exactly two hops; in our example from s_{i+2} to s_i. If s_i receives f_1 originating at one of its neighbors, it will forward f_1 in the upstream direction. If f_1 is not originating at one of its neighbors, it is used but not forwarded.

Since the computation of f_1 is time window based, the frequency with which f_1 gets sent depends on the time window size. Upon receiving f_1 from a two-hop neighbor, the node s_i combines and evaluates it with its own f_1 sample; see Fig. 4(b). Based on this, a behavior classification with respect to the node

s_{i+1} is done. If s_i classifies s_{i+1} as misbehaving, then it computes a sample based on the f_0 feature set. This means, a *co-stimulation* from the f_1 based classification approach is needed in order to activate the energy less efficient f_0 based classification approach; see Fig. 4(c). If misbehavior gets confirmed, the node s_{i+1} is marked as misbehaving. Note that s_i can receive f_1 samples from several two-hop neighbors; see Fig. 4(d) for an example.

The proliferation of f_1 can be implemented without adding any extra communication complexity by attaching this information to CTS or ACK MAC packets (this would require a modification in the MAC protocol). As long as there are DATA packets being forwarded on this connection, the feature set can be propagated. If there is no DATA traffic on the connection (and thus no CTS/ACK packets exchanged), the relative necessity to detect the possibly misbehaving node s_{i+1} decreases. Optionally, proliferation of f_1 can be implemented by increasing the radio radius at s_{i+2}, by broadcasting it with a lower time-to-live value or by using a standalone packet type.

If the node s_{i+1} decides not to cooperate in forwarding the feature set information, the node s_i will switch, after a time-out, to the feature set f_0 computation. In this respect, not receiving a packet with the feature set f_1 can be interpreted as a form of negative co-stimulation. If the goal is to detect a misbehaving node, it is important that the originator of f_1 can be unambiguously identified, i.e. strict authentication is necessary. An additional requirement is the use of sequence numbers for feature sets f_1. Otherwise, the misbehaving node s_{i+1} could interfere with the mechanism by forwarding outdated cached feature sets f_1.

We introduce the following notation in order to keep track of composite feature sets computed at the nodes s_i and s_{i+2}: $\mathcal{F}_1^{s_i} = f_1^{s_i} \cup f_1^{s_{i+2}}$. The feature set $f_1^{s_{i+2}}$ is a co-stimulatory feature set computed by s_{i+2} and sent over to s_i. For simplicity, we will omit the the superscripts indicating the feature set origin. The purpose of $\mathcal{F}_1$ is to provide a co-stimulation to the less energy efficient feature set f_0. The features in the f_0 and f_1 feature sets are to be computed over a (sliding) time window.

5.2   Communication and Learning Complexity Trade-Offs

In [37] it was concluded that as the size of the time window decreases, the classification ability of the feature sets f_0 and $\mathcal{F}_1$ will equalize. That is:

$$\lim_{win.\ size \to 0} class.\ error(\mathcal{F}_1) \approx class.\ error(f_0) \qquad (2)$$

where *class. error* is defined as the sum of false positives and false negatives divided by the number of samples. The above equation is a natural consequence of the fact that instead of observing a data packet's delivery in promiscuous mode by the node s_i, it can be equally well done in a cooperative way by s_i and s_{i+2}, if the window size (sampling frequency) is small. In other words, if the time window is small enough that it always includes only a single event, the relationship between events at s_i and s_{i+2} becomes explicit. This is however connected with a high communication cost (each packet arrival at s_{i+2} must be

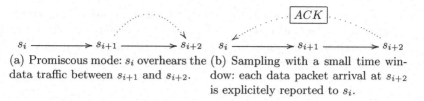

(a) Promiscous mode: s_i overhears the data traffic between s_{i+1} and s_{i+2}.   (b) Sampling with a small time window: each data packet arrival at s_{i+2} is explicitly reported to s_i.

Fig. 5. Promiscous mode vs sampling with a small time window.

explicitly reported to s_i); see Fig. 5. Since both $f_1^{s_i}$ and $f_1^{s_{i+2}}$ consist of several features, a $\mathcal{F}_1$ based misbehavior detection can also be viewed as multiple input-output *black box identification*. In our case, only a single node is part of the observed black box. Notice that, in general, ad hoc wireless networks are an inherently non-linear system [50].

The following observations can be formulated:

– If using f_0, the threshold for a watchdog feature can be set directly. For example, the desired data packet delivery can be set to 0.90 (90% data packets must be correctly forwarded) in order to classify s_{i+1} as misbehavior free.
– If using $\mathcal{F}_1$ with *win. size* $\gg 0$, learning based on data traffic at both s_i and s_{i+2} must be done. Feature averaging over a time window increases the classification task complexity. Frequency of the extra communication between s_i and s_{i+2} depends on the time window size.

The above two observations and Eq. 2 offer a rough characterization of the trade-offs between detection approaches executed by a single node and by several nodes in cooperation.

5.3 Two Immuno-inspired Algorithms

The rule formulated in Eq. 2 motivates the following strategy: compute an approximation of f_0 using $\mathcal{F}_1$. In Fig. 6 an algorithm that builds around this idea is introduced. The feasibility of this algorithm is based on the assumption that the classification error increases monotonically with the time window size.

---

1. *Pre-processing:* each node collects f_0 and $\mathcal{F}_1$ based samples. Label these samples using pre-existing misbehavior signatures. Based on these two labeled sample sets, compute two distinct classifiers.
2. Use the $\mathcal{F}_1$ based classifier computed in the above step to classify any fresh $\mathcal{F}_1$ samples.
3. *Co-stimulation:* if the $\mathcal{F}_1$ based classification hints a possibility of misbehavior then employ the f_0 based classifier to get a more reliable prediction. A computation of a fresh f_0 based sample is necessary.

Fig. 6. Co-stimulation algorithm

There are two basic reasons for applying this strategy: (i) computing $\mathcal{F}_1$ is more energy efficient and (ii) it delivers the desired co-stimulatory effect. Unlike when f_0 gets exclusively applied, in this approach f_0 will get used only if (i) a true positive was detected by $\mathcal{F}_1$ or (ii) a false positive was mistakenly detected by $\mathcal{F}_1$. This means, for a misbehavior free ad hoc network, the energy saving over an exclusive f_0 approach is proportional to $1 - FP\ rate$, where $FP\ rate$ is the false positives rate.

According to [16], the energy consumption when using f_0 (promiscuous mode) grows linearly with the number of data packets that need to be overheard. The energy consumption when using $\mathcal{F}_1$ grows, on the other hand, linearly with the number of $f_1^{s_i+2}$ sent over two hops to s_i. This must, however, be followed by at least a 50-second f_0 based co-stimulation. Such a co-stimulation happens with probability $FP\ rate$ in a misbehavior free network. For a 500-second time window used for $\mathcal{F}_1$ computation, 0.5 packet/second injection rate and 1kB data packet size, the potential energy saving effect is nearly two orders of magnitude. For a smaller data packet size of 64 bytes, more typical for wireless sensor networks, the potential energy saving is about one order of magnitude. The results reported in [37] also document that the final false positives rate could be dramatically reduced with this two-stage algorithm.

The Co-stimulation algorithm assumes that each node is able to acquire enough traffic samples, compute all the features and employ a learning algorithm to build a classifier. In practical terms, it means that a sufficient number of traffic samples labeled as either representing normal network operation or misbehavior must be available. This implies either an existing misbehavior signature knowledge base or the ability *to prime* the network with (generic) examples of the normal behavior and misbehavior.

Drozda et al. [51] proposed an algorithm inspired by the tissue injury signaling role of the innate immune system. The goal of the algorithm is to enforce a network operational strategy in an *energy efficient* way. An operational strategy, or priming, can impose performance limits in the form of e.g. a maximum data packet loss at a node. Unlike in the Co-stimulation algorithm where f_0 and $\mathcal{F}_1$ are learned in an independent way, this algorithm uses priming to assist in initial sample labeling. Then it takes advantage of Eq. 2 in order to propagate the associated classification error induced by this procedure onto the energy efficient $\mathcal{F}_1$ feature set.

Definition 7 (Priming). *Priming is an initial stimulus that induces a minimum level of systemic resistance against misbehavior.*

The Error propagation algorithm is described in Fig. 7 and schematically depicted in Fig. 8(a). The priming step reflects the desired capability to detect any node operation that deviates from the prescribed behavior. The error propagation step utilizes the rule described in Eq. 2. The classification error induced by $\overline{thresh}$ gets this way propagated onto the $\mathcal{F}_1$ feature set. Since for practical reasons $win.\ size \gg 0$ must hold, a co-stimulation ($\mathcal{F}_1 \longrightarrow f_0$) is necessary for improving the classification performance. The less energy effcient f_0 based

1. *Priming:* choose desired levels (thresholds) for one or several point detection features. We denote $\overline{thresh}$ to be the vector of such threshold values.
2. The network starts operating. Allow all nodes to use promiscuous mode. Each node computes both f_0 and $\mathcal{F}_1$ based sample sets. Disable promiscuous mode at a node, when the sample set size reaches a target value.
3. Label the $\mathcal{F}_1$ based samples. A $\mathcal{F}_1$ based sample will be labeled as "normal", if in the f_0 based sample from the corresponding time window, no undesired behavior with respect to $\overline{thresh}$ was observed, and as "misbehavior" otherwise.
4. *Error propagation:* compute a classifier by using only the samples based on $\mathcal{F}_1$.
5. Use the $\mathcal{F}_1$ based classifier computed in the above step to classify any fresh $\mathcal{F}_1$ samples. Co-stimulation: if a sample gets classified as "misbehavior", allow s_i to compute a f_0 sample in promiscuous node.
6. *Classification:* apply $\overline{thresh}$ to the fresh f_0 based sample. If the "misbehavior" classification gets confirmed, mark s_{i+1} as misbehaving. Normal behavior can get classified in a similar way.

Fig. 7. Error propagation algorithm

classification is thus used only if a co-stimulation from its $\mathcal{F}_1$ based counterpart is present. In other words, any detection based on $\mathcal{F}_1$ (adaptive immunity) must coincide with some damage being explicitly detected. This damage detection is based on f_0 (innate immunity).

The rule in Eq. 2 implicates that the final classification performance should equal the (initial) classification ability of the $\overline{thresh}$ threshold vector, if a small time window is applied. This implies that the values for $\overline{thresh}$ must be reasonably chosen in order to keep the classification error low.

The error propagation algorithm can be extended with an *optimization phase*; see Fig. 8(b). The classification outcome can be used to find new thresholds for $\overline{thresh}$ with the goal to minimize the classification error. In this case, the $\overline{thresh}$ threshold vector serves as a seed for the whole process. Then it is possible to relabel the $\mathcal{F}_1$ samples and to recompute the related classifier. Co-stimulation and classification phases are then executed. In general, any suitable optimization procedure with several optimization and error propagation cycles could be applied. In order to use this extended approach in a distributed way, an estimation of natural packet losses (noise) at nodes must be done.

Example: For the results published in [51] the following two thresholds were used. Max. data packet loss at a node was set to 2.5% and max. data packet delay at a node was set to 4ms. A sufficient quantity of f_0 and $\mathcal{F}_1$ based samples was computed. In the same time window, a $\mathcal{F}_1$ based sample was labeled according to whether the f_0 based sample would fullfil both of the requirements or not, thus labeling the sample as "normal" or "misbehavior", respectivelly. Based on a labeled $\mathcal{F}_1$ based sample set, a classifier was computed. Any freshly computed $\mathcal{F}_1$ based sample was then classified using this classifier. If it pointed to a misbehavior, the co-stimulation phase was applied in order to make a final decision.

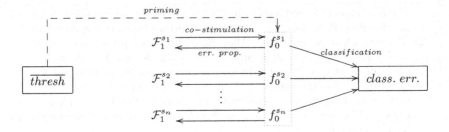

(a) An approach with error propagation and co-stimulation.

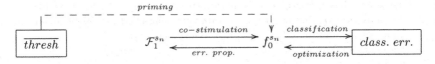

(b) An approach extended with optimization.

Fig. 8. Flow graph of the error propagation algorithm.

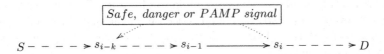

Fig. 9. Safe, danger and PAMP signals for enabling or disabling adaptive learning at s_{i-k}.

As shown in Fig. 4(d), any $\mathcal{F}_1$ computation is also based on the information available at s_{i+2}. This two-hop dependency can be compared to synapses among neurons. Under this comparison, our extended approach bears a certain similarity to the *backpropagation algorithm* for artificial neural networks [52]. Unlike in the backpropagation algorithm where edge weights must be recomputed in order to reflect the classification error, in our case the adjustment is done by a repeated computation of a classifier at each node (wireless device).

As we have already mentioned, co-stimulation seems to have a very positive effect on controlling false positives rate. This is a very good news, since the ambition of detection systems is to provide a solid detection rate while keeping the false positives rate at minimum. The false positives suppressing effect can be of tremendous advantage, if any form of *adaptive learning* should be applied.

Fig. 9 shows how a node could be alerted by means of safe, danger or PAMP signals originating at s_i. Such signals could not only enable or disable adaptive learning at that node but also provide an excellent basis for sophisticated priming. Additionally, it could help identify conditions when the last resort, a human operator, must get notified about adverse conditions reigning in the network.

6   Conclusions

The BIS is an inherently distributed system with functionality that can be very hard to mimic. For example, the success of the negative selection, a learning mechanism applied in training and priming of T-cells in the thymus, rests on the efficiency of the blood-thymic barrier that guarantees that the thymus stays pathogen free at all times. This implies that T-cells being trained in the thymus never encounter any type of pathogen before released into the body. This helps tremendously in detecting foreign cells. Mapping functionality of the BIS to a computational paradigm is a hot topic within the AIS community. Our goal was to review mechanisms with affinity to misbehavior detection in ad hoc wireless networks.

A limiting factor in translating the functionality of the BIS to technical systems remains our narrow knowledge of the BIS. Immunology is a research area with a large number of experimental results published every year. These results are however often very specific to the life form in investigation. As an example, we cite a recent result on priming of plant immunity [53].

The success of the BIS in protecting its host often benefits from the efficiency of underlying chemical reactions at molecular level. Such an efficiency has often no obvious computational parallel. For example, the negative selection process required 15 years until an efficient computational counterpart could be presented [9]. On the other hand, mimicking the somatic hypermutation of B-cells was a less intricate task [11].

Recent immuno-inspired approaches for ad hoc wireless networks [47,37] suggest that the focus of research in this area shifts towards more generic immune principles such as, for example, the regulatory and activating tasks of dendritic cells or the role of co-stimulation in increasing the robustness of the BIS.

Acknowledgments

This work was supported by the German Research Foundation (DFG) under the grant no. SZ 51/24-2 (Survivable Ad Hoc Networks – SANE).

References

1. Szor, P.: The art of computer virus research and defense. Addison-Wesley Professional, Reading (2005)
2. Drozda, M., Schildt, S., Schaust, S., Einhellinger, S., Szczerbicka, H.: A tool for prototyping AIS based protection systems for ad hoc and sensor networks. Cybernetics and Systems 39(7), 719–742 (2008)
3. Murphy, K.P., Travers, P., Walport, M.: Janeway's immunobiology. Garland Pub. (2008)
4. Banchereau, J., Briere, F., Caux, C., Davoust, J., Lebecque, S., Liu, Y.J., Pulendran, B., Palucka, K.: Immunobiology of dendritic cells. Annual review of immunology 18(1), 767–811 (2000)

5. Matzinger, P.: Tolerance, danger, and the extended family. Annual Review of Immunology 12(1), 991–1045 (1994)
6. Whitaker, L., Renton, A.M.: On the plausibility of the clonal expansion theory of the immune system in terms of the conbinatorial possibilities of amino-acids in antigen and self-tolerance. Journal of theoretical biology 164(4), 531–536 (1993)
7. Campbell, N., Reece, J., Markl, J.: Biologie, 6. überarbeitete Auflage (2006)
8. D'haeseleer, P., Forrest, S., Helman, P.: An Immunological Approach to Change Detection: Algorithms, Analysis and Implications. In: IEEE Symposium on Security and Privacy, pp. 110–119 (1996)
9. Elberfeld M., Textor J.: Efficient Algorithms for String-Based Negative Selection. In: Andrews, P.S. (ed.) ICARIS 2009. LNCS, vol. 5666, pp. 109-121. Springer, Heidelberg (2009)
10. Ji, Z., Dasgupta, D.: Applicability issues of the real-valued negative selection algorithms. In: Proc. of the 8th annual conference on Genetic and evolutionary computation, pp. 111–118 (2006)
11. Kelsey, J., Timmis, J.: Immune Inspired Somatic Contiguous Hypermutation for Function Optimisation. In: Cantú-Paz, E., et al. (eds.) GECCO 2003. LNCS, vol. 2723, pp. 207–218. Springer, Heidelberg (2003)
12. Aickelin, U., Bentley, P.J., Cayzer, S., Kim, J., McLeod, J.: Danger Theory: The Link between AIS and IDS? In: Timmis, J., Bentley, P.J., Hart, E. (eds.) ICARIS 2003. LNCS, vol. 2787, pp. 147–155. Springer, Heidelberg (2003)
13. Greensmith, J., Aickelin, U., Cayzer, S.: Introducing Dendritic Cells as a Novel Immune-Inspired Algorithm for Anomaly Detection. In: Jacob, C., Pilat, M.L., Bentley, P.J., Timmis, J.I. (eds.) ICARIS 2005. LNCS, vol. 3627, pp. 153–167. Springer, Heidelberg (2005)
14. Perkins, C.E., Royer., E.M.: Ad hoc On-Demand Distance Vector Routing.. In: Proc. of the 2nd IEEE Workshop on Mobile Computing Systems and Applications, pp. 90–100 (1999)
15. IEEE Std. 802.11: Part 11: Wireless LAN Medium Access Control (MAC) and Physical Layer (PHY) Specifications. IEEE Standard for Information technology
16. Feeney, L.M., Nilsson, M.: Investigating the energy consumption of a wireless network interface in an ad hoc networking environment. In: Proc. of Twentieth Annual Joint Conference of the IEEE Computer and Communications Societies (INFO-COM), vol. 3, pp. 1548–1557 (2001)
17. Iren, S., Amer, P.D., Conrad, P.T.: The transport layer: tutorial and survey. ACM Computing Surveys (CSUR) 31(4), 360–404 (1999)
18. Marti, S., Giuli, T.J., Lai, K., Baker, M.: Mitigating routing misbehavior in mobile ad hoc networks. In: Proc. of International Conference on Mobile Computing and Networking, pp. 255–265 (2000)
19. Kyasanur, P., Vaidya, N.H.: Detection and handling of mac layer misbehavior in wireless networks. In: Proceedings of the International Conference on Dependable Systems and Networks, pp. 173–182 (2002)
20. Jakobsson, M., Wetzel, S., Yener, B.: Stealth attacks on ad-hoc wireless networks. In: Proc. of 2003 IEEE 58th Vehicular Technology Conference (VTC Fall), vol. 3, pp. 2103–2111 (2003)
21. Savage, S., Wetherall, D., Karlin, A., Anderson, T.: Network support for ip traceback. IEEE/ACM Trans. Netw. 9(3), 226–237 (2001)
22. Yang, H., Meng, X., Lu, S.: Self-organized network-layer security in mobile ad hoc networks. In: WiSE 2002: Proceedings of the 1st ACM workshop on Wireless security, pp. 11–20. ACM, New York (2002)

23. Sun, B., Wu, K., Pooch, U.W.: Alert aggregation in mobile ad hoc networks. In: WiSe 2003: Proceedings of the 2nd ACM workshop on Wireless security, pp. 69–78. ACM, New York (2003)
24. Padmanabhan, V.N., Simon, D.R.: Secure traceroute to detect faulty or malicious routing. SIGCOMM Comput. Commun. Rev. 33(1), 77–82 (2003)
25. Hu, Y.C., Perrig, A., Johnson, D.B.: Ariadne: a secure on-demand routing protocol for ad hoc networks. In: MobiCom 2002: Proceedings of the 8th annual international conference on Mobile computing and networking, pp. 12–23. ACM, New York (2002)
26. Douceur, J.R.: The sybil attack. In: IPTPS 2001: Revised Papers from the First International Workshop on Peer-to-Peer Systems, London, UK, pp. 251–260. Springer, Heidelberg (2002)
27. Hu, Y.C., Perrig, A., Johnson, D.B.: Rushing attacks and defense in wireless ad hoc network routing protocols. In: WiSe 2003: Proceedings of the 2nd ACM workshop on Wireless security, pp. 30–40. ACM, New York (2003)
28. Hu, Y.C., Perrig, A., Johnson, D.: Packet leashes: a defense against wormhole attacks in wireless networks. In: Proc. of Twenty-Second Annual Joint Conference of the IEEE Computer and Communications Societies (INFOCOM), vol. 3, pp. 1976–1986 (2003)
29. Juels, A., Brainard, J.: Client puzzles: A cryptographic countermeasure against connection depletion attacks. In: Kent, S. (ed.) NDSS 1999: Networks and Distributed Security Systems, pp. 151–165 (1999)
30. Savage, S., Cardwell, N., Wetherall, D., Anderson, T.: Tcp congestion control with a misbehaving receiver. SIGCOMM Comput. Commun. Rev. 29(5), 71–78 (1999)
31. Aad, I., Hubaux, J.P., Knightly, E.W.: Denial of service resilience in ad hoc networks. In: MobiCom 2004: Proceedings of the 10th annual international conference on Mobile computing and networking, pp. 202–215. ACM, New York (2004)
32. Hofmeyr, S.A., Forrest, S.: Immunity by design: An artificial immune system. In: Proc. of Genetic and Evolutionary Computation Conference (GECCO), vol. 2, pp. 1289–1296 (1999)
33. Sarafijanovic, S., Le Boudec, J.Y.: An artificial immune system for misbehavior detection in mobile ad-hoc networks with virtual thymus, clustering, danger signal and memory detectors. In: Proc. of International Conference on Artificial Immune Systems (ICARIS), pp. 342–356 (2004)
34. Sarafijanovic, S., Le Boudec, J.Y.: An artificial immune system approach with secondary response for misbehavior detection in mobile ad hoc networks. IEEE Transactions on Neural Networks 16(5), 1076–1087 (2005)
35. Drozda, M., Schaust, S., Szczerbicka, H.: Is AIS Based Misbehavior Detection Suitable for Wireless Sensor Networks?. In: Proc. of IEEE Wireless Communications and Networking Conference (WCNC 2007), pp. 3130–3135 (2007)
36. Drozda, M., Schaust, S., Szczerbicka, H.: AIS for Misbehavior Detection in Wireless Sensor Networks: Performance and Design Principles. In: Computing Research Repository (CoRR) (2009) arXiv.org/abs/0906.3461
37. Drozda, M., Schildt, S., Schaust, S.: An Immuno-Inspired Approach to Fault and Misbehavior Detection in Ad Hoc Wireless Networks. Technical report, Leibniz University of Hannover (2009)
38. Schaust, S., Drozda, M.: Influence of Network Payload and Traffic Models on the Detection Performance of AIS. In: Proc. of International Symposium on Performance Evaluation of Computer and Telecommunication Systems (SPECTS 2008), pp. 44–51 (2008)

39. Barr, R., Haas, Z., Renesse, R.: Scalable wireless ad hoc network simulation. In: Handbook on Theoretical and Algorithmic Aspects of Sensor, Ad hoc Wireless, and Peer-to-Peer Networks, pp. 297–311 (2005)
40. Bokareva, T., Bulusu, N., Jha, S.: Sasha: toward a self-healing hybrid sensor network architecture. In: EmNets 2005: Proceedings of the 2nd IEEE workshop on Embedded Networked Sensors, Washington, DC, USA, pp. 71–78. IEEE Computer Society, Los Alamitos (2005)
41. Mazhar, N., Farooq, M.: BeeAIS: Artificial immune system security for nature inspired, MANET routing protocol, BeeAdHoc. In: de Castro, L.N., Von Zuben, F.J., Knidel, H. (eds.) ICARIS 2007. LNCS, vol. 4628, pp. 370–381. Springer, Heidelberg (2007)
42. Kim, J., Bentley, P.J., Wallenta, C., Ahmed, M., Hailes, S.: Danger Is Ubiquitous: Detecting Malicious Activities in Sensor Networks Using the Dendritic Cell Algorithm. In: Bersini, H., Carneiro, J. (eds.) ICARIS 2006. LNCS, vol. 4163, pp. 390–403. Springer, Heidelberg (2006)
43. Kohavi, R., John, G.H.: Wrappers for feature subset selection. Artificial Intelligence 97(1-2), 273–324 (1997)
44. Anantvalee, T., Wu, J.: A survey on intrusion detection in mobile ad hoc networks. Wireless/Mobile Network Security, 159–180 (2007)
45. Mishra, A., Nadkarni, K., Patcha, A.: Intrusion detection in wireless ad hoc networks. IEEE Wireless Communications 11(1), 48–60 (2004)
46. Zhang, Y., Lee, W., Huang, Y.A.: Intrusion Detection Techniques for Mobile Wireless Networks. Wireless Networks 9(5), 545–556 (2003)
47. Davoudani, D., Hart, E., Paechter, B.: Computing the state of specknets: Further analysis of an innate immune-inspired model. In: Bentley, P.J., Lee, D., Jung, S. (eds.) ICARIS 2008. LNCS, vol. 5132, pp. 95–106. Springer, Heidelberg (2008)
48. Atakan, B., Akan, O.B.: Immune system based distributed node and rate selection in wireless sensor networks. In: BIONETICS 2006: Proceedings of the 1st international conference on Bio inspired models of network, information and computing systems, p. 3. ACM, New York (2006)
49. Albergante, L.: Wireless discussion forums: Automatic management via artificial immune systems. In: Proc. of International Symposium on Performance Evaluation of Computer and Telecommunication Systems (SPECTS), pp. 74–81 (2008)
50. Barrett, C., Drozda, M., Marathe, M.V., Marathe, A.: Characterizing the interaction between routing and MAC protocols in ad-hoc networks. In: Proc. of the 3rd ACM international symposium on Mobile ad hoc networking and computing (Mobihoc), pp. 92–103 (2002)
51. Drozda, M., Schaust, S., Schildt, S., Szczerbicka, H.: An Error Propagation Algorithm for Ad Hoc Wireless Networks. In: Andrews, P.S. (ed.) ICARIS 2009. LNCS, vol. 5666, pp. 260–273. Springer, Heidelberg (2009)
52. Alpaydin, E.: Introduction To Machine Learning. MIT Press, Cambridge (2004)
53. Jung, H.W., Tschaplinski, T.J., Wang, L., Glazebrook, J., Greenberg, J.T.: Priming in Systemic Plant Immunity. Science 324(5923), 89–91 (2009)

Immune Decomposition and Decomposability Analysis of Complex Design Problems with a Graph Theoretic Complexity Measure

Mahmoud Efatmaneshnik[1], Carl Reidsema[1], Jacek Marczyk[2], and Asghar Tabatabaei Balaei[1]

[1] The University of New South Wales, Sydney, Australia
[2] Ontonix s.r.l, Como, Italy
{mahmoud.e,reidsema,asghart}@unsw.edu.au, Jacek@ontonix.com

Abstract. Large scale problems need to be decomposed for tractability purposes. The decomposition process needs to be carefully managed to minimize the interdependencies between sub-problems. A measure of partitioning quality is introduced and its application in problem classification is highlighted. The measure is complexity based (real complexity) and can be employed for both disjoint and overlap decompositions. The measure shows that decomposition increases the overall complexity of the problem, which can be taken as the measure's viability indicator. The real complexity can also indicate the decomposability of the design problem, when the complexity of the whole after decomposition is less than the complexity sum of sub-problems. As such, real complexity can specify the necessary paradigm shift from decomposition based problem solving to evolutionary and holistic problem solving.

Keywords: Real Complexity, Design Complexity, Design Decomposition, Decomposability of Design Problems.

1 Introduction

A design problem is a statement of the requirements, needs, functions, or objectives of design. The design problem is a structured representation of the specific question or situation that must be considered, answered, or solved by the designer [2]. Design problem solving is the process of assigning values to the design parameters in accordance with the given design requirements, constraints, and optimization criterion [3]. The design variable sets include subsets of sizing variables, shape variables, topologies and process knowledge and manufacturing variables such as process capabilities [4]. The Design Structure Matrix displays the relationships between components of a system in a compact, visual, and analytically advantageous format as a square matrix with identical row and column labels [5]. Parameter-based Design Structure Matrix (PDSM) is used for modelling low-level relationships between design decisions and parameters, systems of equations, subroutine parameter exchanges which represents the product architecture [5]. For example PDSM of an aircraft at the conceptual level must address how the wing configuration might affect

E. Szczerbicki & N.T. Nguyen (Eds.): Smart Infor. & Knowledge Management, SCI 260, pp. 27–52.
springerlink.com

Table 1. An example PDSM that will be used throughout the paper.

-	V1	V2	V3	V4	V5	V6	V7	V8	V9	V10
V1	0	0.76	0.45	0.16	0.22	0.77	0.12	0.01	0	0
V2	0.76	0	0.11	0.65	0.44	0.78	0	0	0	0.18
V3	0.45	0.11	0	0.64	0.11	0.31	0.02	0	0.15	0
V4	0.16	0.65	0.64	0	0.45	0.34	0	0	0	0
V5	0.22	0.44	0.11	0.45	0	0	0	0.01	0	0.01
V6	0.77	0.78	0.31	0.34	0	0	0	0	0	0
V7	0.12	0	0.02	0	0	0	0	0.2	0.7	0.1
V8	0.01	0	0	0	0.01	0	0.2	0	0.2	0.8
V9	0	0	0.15	0	0	0	0.7	0.2	0	0.9
V10	0	0.18	0	0	0.01	0	0.1	0.8	0.9	0

the cruise speed and fuel consumption of the aircraft. PDSM is the upstream information in the design process about the product and is achieved through the design of experiments, expert suggestions, simulation techniques (in particular statistical Monte Carlo Simulation), and the combination of all/some of these techniques. PDSM can be represented via a graph which, in this paper, is referred to as the self map of the product (Table 1).

A graph G is specified by its vertex set, $V = \{1,...,n\}$, and edge set E. The total number of nodes in G is denoted by $|G|$ and referred to as the order of G. The number of edges in G is the size of G and is denoted by $E(G)$. $G\ (n,\ m)$ is a graph of order n and size m. The most natural matrix to associate with G is its adjacency matrix, A_G, whose entries $a_{i,j}$ are given by:

$$a_{i,j} = \begin{cases} w_{i,j}\ (i,j) \in E \\ 0 \end{cases}. \tag{1}$$

For un-weighted graphs all $w_{i,j}=1$, and in undirected graphs for all $(i,j) \in E$, $a_{i,j} = a_{j,i}$. Another matrix that can be associated with graphs is the Laplacian matrix which is defined as $L(G) = D - A$; where D is $n \times n$ diagonal matrix with entries $D_{i,i} = d_i$ and d_i is the degree of the vertex i defined as the total number of edges (or sum of the weights of the edges) that touch the vertex.

In the domain of the problem solving, every node or vertex of the graph represents a variable of the system/problem and every edge of the graph suggests that two parameters of the system are dependent on each other. The strength of the relationship between the pairs of variables is the corresponding edge weight. An undirected graph as the self map of the system indicates that the variables affect one another mutually and equally. Although this assumption is restrictive and may not always be valid, in this paper it is deemed to hold.

2 Decomposition

Decomposition is a means of *reducing the complexity of the main problem to several sub-problems.* This is a reductionist approach which reduces complex things to their constituent parts and their interactions in order to understand their nature. In contrast holism is an approach to problem solving that emphasizes the study of complex systems as wholes. *"In the design community, decomposition and partitioning of design problems has been attended for the purpose of improving coordination and information transfer across multiple disciplines and for streamlining the design process by adequate arrangement of the multiple design activities and tasks"* [6]. *"For a complex product, such as an automobile, a computer, or an airplane, there are thousands of possible decompositions which may be considered; each of these alternative decompositions defines a different set of integration challenges at the* (design) *organizational level"* [7].

Alexander [8] posed that design decomposition (or partitioning) must be performed in a way that the resulting sub-problems are minimally coupled (least dependent). In the literature this is also referred to as optimal decomposition [6] and robust decomposition [9]. Simon [10] made the strong suggestion that complex design problems could be better explained when considered as *"hierarchical structures consisting of nearly decomposable systems organized such that the strongest interactions occur within groups and only weaker interactions occur among groups"*. More coupled sub-problems usually lead to more process iterations and rework, because conflicts may arise when dependency (edges) in between the subsystems exists. A conflict is when the solution to one sub-problem is in contrast with the solutions to another sub-problem(s). An important conflict resolution technique is negotiation. Negotiation leads to iteration in the design process. Obviously a design process with least number of iterations is more desirable, and to do this decomposition must be performed in way that the entire system after decomposition has less coupling (see Fig. 1).

Problem decomposition and partitioning of the self map of the system fits in the area of graph partitioning. A *k*-partitioning divides the vertices of the graph into k non-empty sets $P = \{P_1, P_2, ..., P_k\}$. A *problem decomposition* is defined as *partitioning the corresponding graph or adjacency matrix of the problem (PDSM)*. As mentioned before a graph can be partitioned in many different ways. The sub-graphs can be regarded as subsystems or sub-problems or agents [11]. The notion of agency implies that the sub-problems are solved more or less independently from each other. Each design team has autonomy to explore parts of the solution space that is of interest to its own assigned sub-problem (agent). The system/problem is fully decomposable if there is no edge in between the sub-systems. In this case the corresponding design process can be made fully concurrent: problems are solved separately and solutions are added together.

A two stage algorithm is usually used to decompose a design problem into sub-problems that are less coupled [11, 12]. These stages are:

1. *To diagonalize the PDSM of the problem (or the adjacency matrix of the corresponding graph). This requires a diagonalization algorithm.*

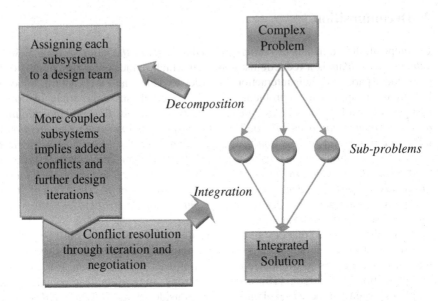

Fig. 1. More coupled sub-problems increase the number of process iterations.

2. *To cut the diagonalized PDSM from the appropriate points. For optimal or robust decomposition, cutting is usually performed to assure the minization of a partitioning quality criterion.*

In this paper we propose a spectral partitioning diagonalization algorithm along with a complexity based and graph theoretic quality partitioning criterion. This measure indicates the fact that decomposition increases the overall complexity, and whilst it is a suitable tool for large scale systems, it might not be overtly employed for highly complex problems. This result is consistent with the recent arguments from complex systems researchers (e.g. Bar-Yam [13]) who argue against the decomposition based techniques for complex systems design. Alternative and holistic techniques (such as evolutionary engineering) for the design of complex systems must be deployed. One important utility of our measure is to capture the essentiality of the paradigm shift from reductionist to holistic problem solving.

The organization of the paper is as follows. In the remaining of the *Section 2.* the spectral diagonalization technique, as well as the available graph theoretic quality partitioning criteria are explained. This section ends with a literature review of complexity measures as partitioning criterion especially in mechanical design context. First part of *Section 3.* contains the main properties that a complexity measure must possess, accompanied with a deep perspective on complexity as a phenomenon. In the next part of the *Section 3.* a graph theoretic complexity measure and its application as quality partitioning criterion (real complexity) are proposed. Also the application of real complexity for overlap decompositions is highlighted. *Section 4.* is a reflection on decomposability analysis with real complexity and some useful interpretations. *Section 5.* closes the paper by some open ended remarks on the utility of decomposition as a main tool for the management of complex and large scale systems engineering.

2.1 Spectral Diagonalization Technique

Several methods exist for diagonalization including integer programing [11], genetic algorithms [14] and spectral methods. Spectral graph theory uses the eigenvalues and eigenvectors of the adjacency and Laplacian matrices. The eigenvectors of adjacency matrix and Laplacians can be used to diagonalize the adjacency matrices of both weighted and un-weighted graphs. Consider A to be the adjacency matrix of an undirected, weighed graph (G). An automorphism of a graph G is a permutation g of the vertex set of G with the property that, for any vertices u and v, we have ug ~ vg if and only if u ~ v. "vg" is the image of the vertex v under the permutation g and (~) denotes equivalence. Automorphisms of graph G produce isomorphic graphs [15].

The first step in spectral partitioning of graphs is to sort the eigenvectors of the adjacency and Laplacian matrices in ascending order, and then to permute G by those indices of the sorted eigenvectors. Some of these permutations (by different sorted eigenvectors) are diagonalized. Initially it was thought that only the eigenvector of the second eigenvalue of the Laplacian (known corresponding as the Fiedler vector and Fiedler value) has the property of diagonalization; but later it was shown that using other eigenvectors (of both adjacency and Laplacian) can outperform the Fiedler

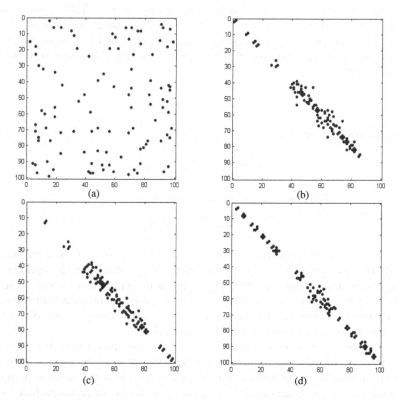

Fig. 2. Shows the adjacency matrices of a graph (a) and its spectral permutations by various eigevectors that are diagonal automorphisms of the graph.

vector in this regard, specifically in case of the weighted graphs [16]. Fig. 2.a shows the graphical representation of the adjacency matrix of an un-weighted randomly generated graph of order one hundred. Each dot points to the existence of a link between two corresponding variables. The automorphisms of this adjacency matrix are also shown which are permuted by Fiedler vector (Fig. 2.b), third eigenvector of the Laplacian (Fig. 2.c), and 98th eigenvector of adjacency matrix (Fig. 2.d).

Traditionally the diagonalization was achieved through lengthy integer programming and hefty branch and bound algorithms [11]. The spectral diagonalization technique is already exploited extensively in the context of discrete mathematics [16, 17], circuit design [18] data mining [19, 20] and image segmentation [21]. This algorithm is very fast compared to the traditional integer programming and branch and bound algorithms that were iteration based as those reported in Kusiak [11]. The spectral diagonalization has not been reported in the engineering design literature.

2.2 Partitioning Quality Criteria

After diagonalization, the cutting points must be determined. There must be a metric that enables comparison between possible and distinct decompositions of a given graph. Such metric is referred to as partitioning quality criterion. Table 2 shows some of these metrics and their characteristics. In Table 2 k denotes the number of sub-graphs and n is the cardinality (order) of the original graph, and λ_i is the i_{th} eigenvalue of the Laplacian matrix, E_h is sum of the weights of all edges that have only one end in sub-graph P_h:

$$E_h = \sum_{i \in P_h, j \notin P_h} a_{i,j}. \tag{2}$$

Also the cut size is defined as:

$$\mathbf{cut}(P_r, P_s) = \sum_{i \in P_r, j \in P_s} a_{i,j}. \tag{3}$$

And finally the total edge weights in the sub-graph P_h is:

$$E(P_h) = \sum_{i \in P_h, j \in P_h} a_{i,j}. \tag{4}$$

For a more detailed comparison in between the performance of these metrics see [18, 24]. The minimization of the quality partitioning criterion is an optimization problem and requires employing the appropriate optimization techniques. There are various spectral methods to determine the indices of cut points that can minimize different partitioning criteria however their accuracy is disputed [16, 24]. We suggest using exhaustive search algorithm after diagonalization.

To reduce the computational costs it is desirable to have an estimate of the number and order of the subsystems. As a general rule, a higher number of sub-problems is better [6]. In Section 3 a partitioning quality criterion of decompositions is presented that amongst other advantages explicitly suggests the number of partitions that should not be used.

Table 2. Several partitioning quality criteria.

Name	The measure	Proposed by	General remarks
Cut Ratio	$\dfrac{cut(P_1,P_2)}{min(\lvert P_1\rvert,\lvert P_2\rvert)}$	Spielman [17]	Lower bound = $\dfrac{\lambda_2}{2}$
Cut Ratio	$\dfrac{cut(P_1,P_2)}{\lvert P_1\rvert \times \lvert P_2\rvert}$	Cheng and Hu [22]	Lower bound = $\dfrac{\lambda_2}{n}$
Min-Max Cut Ratio	$\dfrac{cut(P_1,P_2)}{E(P_1)}+\dfrac{cut(P_1,P_2)}{E(P_2)}$	Ding et al [19]	Leads to balanced sub-graphs in terms of size
Normalized Cut Ratio	$\dfrac{cut(P_1,P_2)}{E_1+E(P_1)}+\dfrac{cut(P_1,P_2)}{E_2+E(P_2)}$	Shi and Malik [21]	Leads to balanced sub-graphs in terms of size
Min Cut	$\displaystyle\sum_{i=1}^{k} E_i$	Alpert et al [16]	Can lead to unbalance sub-graphs in terms of size
Cost	$\dfrac{\sum_{i=1}^{k} E_i}{2\sum_{i=1}^{k-1}\sum_{j=i+1}^{k}\lvert P_i\rvert \times \lvert P_j\rvert}$	Yeh et al [23]	Leads to balance sub-graphs in terms of order
Scaled Cost	$\dfrac{\sum_{i=1}^{k}\dfrac{E_i}{\lvert P_i\rvert}}{n(k-1)}$	Chan et al [18]	Lower bound = $\dfrac{\sum_{i=1}^{k}\lambda_i}{n(k-1)}$
Modality Function	$\displaystyle\sum_{i=1}^{k}\left(\dfrac{E(P_i)}{E(G)}-\left(\dfrac{E_i+E(P_i)}{E(G)}\right)^2\right)$	White and Smyth [20]	Strong cluster identification metric for very large networks. Maximizes at $k=3$.

2.3 Complexity Measure as Quality Partitioning Criterion

In engineering design, there are three elements that may be externally represented by a designer and for which complexity can be measured: the design problem, the design process, and the design artefact [2]. Within the problem domain, measuring complexity has been regarded useful because it can give a quantitative estimation of the problem solving difficulty, required problem solving effort (design effort), design lead time, cost and risk [25]. Design risk is the probability of not satisfying the functional requirements at the end of the design cycle. Measuring the complexity of a design problem allows for planning by using the results (design process or design artefact) from previous comparable complex design problems to predict necessary resources, time, or commitment required for the new design problem [2]. It is important to notice that metrics in general can be either result or predictor oriented [25]. A result metric is an observed characteristic of a completed system such as development time and design effort [25]. A predictor metric has a strong correlation to some further result, such as product complexity, design difficulty, etc., as they relate to design effort or duration [25]. Complexity measures that apply to the design problem are predictors whereas measures that apply to the design artefact are result.

Summers and Shah [2] provided an extensive literature review on the several measures of design complexity, and process complexity. Determining the complexity of a design process may be useful for selecting between two or more design processes for the same design problem. The design problem's complexity does not change, but the complexities of the available design processes may be different. The process complexity measures have applications in design automation where the machine needs to decide between several available processes. It is important to note that design problem complexity and the design process complexity are interrelated through decomposition. As such a complexity index presented by Chen and Li [1] that was associated with the process efficiency was the ratio of the complexity of the problem after decomposition to that of the problem before decomposition. Obviously the index is also a quality partitioning criterion. They claimed originality in their approach to mitigate the design process complexity.

In Chen and Li [1] the problem was modelled as the set of design components or physical constituents of a design (design variables), and the design attributes that described the behavioral properties of a design (functional response). They studied the decomposition of the incident matrix of components-attributes pair that reflects how components influence attributes. The immediate observed drawback with this approach is that it does not include the interactions of components-components or attributes-attributes. They presented the following measure for the problem complexity (before decomposition):

$$\text{Com}_o = m \times \ln(2^n).$$ (5)

Where m is the number of attributes and n is the number of components. The complexity after decomposition had two sources contributing to the total complexity of the interaction-involved matrix: the interaction part and the blocks (the resulting, subsystems, or sub-problems):

$$\text{Com} = m_a \ln(2^{na}) + \sum_{i=1}^{nb} m_i \ln(2^{nc_i}).$$ (6)

where m_a is the number of attributes present in the interaction part that is a number between 2 and m, nb the number of blocks, m_i the number of attributes inside the blocks, and nc_is the number of components inside each block. The first term in Equation (6) is the complexity of the interactions and the second term is the sum of the complexity of the blocks (Fig. 3). Although Chen and Li [1] also presented coupling as a original but separate measure of decomposition, their complexity measure is not an indicator of coupling. Obviously (5) and (6) only regard the size in measuring the complexity.

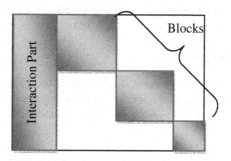

Fig. 3. Shows the block and the interaction part of a decomposition of an incident matrix. From Chen and Li [1].

Our approach to measurement of the process complexity and problem complexity have similarities to that of [1]; although with the distinction that our partitioning quality criterion is more holistic and does not sum up the complexity of the blocks and interactions to arrive at the complexity of the problem after decomposition. Our measure of complexity is a function of coupling as well as size. The importance of this last point is demonstrated in the next section. Perhaps as the consequence of their reductionist design complexity measure, Chen and Li [1] incorrectly argue that decomposition can decrease the complexity. We will show that regardless of the decomposition type, the overall complexity of the process cannot be reduced, which is utterly rational when the 'no free lunch theorem'[1] is considered.

3   Graph Theoretic Complexity Measure of Design Problem

Summers and Shah [2] posed that a design complexity measure must reflect three intrinsic characteristics and components: solvability, size, and coupling. They presented the following definitions:

1. *Solvability is whether the design artefact may be predicted to satisfy the design problem.*
2. *Size of several elemental counts including the number of design variables, functional requirements, constraints, sub-assemblies, etc.*
3. *The coupling between elements.*

They extended their definition to the design problem, the design process, and the design artefact. We believe that solvability can be represented with cyclic dependencies or cyclomatic complexity because cycles produce emergent effects that cannot be predicted: by presenting empirical results Watson and McCabe [26]

---

[1] No free lunch theorem is discussed in the context of optimization theory and states that if a search algorithm achieves superior results on some problems, it must pay with inferiority on other problems. We argue that decomposition makes a problem tractable at the price of more overall complexity.

reported that the number of errors in the implemented (software) systems has been in direct proportion with the cyclomatic complexity.

3.1 A Note on Emergence

A property of a system is emergent if and only if the property is present in global scales and cannot be traced to the local properties of parts of the system, an emergent property is thus the global effect of local interactions. Edmonds [27] defines complexity as:

> ... that property of a model which makes it difficult to formulate its overall behaviour in a given language, even when given reasonably complete information about its atomic components and their inter-relations.

This definition couples complexity with emergence. Complex systems have interconnected elements and are characterized by [28]:

1. *Circular causality, feedback loops, logical paradoxes, and strange loops*
2. *Small change in the cause implies dramatic effects*
3. *Emergence and unpredictability.*

According to Erdi [28] circular causality in essence is a sequence of causes and effects whereby the explanation of a pattern leads back to the first cause and either confirms or changes that first cause. Complexity as structure gives birth to emergent properties that are hard to predict, if not impossible. Although we may not be able to exactly describe the emergent property; we can argue about the potential existence of emergent properties. Therefore a comprehensive definition of complexity would be *the intensity of emergence.* This is a fundamental notion and is in accordance with the concept introduced by Marczyk and Deshpande [29] that complexity is a *potential* for creating top level properties and overall functionalities. For example a car is relatively more complex than a bike and it has also more functionalities. So is a human community (more complex) relative to an ant community and has relatively more functionalities. Complexity allows the potential for emergence be it desirable emergent properties (functionality) or catastrophic ones (surprise failure).

3.2 A Graph Theoretic Complexity Measure

With regards to the complexity of a system/problem and its structure being represented via a graph, there are different perspectives amongst system researchers as to what represents a system's complexity. The general belief is that the complexity can be fully represented by size, coupling and cyclic interactions. With regards to size there is clearly a sense in which people use "complexity" to indicate the number of parts [27].

Definition 1. The *size* of a system is the minimum number of variables that the system can be described with. Thus the size is the order of the system's graphical representation.

The notion of minimum size overcomes some of the inadequacies of mere size as a complexity measure [27]: it avoids the possibility of needless length.

Definition 2. Coupling (or connectivity) is the sum of the densities of dependencies between the system's variables. The coupling is in fact the size of the system's graphical representation.

The coupling of a system is a strong indicator of its decomposability: it is difficult if not impossible to decompose a system with densely interconnected elements/components without changing the overall characteristics of the system [27].

Definition 3. The number of *independent cycles* is a basic graph measure sometimes referred to as cyclomatic complexity and is the number of independent cycles or loops in a graph [26]. Number of independent cycles is easily calculated by the formula $c(G) = m - n + p$ where m is graph size, n is graph order, and p is the number of independent components determined by multiplicity of zero in eigenvalue spectrum of Laplacian matrix.

Since complex systems are characterized by circular causality, thus a graph theoretic measure of complexity must point to circularity of dependencies between the system's variables. In general there is no direct relation between the order (number of vertices) and the cyclomatic complexity: the number of vertices will limit the cyclomatic complexity but this effect is only significant with very few vertices as the number of possible edges goes up exponentially with the number of vertices [27].

Proposition 1. The *complexity* of the system S, $C(S):R^n \rightarrow R$ with n as the size of the system is the largest eigenvalue of Laplacian matrix of the system's graph.

This measure pronounces all the previously stated characteristics of complexity. On average, the measure increases with three components of complexity that are size, coupling, and cycles (Fig. 4). In Fig. 4 each dot represents a randomly generated graph. The notion of a randomly generated graph is quite different from the common term random graph. A randomly generated graph is generated by random number generators with a know probability of existence of each edge. This probability determines and is roughly proportional to the size of the graph.

The notion of emergence thus has very close relationship with the introduced measure. The emergence of connectivity in random graphs (Erdős–Rényi graphs) is usually analysed in terms of the size of the largest connected subgraph (LCS) [30]. A subgraph of a graph G is a graph whose vertex set is a subset of that of G, and whose adjacency relation is a subset of that of G restricted to this subset. A subgraph is connected if only if it cannot be any further partitioned into two or more independent components (or subgraphs with no edge in between them). It is simple to see that $C(G)$ is equal to $C(LCS)$. This property renders the measure more attractive since it illustrates the reasonable argument that *decomposition cannot reduce the overall complexity*.

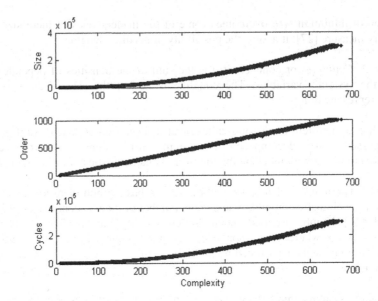

Fig. 4. Maximum eigenvalue of LCS is an ideal graph theoretic complexity measure, and on average increases with size, order and cycles.

One last point before proceeding is that central to any measurement of complexity is the employment of the entropy based mutual information as a measure of dependency between two variables [27]. The main property of mutual information is the ability to capture both linear and non-linear relationships, making it more attractive than the covariance based correlation coefficient. Equation (7) shows the entropy based mutual information between two random variables X and Y. This requires estimation of the probability distribution of every random variable ($P_X(x)$, $P_Y(y)$) in the data set and also the mutual probability distribution of all pairs of random variables ($P_{X,Y}(x,y)$).

$$I(X, Y) = \iint P_{X,Y}(x, y) \times \log \frac{P_{X,Y}(x,y)}{P_X(x) \times P_Y(y)} \, dxdy \qquad (7)$$

In the section 3.3 a complexity based partitioning quality measure is introduced that would also serve as a problem complexity measure after decomposition.

3.3 Real Complexity of Design Problems

Let S be the graphical representation of a problem with the adjacency matrix $A = [a_{i,j}]$ and its complexity $C(S)$ (the problem or self complexity) measured by the presented graph theoretic complexity measure. Consider k-partitioning $P = \{P_1,..., P_k\}$ of the graph S. Each of these sub-graphs is a block of the system. Let $C(P_i)$ be the complexity of each sub-graph determined by the complexity measure.

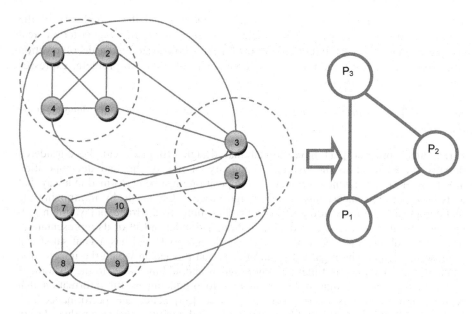

Fig. 5. The partitioning and block diagram of graphs.

Definition 4. A *block diagram* is the graph representation of partitioned graph i.e. it is more abstract than the self of the system. As such a block diagram is the graphical representation of the decomposed system (Fig. 5).

Definition 5. We define the k dimensional square matrix B as the *Complexity Based Adjacency Matrix of the Block Diagram* with the diagonal entries as the complexity of the sub-graphs (or blocks), and the off-diagonal entries as the sum of the weight of the edges that have one end in each of the two corresponding sub-graphs:

$$B = \begin{bmatrix} C(P_1) & \cdots & \sum_{i \in P_1} \sum_{j \in P_k} a_{i,j} \\ \vdots & \ddots & \vdots \\ \sum_{i \in P_1} \sum_{j \in P_k} a_{i,j} & \cdots & C(P_k) \end{bmatrix}. \tag{8}$$

Definition 6. The *real complexity* of the block diagram $C(B)$ is achieved by deriving the maximum eigenvalue of the Laplacian of B. The Laplacian of B has on its diagonal entries the sum of nodes (here subsystems) degrees and their complexities.

Other names such as *design complexity* or *process complexity* are also appropriate for this notion. By indicating the three components of complexity, this measure is a better measure of the design problem complexity after decomposition than the measure of Chen and Li [1] that was only a function of size. Real complexity $C(B)$ is a *subjective* measure of the *system's complexity* and is relative to how one might decompose the system. Conversely the self complexity $C(S)$ is an *objective* measure of the system and is *absolute* in being *independent* from the *type* of decomposition P. The purpose of decomposition is to reduce the initial problem complexity $C(S)$ to a number of sub-problems with complexity $C(P_i)$ less than self complexity $C(S)$.

The real complexity represents the *overall complexity* and the complexity of the whole system of subsystems. While being a *quality of partitioning* criterion, real complexity represents the integration effort for the whole system after decomposition since real complexity in addition to the complexity of each sub-problem, expresses the coupling of the system of sub-systems.

Remark 1. The *integration efficiency* and *risk* of a design system are dependent on *real complexity* as much as they depend on *self complexity*.

Ulrich and Eppinger [31] have argued that design efficiency can be considered directly proportional to the overall complexity of the system and decomposition affects the design efficiency. Problem decomposition must be performed in a way that adds least possible uncertainty to the design process. Obviously by minimizing the real complexity the integration effort and risk is minimized. Braha and Maimon [32] defined design *information* as a distinct notion, and independent of its representation; information allows the designer to attain the design goals and has a goal satisfying purpose. Design can be regarded as an information process in which the information of the design increases by time: making progress and adding value (to the customer), which in a product design system compares to producing useful information that reduces performance risk [33]. Browning et al [33] states the performance risk increases with product and problem complexity and complex system product design involves enormous risk. Since decomposition increases the overall complexity, it must, as a result, reduce design information and increase the design risk. Fig. 6 demonstrates that the Process (1), by employing a better decomposition, adds less risk to the design process and reduces less information from it; and thus Process (1) is likely to have higher design efficiency.

Definition 7. A decomposition with minimum real complexity is an immune decomposition.

Remark 2. Indeed robust decomposition might be a more appropriate name and while the immunity notion, here, conveys the intended meaning, it is not directly connected to the notions in Natural Immune Systems. However, we believe that extremely complex natural and organic systems are robust because they have immune systems. Thus the complex systems researchers must have a firmer eye on immunization procedures than before. It should also be noted that, contrary to the general belief, Artificial Immune Systems do not necessarily need to mimic the Natural Immune Systems.

The immune decomposition is likely to lead to better, cheaper and faster production of complex products and to enhance the design process efficiency with respect to:

1. *Better products in terms of quality and robustness:* Less complexity means lower risk or rework during the design process [33] and this implies that higher quality can be achieved with less effort. Furthermore, in terms of the robustness of the product itself, immunity from a chaotic and overly sensitive response to stochastic uncertainties in the manufacturing process capabilities and during the performance cycle of the product is the direct influence of lower complexity.

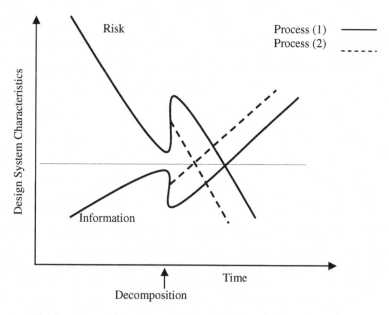

Fig. 6. Decomposition increases risk and reduces information.

2. *Cheaper product and process design costs:* The lower complexity structure implies less coupling between subsystems and that means less number of design iterations amongst various engineering tasks in a large problem [34] easier coordination, less conflicts arising in the integration of the product, and all these suggest a cheaper design process.

3. *Faster design process:* Designing a product is a value adding process, and that involves an accumulation of the required information to meet the design requirements [33]. Obviously for less complex products this accumulation happens faster because the design process system of a less complex product would be less fragile (at lower risk) to uncertainties available in the early stages of the design process.

Fig. 7 shows the immune decompositions of the system example in Table 1, for various numbers of subsystems. The figure shows that amongst all decompositions 2-partitioning has had the minimum real complexity. The spectral diagonalization method combined with random search algorithm was used. Fig. 8 compares the performance of real complexity against three other cut quality measures for *20000* randomly chosen and distinct decompositions of a randomly generated graph of order *100* and of the size *1500*. This figure shows that real complexity has responded differently to the number of subsystems than other measures. For this randomly chosen graph, the minimum real complexity is the global minimum (amongst different partition numbers) at bi-partitioning ($k=2$). The minimum real complexity maximizes at the number of subsystems equal to 4. After 4-partitioning, the minimum real complexity is inversely proportional to the number of subsystems. Other cut quality measures show strictly decreasing linear relationship between the measures of

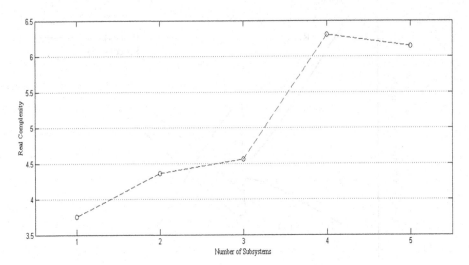

Fig. 7. Immune Decompositions of the example PDSM at various numbers of partitions.

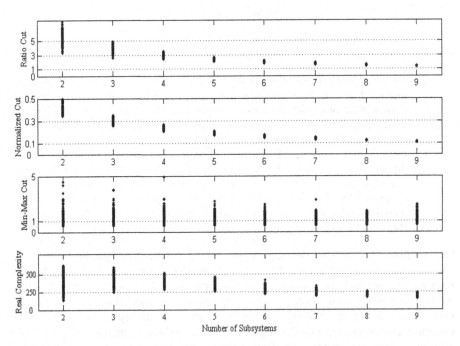

Fig. 8. Comparison between the performance of real complexity and other cut quality measures.

minimum cut and number of partitions. Another observation and interesting characteristic of real complexity is that its minimum for a particular number of subsystems maximizes at a certain number of subsystems which is different for different graphs. Amongst the quality partitioning criteria presented in Table 2 only modality function had this property which always maximizes at 3-partitioning. This sophisticated and nonlinear behavior of real complexity helps in ruling out the number of partitions that should be avoided.

3.4 Real Complexity of Overlap Decompositions

Several operation researchers have addressed the overlapping decomposition problem and have emphasized its quality improving, lead time and cost reducing benefits [35, 36] to the extent that design process efficiency can be regarded as proportional to the amount of the overlap between the subsystems [37]. As it will be clarified later in this section, this statement implies that the parts of the problem that are shared between design groups must preferably have denser connectivity (i.e. higher complexity); otherwise perfectly overlapped sub-systems lead to identical sub-problems. Taking the information processing view of the product design process, Krishnan et al. [36] argued that overlap decomposition of the problems can lead to faster information acquisition and more frequent information exchange between the subsystems enabling the concurrent execution of coupled activities in an overlapped process. However, the design teams can be collaborative only when the design space or problem space is decomposed in an overlapped manner, so that the design teams share some of their parameters, problems, and tasks. Krishnan et al. [36] however noted that the overlap decomposition of the system must be performed very carefully as without careful management of the overlapped product design process, the development effort and cost may increase, and product quality may worsen. It should be noted that no appropriate measure has yet been proposed to distinguish the overall behaviour of the systems under a range of possible overlap decompositions.

An overall complexity measure for overlap decomposition can be readily obtained by exploiting the real complexity. In overlap decomposition of graphs, vertices are allowed to be shared between the sub-graphs. The measurement of the overall real complexity of the overlap decompositions can be gained based on the formulation of the decentralized control strategies for overlapping information sets [38]. Ikeda et al [38] stated that:

> *The simple underlying idea is to expand the state space of the original system* (design space in case of the product design) *so that the overlapping subsystems appear as disjoint. The expanded system contains all the necessary information about the behaviour of the original system which then can be extracted using conventional techniques devised for standard disjoint decompositions.*

Fig. 9 shows the extraction of the Complexity Based Adjacency Matrix of Block Diagram with overlapped subsystems. It can be tested that the dimension of this matrix is four whereas the number of subsystems are two.

In general the extended Complexity Based Adjacency Matrix of Block Diagram with elements $B=[b_{ij}]$ where sub-graphs share some vertices can be defined as:

$$b_{i,j} = \begin{cases} C_i & i = j, \text{ for nonoverlapping blocks} \\ C_{lap} & i = j, \text{ for overlapping blocks} \\ 0 & i \neq j, \text{ for overlapping blocks} \\ \sum_{k\in P_i} \sum_{l\in P_j} a_{k,l} & i \neq j, \text{ for nonoverlapping blocks} \end{cases} \quad (9)$$

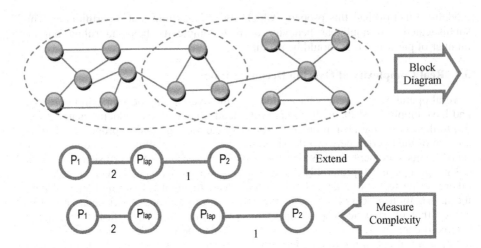

Fig. 9. Real structural complexity measurement for overlapping subsystems. All link weights are equal to *1*.

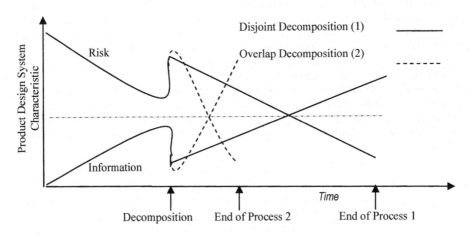

Fig. 10. Decomposition increases risk and reduces information. Overlap decomposition makes the system to converge faster.

The effect of overlapping decomposition on the design efficiency is a very subtle one. The integration phase of the design process is often accompanied by inadvertent information hiding due to the asynchronous information exchanges between the design teams, referred to as *design churn effect* [39]. Design churn delays the design process convergence to a global solution. The remedy to this effect lies in overlapping the design tasks. Overlapping leads to faster and *in time* information transfer between the design teams. This is to say that overlapping can increase the design process response and sensitivity to new information reducing design lead time and increasing design process efficiency (Fig.10).

Overlapping leads to more real complexity in comparison with the disjoint decomposition of the design space, since overlapping virtually increases the dimensionality of the problem space. Thus, it is not recommended to simply overlap the subsystems as much as possible because it may lead to high overall complexity. We propose two seemingly conflicting objectives when overlapping subsystems:

1. *To minimize the real complexity of the whole (extracting immune decompositions).*
2. *To maximize the sum complexity of the overlapped parts. The complexity sum of the overlapped parts (C_{lap}) is representative of how much the system is overlapped.*

Fig. 11 demonstrates the real complexity and the overlapping degree (complexity sum of the overlapping parts) of many random decompositions of the example PDSM in Table 1. This figure shows the desired and compromised region for the decompositions, the characteristic of which is minimum overall real complexity (maximum efficiency, minimum fragility of the corresponding design process or immunity) and maximum overlapping complexity (high sensitivity of the design process to new information).

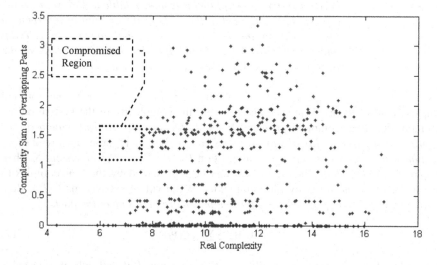

Fig. 11. Extracting the desired overlap decompositions of the example PDSM.

4 On Decomposability

Edmonds [27] provided a comprehensive literature review on the general area of decomposability. The complexity based approach to decomposability has also been considered in [27]. Edmonds used the analytical complexity measure to determine the decomposability of the syntactical expressions in formal languages. The ease with which a system can be decomposed into sub-systems has close connections with the real complexity which provides a medium for decomposability testing. A system that is not decomposable is irreducible. Before proceeding lets introduce the lower bound for the real complexity that will be used in the subsequent discussion.

Observation 1: Given system S with graph G_S as its graphical representation and for all k-partitioning $P=\{P_1,P_2,...,P_k\}$ of S with $i=\{1...k\}$, the following is valid:

$$C\,(B) \geq C\,(S) \geq C\,(P_i). \qquad (10)$$

B is the Complexity Based Adjacency Matrix of the Block Diagram of decomposition P on G_S. Equation (10) means that the lower bounds for real complexity are the self complexity, and also the complexity of each of the sub-systems. The equality happens when the system can be fully decomposed. This observation is important since it indicates that decomposition cannot decrease the overall complexity of a problem, and the perceived complexity after decomposition: $C(B)$ is never less than the complexity before decomposition $C(S)$. Note that in the block diagram the information that indicates which nodes in different subsystems have been linked is lost. Similarly when a system is decomposed the information indicating which vertices are linked to which ones in other subsystems is also lost. Klir [40] states:

> *When a system is simplified it is unavoidable to lose some of the information contained in the system; the amount of information lost results in the increase of an equal amount of relevant uncertainty. Any kind of simplification including break of the overall system to subsystems can increase uncertainty.*

More uncertainty implies more complexity and thus the lower bound for real complexity *must* be and cannot be anything but the complexity of the system before decomposition (self complexity); in other words *facing more overall complexity as the price for the tractability of sub-problems* can, in fact, be regarded as a read of *no free lunch theorem*. Considering decomposition $P=\{P_1, P_2,...,P_k\}$ of system S. Then complexity of *the whole* for the system can be represented by the real complexity $C(B)$ and that of *the parts* by the complexity of individual subsystems $C(P_i)$. Real complexity can then explain whether *the whole is more than sum of the parts*:

$$C(B) \geq \sum_{i=1}^{k} C(P_i). \qquad (11)$$

Proposition 2. A system is *irreducible by decomposition P* if and only if (11) holds for that system. If (11) holds for every possible P, then the system cannot be reduced to the sum of its constituents and therefore is *irreducible*.

It should be reminded that the graph theoretic complexity measure is a measure of the *intensity of emergence* and possibility of *emergent surprise* (characteristics). Therefore, the integration of a system, being decomposed in a way that (11) holds, is prone to a significant amount of risk; it is very likely that the whole system of sub-systems shows emergent properties that do not exist in the sub-problems. The integration, in such conditions cannot be implicitly performed while the sub-problems are being solved concurrently because as much attention must be given to communication between the design teams. Where all viable decompositions of a system have the property of (11), then decomposition is not a valid and robust methodology for problem solving (however it may not be impossible). In such cases the whole is more than sum of the parts (regardless of what is deemed as parts), and distributed problem solving must be forsaken; the problem may be tackled satisfactorily as a whole utilizing several design teams working and competing parallel to each other (with no collaboration).

When the opposite of (11) is true and a decomposition P can be found in a way that the real complexity is less than the sum of the subsystems complexities then the system is reducible. Under such circumstances, the complexity of the whole *can be* reduced to the complexity of the individual components. The main result from here is that the reducibility depends not only on the self complexity but on the decomposition restrictions namely the partition size and order. Fig. 12 compares the whole and the sum of the parts for many decompositions of the example PDSM. This system was decomposed in *200* different ways all of which had relatively balanced subsystems (equally ordered). The system was reducible for only a few of the viable decompositions (i.e. the system is nearly irreducible). The systems that are contrary to this (irreducible under only a few decomposition) can be regarded as the nearly decomposable (reducible) systems. For strictly irreducible systems decomposition does not lead to higher tractability. Decomposition for such systems can further complicate the process of collaboration between design teams: the overall effort of integrating the system of subsystems amounts to more than sum of the efforts spent on integrating each subsystem. This notion indicates substantial amount of rework

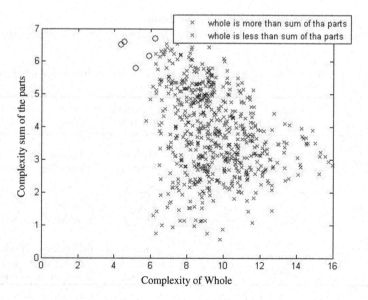

Fig. 12. Using real complexity to test decomposability.

and design iteration in the design of complex systems by means of decomposition. As such the application of concurrent engineering for some highly complex systems can be questioned.

5 Concluding Remarks

The PDSM can be decomposed in different modes as demonstrated in Table 3, which can, for example, be decided upon based on PDSM connectivity level.

Definition 8. Problem connectivity is the total number of edges in the self map of the product/problem divided by the total number of possible edges (which is a PDSM's order function).

The definition introduced for connectivity in section 3.2 basically implies the same notion. The importance of choosing the right decomposition pattern is critical for the efficiency of the design process and the successful management of the integration process. Obviously choosing an integration scheme is critical in determining how efficient or how flexible a resulting problem solving architecture will turn out to be [4]; and depending on the flexibility of design organization, the integration schemes can be determined by the decomposition patterns (see Table 3). This is so since the information exchange and interactions between the design teams are more or less determined by the problem structure.

Bar Yam [13] stressed that there are two effective strategies for achieving quality criteria in the design of large scale systems:

1) *restricting the complexity of the problems.*
2) *adopting an evolutionary paradigm for complex systems engineering.*

Table 3. Decomposition modes and their corresponding integration schemes.

Connectivity→ Suggestion↓	Very Low (0-0.02)	Low (0.02-0.1)	Intermediate (0.1-0.2)	High (0.2-0.3)	Very High (0.3-1)
Possible or best decomposition strategy	Full decomposition	Integrative clustering	Modular clustering	Overlap Clustering	No decomposition
Illustration					
Corresponding integration scheme	Concurrent Engineering	Low level integration scheme	Unsupervised multi agent system	Information intensive architecture	Evolutionary architeture (Holistic)

The holistic approach to design relies on radically innovative solutions provided by several design groups working parallel to each other on the same problem [13]. The evolutionary process advocates competition (rather than collaboration) in between the design groups for achieving higher fitness of their design solutions. The integration scheme for routine/traditional design processes, on the other hand, uses decomposition and collaboration at the design organization level. The routine processes may have a *high level integration team* to centrally supervise the problem solving process and through that seek the overall integration of the solutions. Another routine process employs *low level integration team:* this is to utilize coordination based decompositions that aim at decomposing the system into several relatively independent subsystems and only one severely connected subsystem namely the coordination block. The design team in charge for design of integrative subsystem is then regarded as low level integration team. Since the design of the integrative subsystem must be much more complex than the other subsystems, the integration of the complex systems with more than a certain amount of coupling is not desirable by coordination based decomposition. *Multi agent design systems* are in essence unsupervised problem solving systems. The interaction between the design teams in these systems are autonomous and based on agents social knowledge. The solution to the design problems in multi agent systems is formed in a *self organizing* fashion and as the emergent result of autonomous interaction of the agents; multi agent systems respond to the modular problem decomposition where the connectivity distribution in between the subsystems is relatively uniform. For large scale and severely coupled problems collaborative problem solving is possible when the design space or problem space is decomposed in an overlapped manner: the design teams explicitly share some of their parameters, problems, and tasks. The main characteristic of this process model is the intense collaboration between coalitions of agents making this mode an information and knowledge intensive process. The impact of new information on the design process in this integration scheme is relatively high; as such overlap decomposition and its corresponding integration scheme are suitable for problems of high complexity and self connectivity.

The fixed organization structures are sometimes hinged upon to determine the number of subsystems in decomposition process. The computational resources and the number of available design teams, their size and the hierarchies in the organization, only reflect some aspects of the organizational structure. It is sometimes feasible to form the teams after decomposition takes place. In any case, the subjective measure of decompositions as real complexity can be employed for all the decomposition modes and even comparison between them.

5.1 Summary and Future Work

A measure of complexity for decomposition was introduced which was based on an spectral measure of LCS. The minimum bound for this measure is the self complexity (complexity before decomposition). The minimization of the real complexity as a quality partitioning criterion does not necessarily lead to balanced subsystems which may be regarded as a disadvantage. Regardless of this, real complexity for a various number of subsystems can be compared and the optimal decomposition can be chosen, whether the number of subsystems is known or not. Depending on the complexity of the problem, real complexity can be more than the sum of the complexities of the sub-problems. In this case decomposition cannot reduce the problem into the sum of its parts and the problem is strictly irreducible alternative problem solving techniques such as evolutionary design process may be used.

There are two distinct directions for this research:

1) To further investigate the emergence and some characterizations of emergent properties, by real complexity. This needs to be a rigorous mathematical endeavor.

2) To utilize the presented knowledge in design decision support systems. This trend has already been to some degrees pursued by the authors. However, in our opinion, the most interesting application of this (presented) type of design domain knowledge should be sought in design automation systems.

3) The measure was only demonstrated for two level hierarchies; however the real complexity can be applied to multiple hierarchal levels.

References

1. Chen, L., Li, S.: Analysis of Decomposability and Complexity for Design Problems in the Context of Decomposition. Journal of Mechanical Design 127, 545–557 (2005)
2. Summers, J.D., Shah, J.: Developing Measures of Complexity for Engineering Design. In: ASME Design Engineering Technical Conferences DETC 2003, Chicago, Illinois (2003)
3. Zdrahal, Z., Motta, E.: Case-Based Problem Solving Methods for Parametric Design Tasks. In: Smith, I., Faltings, B.V. (eds.) EWCBR 1996. LNCS, vol. 1168, pp. 473–486. Springer, Heidelberg (1996)
4. Prasad, B.: Concurrent Engineering Fundamentals: Integrated Product and Process Organisation. Prentice Hall, New Jersey (1996)
5. Browning, T.R.: Applying the Design Structure Matrix to System Decomposition and Integration Problems: A Review and New Directions. IEEE Transactions on Engineering Management 48 (2001)
6. Michelena, N.F., Papalambros, P.Y.: A Hypergraph Framework for Optimal Model-Based Decomposition of Design Problems. Computational Optimization and Applications 8, 173–196 (1997)
7. Pimmler, T.U., Eppinger, S.D.: Integration Analysis of Product Decompositions. In: ASME Design Theory and Methodology Conference, Minneapolis, MN (1994)
8. Alexander, C.: Notes on the Synthesis of Form. Harvard University Press, Cambridge (1964)
9. Browning, T.: Designing System Development Projects for Organizational Integration. Systems Engineering 2, 217–225 (1999)
10. Simon, H.: Sciences of the Artificial. M.I.T. Press, Cambridge (1969)
11. Kusiak, A.: Engineering Design: Products, Processes, and Systems. Academic Press, London (1999)
12. Chen, L., Ding, Z., Simon, L.: A Formal Two-Phase Method for Decomposition of Complex Design Problems. Journal of Mechanical Design 127, 184–195 (2005)
13. Bar-Yam, Y.: When Systems Engineering Fails – Toward Complex Systems Engineering. In: International Conference on Systems, Man & Cybernetics, vol. 2, pp. 2021–2028. IEEE Press, Piscataway (2003)

14. Altus, S., Kroo, I.M., Gage, P.J.: A Genetic Algorithm for Scheduling and Decompostion of Multidisciplinary Design Problems. Journal of Mechanical Design 118, 486–489 (2003)
15. Cameron, P.J.: Automorphisms of Graphs. In: Beineke, L.W., Wilson, R.J. (eds.) Topics in Algebraic Graph Theory, pp. 137–155. Cambridge University Press, Cambridge (2004)
16. Alpert, C.J., Kahngb, A.B., Yao, S.-Z.: Spectral partitioning with multiple eigenvectors. Discrete Applied Mathematics 90, 3–26 (1999)
17. Spielman, D.A., Teng, S.-H.: Spectral Partitioning Works: Planar Graphs and Finite Element Meshes. In: IEEE Symposium on Foundations of Computer Science (1996)
18. Chan, P., Schlag, M., Zien, J.: Spectral K-Way Ratio-Cut Partitioning and Clustering. IEEE Transactions on Computer-Aided Design of Integrated Circuits and Systems 13, 1088–1096 (1994)
19. Ding, C., He, X., Zha, H.: A Min-Max Cut Algorithm for Graph Partitioning and Data Clustering. In: IEEE International Conference on Data Mining, San Jose, CA (2001)
20. White, S., Smyth, P.: A spectral Clustering Approach to Finding Communities in Graphs. In: 5th SIAM International Conference on Data Mining. Society for Industrial and Applied Mathematics, Newport Beach, CA (2005)
21. Shi, J., Malik, J.: Normalized Cuts and Image Segmentation. IEEE Transactions on Pattern Analysis and Machine Intelligence 22, 888–905 (2000)
22. Cheng, C., Hu, T.: The Optimal Partitioning of Networks: Technical Report. University of California, San Diego (1989)
23. Yeh, C., Cheng, C., Lin, T.: A Probabilistic Multicommodity-Flow Solution to Circuit Clustering Problems. In: IEEE Int. Conf. Computer-Aided Design ICCAD 1992, Santa Clara, CA (1992)
24. Verma, D., Meila, M.: A Comparison of Spectral Clustering Algorithms. Technical Report. University of Washington (2003)
25. Bashir, H., Thomson, V.: Metrics for Design Projects: A Review. Design Studies 20, 263–277 (1999)
26. McCabe, T.: A Complexity Measure. IEEE Transactions on Software Engineering 2, 308–320 (1976)
27. Edmonds, B.: Syntactic Measures of Complexity. PhD. University of Manchester, Manchester, UK (1999)
28. Erdi, P.: Complexity Explained. Springer, Heidelberg (2008)
29. Marczyk, J., Deshpande, B.: Measuring and Tracking Complexity in Science. In: 6th International Conference on Complex Systems, Boston, MA (2006)
30. Boschetti, F., Prokopenko, M., Macreadie, I., et al.: Defining and Detecting Emergence in Complex Networks. In: 9th International Conference on Knowledge-Based Intelligent Information and Engineering Systems, Melbourne, Australia, pp. 573–580 (2005)
31. Ulrich, K.T., Eppinger, S.D.: Product Design and Development. Mc Graw-Hill/Irwin (2004)
32. Braha, D., Maimon, O.: The Measurement of a Design Structural and Functional Complexity. IEEE Transactions on Systems, Man, and Cybernetics-Part A 28, 527–535 (1998)
33. Browning, T.R., Deyst, J.J., Eppinger, S.D.: Adding Value in product Developement by Creating Information and Reducing Risk. IEEE Transactions on Engineering Management 49, 443–458 (2002)
34. Smith, R.P., Eppinger, S.D.: Identifying Controlling Features of Design Iteration. Managment Science 43, 276–293 (1997)
35. Roemer, T., Ahmadi, R.: Concurrent Crashing and Overlapping in Product Development. Operations Research 52, 606–622 (2004)

36. Krishnan, V., Eppinger, S., Whitney, D.: A Model-Based Framework to Overlap Product Development Activities. Management Science 43, 437–451 (1997)
37. Clark, K.B., Fujimoto, T.: Overlapping Problem Solving in Product Development, pp. 127–152. Managing International Manufacturing, Amsterdam (1989)
38. Ikeda, M., Siljak, D.D., White, D.E.: Decentralized Control with -Overlapping Information Sets. Journal of Optimization Theory and Applications 34, 279–310 (1981)
39. Eppinger, S.D., Whitney, D., Yassine, A., et al.: Information Hiding in Product Development: The Design Churn Effect. Research in Engineering Design 14, 145–161 (2003)
40. Klir, G.J. (ed.): Facets of Generalized Uncertainty-Based Information. Springer, Heidelberg (2003)

Towards a Formal Model of Knowledge Sharing in Complex Systems

Nadim Obeid[1] and Asma Moubaiddin[2]

[1] Department of Computer Information Systems,
King Abdullah II School for Information Technology,
The University of Jordan
obein@ju.edu.jo
[2] Department of Linguistics,
Faculty of Foreign Languages,
The University of Jordan

Abstract. Knowledge sharing between various components of a system is a prerequisite for a successful knowledge management system. A knowledge sharing model includes providing knowledge workers with the knowledge, experiences and insights which they need to perform their tasks. We propose a multi-agent system that assists in the process of knowledge sharing between concerned knowledge worker groups. Each agent is a knowledge broker and organizer for a specialized knowledge worker group involved in the operations of a sub-system. In addition to timely access to knowledge, it should help in understanding the motivation which underlies decisions made by other groups and/or the information/knowledge bases for such decisions. Each agent is expected to learn about the activities, contexts of decisions, knowledge employed and experiences of other knowledge workers groups whose activities are considered to be relevant to the group it represents. We shall employ Partial Information State (PIS) to represent the knowledge of the agents. We shall employ a three-valued based nonmonotonic logic for reasoning about PISs. We present a multi-agent based model of argumentation and dialogue for knowledge sharing.

Keywords: Knowledge Management, Knowledge Sharing, Multi-Agent Systems, Three-Valued Logic, Nonmonotonic Reasoning, Dialogue, Argumentation.

1 Introduction

Knowledge Management (KM) allows knowledge workers and decision makers to learn what they need, when they need it and to improve organizational performance. It should provides relevant knowledge to assist knowledge workers in performing knowledge intensive tasks. Knowledge is considered as a source of competitive advantage for an organization. It can be regarded as a integrated array of processes that capture, store, and disseminate knowledge in an organization. The term organizational knowledge refers to more than just databases and information systems involved in the organization. It also encompasses processes and people. Knowledge is intrinsic to

E. Szczerbicki & N.T. Nguyen (Eds.): Smart Infor. & Knowledge Management, SCI 260, pp. 53–82.
springerlink.com   © Springer-Verlag Berlin Heidelberg 2010

people. KM takes place when there are mechanisms that structure the entire knowledge and build the corporate memory. KM is associated with constructs such as: knowledge acquisition, information distribution, information interpretation. Extending research prototypes to real-world solutions may requires an integration of some basic technologies such as ontologies for Knowledge Sharing (KS) and reuse, and collaboration support.

The notion of knowledge is a critical organizational asset integrated within the organizational learning process. It is widely agreed that knowledge is an essential imperative of an organization regarding growth, development, goals achievement and having a competitive advantage [46]. Knowledge is essential for effective decision making where there is a need for full understanding of the related domain(s). Knowledge is classified into: (a) explicit knowledge [1] that can be seen, shared and communicated with others and (b) tacit knowledge [17] which is embedded in a person's memory and thus, difficult to extract and share with others such as when a senior manager uses a particular decision theory to solve certain problems.

The study of "Organisational Knowledge" seeks to understand how knowledge is created, stored, accessed, transformed (e.g. from tacit to explicit and vice versa), handled and used effectively. Knowledge tasks have a collaborative feature in the sense that an agent/process can attain and use knowledge, which it needs, by reusing information already collected and interpreted by other agents/processes [15]. Thus, the efficient and effective handling and communication of knowledge, experiences and insights between Knowledge Workers (KW), such as domain experts and decision makers is a requirement for high-quality decision-making and coordinated organizational action [73] in organizations that deal with complex issues.

A KM system must be able to adapt to changes in the environment, to the different needs and preferences of individual agents and work groups, and to integrate naturally with existing methods and processes. That is, KM systems must be both reactive (e.g., able to respond requests or environment changes) and proactive (e.g., able to take initiatives to attend to agents' needs).

We make the assumptions that deep knowledge based reasoning and analytical capability are necessary to effectively deal with various issues related to complex systems operations that transcend the boundary of a single component/unit of the system. A KM system should involve providing knowledge workers with the knowledge, experiences and insights which they need to perform their tasks. These features characterize intelligent agents, what seems to indicate the applicability of agent technology in the KM area. With agents, it is possible to capture the representation, coordination, and cooperation between heterogeneous processes and other agents.

In this paper, we aim to emphasize the role of agents in the development of a collaborative KS system. We shall employ Partial Information State (PIS) to represent the knowledge of the agents. We employ a three-valued based nonmonotonic logic for reasoning about PISs. We present a multi-agent based model of argumentation and dialogue for knowledge sharing. The paper is organized as follows: In section 2 we discuss the basic concepts of data, information and knowledge. In Section 3 we provide some background on KS and communication. In section 4, we present multi-agent systems. In sections 5 and 6, we discuss the role of agents in KM and the need for knowledge sharing in KM. Section 7 is dedicated for reasoning with incomplete information, context and dialogue. In section 8, we present a multi-agent based model of argumentation and dialogue for KS.

2 Data, Information and Knowledge

To grasp the role of communication in KS, there is a need to understand the relationships that exist among data, information and knowledge. Data can be defined as a "set of discrete , objective facts and event ... and is most usefully described as structured records of transaction" [18]. Information represents an understanding of the relationships between pieces of data, or between pieces of data and other information. Information relates to different forms of descriptions and definitions. Information is used to make decisions. For Data to be transformed into information, some processing must be performed while considering the context of a decision. We often have abundance of data but lack relevant information. Information can be considered as knowledge that can be communicated or conveyed.

Unlike information, *knowledge* is less tangible. It is inferred from information (e.g., by humans) using methods of comparison, consequences, connections, and conversation. According to [74]:

> "Knowledge is a fluid mix of framed experience, values, contextual information, and expert insights that provides a framework for evaluating and incorporating new experiences and Information. It originates and is embedded in the minds of knowers. In organizations, it often becomes embedded not only in documents or repositories but also in organizational routines, processes, practices, and norms."

Knowledge represents a self-contextualizing pattern relation between pieces of information and/or data. It may inform us of how the pattern it represents will evolve. It encompasses what we may call strategies, practices, methods and principles. Knowledge can be defined as the application of the relevant information to address specific problems.

There are several types of *knowledge* and some of these types are difficult to distinguish from *information*. To appreciate the degree of overlap, we only need to consider the scope of the notion of an information system that spans database systems and intelligent knowledge base systems. Knowledge can inform us of how the pattern it represents will evolve. Explicit knowledge can be formally articulated, although removed from the original context of creation or use [79]. The key aspect of explicit knowledge is that expressed and shared. Tacit knowledge is rich in context and difficult to articulate. It is acquired through experience and action and can be shared through interactive conversation. Context and experience are central to the concept of tacit knowledge. Therefore, sharing tacit knowledge is difficult due to context-dependency. There is a third type of knowledge, implicit knowledge, which may be inferred from the behaviors of agents or individuals and in such cases KS may be performed through "learning by doing" [8].

It is clear that the notion of *knowledge* encompasses *information* and *data*. It also seems incomplete to organize and communicate knowledge without taking into consideration the possible ways it can be manipulated in order to satisfy a goal. Thus, KS cannot disregard embedded intelligence which is associated with the principles of generating, manipulating and applying knowledge. Furthermore, it cannot dispense with data and information because they form the basic constituents of knowledge. Other definitions of *data*, *information* and *knowledge* are found in [13].

3 Knowledge Sharing and Communication

AI researchers have recognized that capturing knowledge is the key to building large and powerful AI systems. On the other hand, it was realized that the issue of knowledge representation could be difficult and time consuming. For instance, it is straightforward to exploit a previously developed Knowledge Base (KB) in order to develop a closely related AI system as one needs to understand from the KB's developer many of the decisions she/he has made and the conventions that she/he has employed. Even if the documentation clarifies many of these decisions and conventions, there are always implicit conventions which are not documented but must be observed when trying to make use of a previous KB.

The prospect of building larger systems may be hampered by the high cost of knowledge transfer/sharing. Hence, it was necessary to find ways to overcome this barrier and to address some important problems such as:

(1) Heterogeneous Representation Languages: Allowing for diversity in representing the same situation is one of the keys to success in properly managing and making an optimal use of the knowledge available. The diversity of representations allows for a deeper and a better understanding of the different patterns and characteristics of a situation, and, if properly handled, supports cooperative work. We may distinguish between static (possibly timeless) representations of knowledge, which are particularly useful for knowledge re-use and dynamic representations of knowledge that are needed for knowledge creation. In the presence of static diverse knowledge representation formalisms, sharing and reusing knowledge may involve translating from one representation to another. The dynamic views are more challenging. They are based upon ongoing re-interpretation of data, information and assumptions while pro-actively deciding how the decision-making process should be adjusted to deal with future possibilities. They also allow for diversity of interpretations of the same information across different contexts and at different times.

(2) Common Ontology: Even if the KR formalism-level problems are resolved, it can still be difficult to combine the knowledge of two process/agents or establish effective communications between them unless they share a common ontology. For a shared ontology to be useful, the definitions provided must include declarative constraints that specify the semantics of the terms being defined, and the ontology must provide mechanisms that impose those constraints when the terms are used in an application.

Individual workers are the basic entities for knowledge creation in organizations [53]. The process of knowledge creation and transfer is captured in a model where tacit knowledge is transferred to explicit and then back to tacit knowledge in a dynamic and on-going process depicted as an expanding spiral. An alternative model, suggested by [36], is based on communication practices in communities and therefore is highly context dependent where communication represents a continuing process by the participants. Thus, communication is an integral part of KS. Therefore, understanding how information is exchanged through communicative process and transformed into knowledge in organizations and in the task-directed teams may help in developing approaches for its more effective use and potential benefits.

A proper handling of (1) and (2) enhance the usability of MAS in addressing many real life problems: agents' knowledge could easily be shared and agents with specialized reasoning capabilities that embody problem solving methods could easily be developed. A new agent should have the ability to interoperate with existing systems and communicate with them in order to perform some of their reasoning tasks. In other words, declarative knowledge, problem solving techniques, and reasoning services could all be shared among agents.

4 Multi-Agent Systems

Agents are specialized problem solving entities with well-defined boundaries with the ability to communicate with other agents. They are designed to fulfill a specific purpose and exhibit flexible and pro-active behaviour. They are autonomous in the sense that they can operate on their own without the need for guidance. They have control both over their internal state and over their actions. An agent can decide on its own whether to perform a requested action. They are capable of exhibiting flexible problem solving behavior in pursuit of their objectives which they are designed to fulfill.

Autonomous agents have the possibility to interact with other agents using a specific communication language which enable them to be involved in some larger social networks, respond to changes and/or achieve goals by simply with the help of other agents.

A Multi-Agent System (MAS) can be defined as: a collection of agents with their own problem solving capabilities and which are able to interact among them in order to reach an overall goal [27].

Agents usually operate in a dynamic, non-deterministic complex environment. In MAS environments, there is no assumption regarding global control, data centralization and synchronization. Thus, agents in a MAS are assumed to operate with incomplete information or capabilities for solving the problem. Communication is then the key for agents to share the information they collect, to co-ordinate their actions and to increase interoperation.

The interactions between the agents can be requests for information, particular services or an action to be performed by other agents, issues that concern cooperation, coordination and/or negotiation in order to arrange interdependent activities.

Agent will need to interact with one another, either to achieve their individual objectives or to manage the dependencies that result from being situated in a common environment [15, 37]. However, such interaction is conceptualized as taking place at the *knowledge level* [52]. That is, it can be conceived in terms of which goals should be followed, at what time, and by which agent. Furthermore, as agents are flexible problem solvers that have only a partial control and partial knowledge of the environment in which they operate, interaction has to be handled in a flexible manner and agents need to be able to make run-time decisions regarding whether or not to initiate interactions and regarding the nature and scope of these interactions.

Organizations are considered to be real world complex problems that involve lots of interactions. An essential part of a design process is finding the right models for viewing the problem. There are many well known techniques and tools for tackling complex, real-world problems in software. [12] identifies three such tools:

(1) Decomposition: The idea is to divide large problems is to divide them into smaller, more manageable modules each of which can then be dealt with in relative isolation.
(2) Abstraction: This process allows the designer to work with a simplified model of the system that emphasizes the more relevant properties, while ignoring the less relevant details.
(3) Organization: In this process the interrelationships between the various problem components are identified. Some of the basic components, that provide a particular functionality, may be grouped together and considered as one unit at a higher level.

One natural way of viewing complex systems is as subsystems, subsystem components, interactions and organizational relationships. Agent organizations are very much like complex systems in that they involve a number of constituent components that act and interact according to their role within the larger enterprise. The interplay between the subsystems and between their constituent components can naturally be viewed in terms of high level social interactions.

Complex systems require collections of components to be treated as a single entity when viewed from a different level of abstraction and are dynamic in terms of relationships between their various components. MAS have been proposed as a suitable model for engineering complex, distributed systems [38] because:

(1) They can naturally represent the decentralized nature of such problems
(2) They allow for decentralized control.
(3) They allow for multiple perspectives on a situation.
(4) They provide an adequate and flexible mechanism for interaction between the various computational entities.
(5) They allow us to explicitly represent the system's inherent organizational structure

In organizations, agents are expected to behave so that they achieve the objectives of the work groups with which they are associated or as part of a more general problem solving initiative. Interaction between agents take place in a context that defines the nature of the relationship between them [27, 30] and reflect their functionality. Thus, such social/functional role has to be specified explicitly. Since such relationships may evolve with time, it is important to take into consideration the temporal aspect.

Due to the dynamic and unpredictable nature of the environment, in which organizations operate, it is not possible to give, at design time, a complete specification of all the activities that need to be performed in an organization. It is also difficult to predict how these activities must be ordered or what knowledge they may need. Decentralization of the tasks and information and flexibility seem to be a reasonable way to keep control within large complex systems and organizations. Different work groups may be allowed relative autonomy, in the sense that they could control how their activities are handled. This may be necessary in cases for physically distributed organizations where attempts to maximize its capabilities and competence in line with the overall organizational goal(s). These characteristics require an environment that integrates the different knowledge groups, though heterogeneous, of the organization with active support of accessing and exchanging the information, expertise and knowledge they need. At knowledge level, integration must be based on the semantics and the context of the problem at hand. A knowledge-level integration framework must be able to create dynamic relationships between knowledge workers that do not compromise the autonomy of any entity. Furthermore, there may be a need to deal with various heterogeneous knowledge representations. To be able to support the collaboration of the various knowledge groups, there is a need for ontology; a common knowledge description.

In this context, a KM system has to provide uniform access to a diversity of knowledge and information sources and the environments must be both *reactive* and *proactive* to adapt the different needs.

5 Agents in Knowledge Management

KM environments can be described as distributed systems where different autonomous agents act on behalf of a knowledge group/user while pursuing their own goals. In such environments, uniform access to a diversity of knowledge and information sources and the environments must be both *reactive* and *proactive* to adapt the different needs and the ability to communicate and negotiate is essential. MAS is suitable for KM for many reasons. Among these are:

(1) KM domains are inherently distributed regarding the available knowledge and problem solving capabilities and responsibilities.
(2) The structure of the organization and the relative autonomy of its components need to be preserved.
(3) Interactions in KM environments are fairly sophisticated, including negotiation, information sharing, and coordination.
(4) In KM domains, agents can act as mediators within and between the different knowledge groups.
(5) In KM domains, problem solvers are expected to appropriately react to changes in the environment, handle unpredictability and to proactively take initiatives when the need arises.

To cope with such an inherent complex process, we propose a multi-agent system that assists in capturing the representation, coordination, collaboration and KS between heterogeneous processes and other agents.

In KM environments, agents are expected to monitor the changing conditions of the environment, to decide on what is needed using their reasoning ability to draw inferences, to be involved in solving problems and in determine actions, and finally, act accordingly.

The employment of MAS serves two roles: (1) agents can model the organizational environment where the KM system has to operate and (2) agents can be used to implement the functionality of KM systems.

6 Need for Knowledge Sharing in KM

Organizations rely on many kinds of work groups to develop products, improve services, and manage operations. For these groups to be effective, structures and processes must be in place to encourage them working together [16]. Numerous studies have demonstrated benefits for work groups that engage in information exchange and task-related communication within the group [3, 75].

Previous research has shown that KS with other groups is positively related to performance [5]. It is increasingly clear that knowledge transfer, both within and outside of groups, plays a fundamental role in the effectiveness of organizations [6, 7].

The knowledge necessary for high performance in work groups can be tacit [62], codified [80], or embodied in routines [51]. KS is defined here as the process that involves sending and/or receiving information or know-how concerning a task or an issue of concern. In addition to communication and the exchange of information, KS includes the implicit coordination of expertise [24] and information about who knows what in the group [70]. Since different agents and/or agent groups may know different things, exchange of information may help in eliciting new ideas and ideas and insights from different sources of knowledge [34, 35].

The structural diversity in agent groups such as diversity in location, expertise, role and functional responsibility of the agent could play a positive role in the value of information exchange and collaboration. For instance, agent groups members in different locations also likely to need more to share knowledge and to communicate in order to handle a task. KS in the presence of diverse expertise is essential for the integration of expertise. Advantages are realized in the case of diverse functional roles and responsibilities, because the collaboration may draw on differences in the agents' approaches, methods and experience [14]. Furthermore, such diversity will enrich the communication with other agent groups. Within a group, KS may be enhanced when an agent has to collaborate with or report to more than one agent, up the social hierarchy, in the group.

A KM system has to allow some flexibility to the agents. An agent should be able to initiate a dialogue; ask for information about an issue, as how to solve/deal with an issue, discuss replies, make proposals, challenge proposals and reply to challenges and other proposals.

KS can be considered as an ongoing collaborative process. It is a process that may involve determining and generating acceptable arguments and lines of reasoning underlying assumptions and bodies of knowledge. KS can be regarded as a process that encourages agents to be involved in activities such as reasoning, belief revision and conceptual change.

We believe that there is a relation between knowledge construction and argumentation in KS situations. Collaborative argumentation allows agents to articulate and negotiate alternative perspectives regarding a particular task. In KS contexts, agents may need to assess the information they receive critically, considering the problem or question under discussion. Various perspectives can be discussed and/or elaborated upon by the use of critical argument. Agents can verify information when they do not fully understand information which they have received.

When agents disagree regarding previously stated information by another agent, they can use challenges. Challenging information means that the agent poses questions which are aimed at triggering justifications.

Communication involves [76] an ability to fully, if possible, understand the content of what is being communicated from the perspective of the agent that has proposed it and an ability to integrate it into the receiver's knowledge to make a proper use of it. The competencies, which determine how well tasks are performed and decisions are made, are a function of the knowledge being employed, including understanding, expertise, experiences and skills.

Problem-solving involves activities such as communicating insights, assessments, experiences or skills. Communication can be employed to transfer/exchange various types of knowledge/information such as: (1) simple facts; (2) proof/recipe specifying the steps to accomplish a task (know-how) or reach a conclusion; (3) the cause effect relationships that concern a phenomenon and other types of knowledge. In addition to relevant information, there may be a need to exchange contextual information and other constraints associated with the application of the piece of knowledge being exchanged. However, it must be to emphasized that the usefulness of the process is reflected by the extent to which the learner acquires potentially useful knowledge and utilizes this knowledge in its own operations.

To be able to organize the knowledge of a KM system, there is a need:

(K1) To identify, model and explicitly represent each of the agents and agent-groups knowledge. This entails modeling its processes, together with its control mechanism, and its decision-making.

(K2) For the ability to handle the computational aspects of multi-agent systems such as task allocation, interaction, coordination, process and organization representation, consistency management, protocol, adaptation and evolution of knowledge.

(K3) For the ability to assess the performance parameters of the system in real time.

Some of the major problems that face KM activities and/or are associated with immediate knowledge transfer between the knowledge groups are:

(D1) Agents' ability to clarify a message or to find a weakness in an argument outside their specialty is rather limited.

(D2) Cooperation is necessary between the different agents in the various work groups.

(D3) Constraints and contextual factors: There is a need for shared knowledge and shared understanding of the context and constraints of the working environment.

7 Reasoning with Incomplete Information

The basic idea that underlies the use of PIS is that it is useful to view dialogues in terms of the relevant information that the dialogue participants have at each stage in an on-going dialogue. Thus, it is important to note that the way a move affect an agent's information depends on how the agent reacts to the exchanged information. Agents are expected to have the ability to acquire and manipulate (modify, derive), through reasoning, knowledge [60, 68, 61, 78]. Each agent/participant in a dialogue has a certain well-defined role, determined by the type of dialogue, the goal of that type of dialogue, and the rules for making moves in it. Furthermore, agents are cooperative, abide by the rationality rules, such as rules of relevance [32], and they fulfill their commitments and obligations in a way that truthfully reflects their beliefs, intentions and/or desires.

An information state of an agent G is akin to a default theory $\Delta(G) = <W(G),D(G)>$ where $W(G)$ is a set of facts that are unquestionably accepted by G and $D(G)$ is a set of defaults [69] that reflects G's view of how things are related in the domain of concern. $\Delta(G)$ may have multiple extensions. If G is cautious as a reasoner, it will sanction only those propositions that are derivable from the intersection of all the extensions of $\Delta(G)$ whereas if it is a credulous reasoner, G will select one extension and it can switch to another if the need arises. Reiter was more in favor of a credulous reasoner.

Let KB(G) represents the set of propositions which G accepts as true, i.e., KB(G) is a default extension of $\Delta(G)$. If $\Delta(G)$ has multiple extensions, then G has to select one.

7.1 Temporal First Order Nonmonotonic Logic (TFONL)

The time structure employed in this paper is an interval-based theory of time. The agent's knowledge and reasoning capability are expressed in a Temporal First Order Nonmonotonic Logic (TFONL) [54]. The system is based on the quantified version of the non-temporal system T3 [57].

The language, L_{T3}, of T3 is that of Kleene's three-valued logic extended with the modal operators M (Epistemic Possibility) and P (Plausibility). In T3, L is the dual of M and N be the dual of P, i.e., $LA \equiv \sim M \sim A$ and $NA \equiv \sim P \sim A$. A truth-functional implication $\supset$ that behaves exactly like the material implication of classical logic is defined in [57] as follows: $(A \supset B = M(\sim A \,\&\, B) \,V \sim A \,V\, B$. Non-monotonic reasoning is represented via the *epistemic possibility operator* M and the *plausibility operator* P. Informally, MA states that A is not established as false. Using M, we may define the operators $UA \equiv MA \& M \sim A$ (*undefined*), $DA \equiv \sim UA$ (*defined*) and $\neg A \equiv DA \& \sim A$ (*classical negation*).

TFONL employs an explicit temporal representation with a theory of time that takes both points and interval as primitives. It employs an explicit representation of events. We may embody default logic into TFONL [55]. Furthermore, it is suitable for argumentation and dialogue frameworks cf. [47, 48, 49, 56] for more details).

We have implemented a proof system for T3. One of its essential features is that it allows free and complete access to all stages of the proof process. The proof method proceeds by the construction of a tableau. This is a tree-structure in which all the possible models allowed by the premises and negated conclusion are set out and examined for consistency. The construction of the tree is governed by rules for each logical connective in the language. These rules are closely related to the semantics of the language. The method performs a case-wise analysis of all models in which the premises might be true while contradicting the conclusion, if no such models are found to exist, the theorem is proven. We employ this method because it allows an agent absolute access to every stage of the proof process. We then exploit this access in order to find a cooperative and/or informative answer. If a proof succeeds monotonically, the agent's answer is simply *Yes*. If it succeeds by means of one or more assumptions, the answer is of the form *Yes, if* .... where the body of if-clause is the information that was assumed. Where a proof fails, we have the task of determining the reason why it failed - i.e. which assumptions should be made to yield a yes-answer and what additional knowledge is needed. This is essentially how agents can determine what knowledge/information they need. A failed proof has one or more models, which are consistent, and therefore counterexamples to our intended inference. We are able to compare these consistent models with closely related inconsistent ones. We can then identify the contradiction, which is in some sense missing. In other words, we point to the particular premise or premises which are too weak to support the inference. A cooperative and/or informative answer in this case takes the form *No, unless* ... and the body of the *unless*-clause is composed of the strengthening required in a premise or premises so that the counterexamples would no longer arise.

7.2 Context and Dialogue

Contextual knowledge representation and reasoning are challenging issues for interdisciplinary research. The roots of the problem lie in the ambivalence of the notion of context across and within different research specializations such as philosophy, psychology, pragmatics, linguistics, computer science, and Artificial Intelligence(AI). Indeed, there are many different notions of context that can be found in the AI literature [2, 31, 33, 43].

There are many definitions of context and many dimensions along which representations of contextual knowledge may vary [10, 11]. However, most researchers in the different disciplines seem to agree on the following fundamental properties.

(1) Partiality: A context is partial – it describes only a subset of a more comprehensive state of affairs. Furthermore, there may be various kinds of relationships between different partial representations (such as overlap or inclusion).

(2) Approximation: Approximation is closely tied to the notion of relevancy. Contextual information which is not required by the user must be abstracted away, rendering the remaining information partial.

(3) Perspective: A given state of affairs can be considered from several independent perspectives. The perspective dimension indicates how information must account for various perspectives such as a spatial perspective (e.g., a user's location), temporal perspective and logical perspective.

There are no clear rules for determining what will count as the context of a Topic of Concern (ToC). However, something is generally more likely to be considered part of the context if it is close to the ToC along some particular dimension or for some particular purpose. In short, context is a set of things, factors or attributes that are related to a ToC in important ways (e.g. operationally, semantically, conceptually, pragmatically). We believe that an agent never considers all its knowledge but rather a very small subset of it. This subset is usually determined by what an agent considers to be of relevance to a state of affairs based on what it has in its knowledge base and the input it receives concerning the state of affairs. This subset, however, is what determines the context of reasoning. We therefore take a context C to be a set of conditions that helps in extracting a local theory; a subset of the knowledge base together with the inference machinery needed to reason with that is used locally to prove a given goal.

In knowledge sharing situations, Context can be defined as the set of all conditions are related to the agent functionality, together with those that influence the understanding and generation of appropriate locutions in a dialogue. It includes the agent's beliefs about various types of information and about the other participant's current beliefs, related to the underlying task/topic and to each other's information state. The definitions of the basic types of locutions and their effect rules (cf. Subsection 8.3.4) specify the way in which certain context information is to be updated when the locution is properly understood by the receiver. Therefore a model of context should include at least the following information:

(C1) Information that is needed for the interpretation (and generation) of appropriate locution such as information about goals and tasks, and about progress made up to that stage in the dialogue toward achieving particular goals.

(C2) Information about participants' goals and beliefs

(C3) Information about the interaction, needed for the interpretation and generation of dialogue control actions such as:

   (C3.a) information about the interpretation, evaluation and application of previous utterances.

   (C3.b) information about the participant's interpretation, evaluation and application of previous utterances.

   (C3.c) information about protocols and turntaking.

The essential use of context is for the agents to judge the relevance of moves and the continual change to their information states throughout the different stages of the dialogue.

8 A Multi-Agent Based Model of Argumentation and Dialogue

The primary purpose of a KM system should be to make knowledge accessible and reusable by its different components whether human or software agents [30]. The core of KS is a dialogue and argument framework for collaborative autonomous agents which, should allow dialogue participants to communicate effectively; convey information, generate appropriate questions that express the needs of the represented groups, annotate responses (e.g., in the form of arguments) and judge their suitability and quality [47, 48, 49, 56]. The participating agents are expected to recognize their limitations, determine when they should seek help. This has the effect of allowing agents to make use of other agents' available knowledge and expertise for tasks that are outside the knowledge scope and expertise of their represented groups.

8.1 Argumentation and T3

Agents, in a KM system, need to interact in order to fulfill their objectives or improve their performance. Different types of interaction mechanisms suit different types of environments and applications. Thus, agents might need mechanisms that facilitate information exchange [19, 44], coordination [20, 50] where agents arrange their individual activities in a coherent manner, and collaboration [65] where agents work together to achieve a common objective. Various interaction and decision mechanisms for automated negotiation have been proposed in the MAS literature. These include: game-theoretic analysis [41, 71], heuristic-based approaches [25, 26, 40] and argumentation-based approaches [42, 60, 72, 66].

In this paper we are mainly interested the argumentation-based approaches. Argumentation is gaining more significance as an essential concept in multi-agent interaction as it enables richer forms of information exchange and rational dialogue. Argumentation is a process of reasoning. Its ultimate goal is to resolve any controversy in the sense that the available information provides both a "justification" and refutation for a given proposition expressing a belief or a goal. Argumentation can be used by an agent to reason about what to believe and about what to do. Arguments have an essential role to play in situations of conflict between communicating agents. They allow an agent to critically question the validity of information presented by another agent, explore multiple perspectives and/or get involved in belief revision processes.

In [21, 22], a theory of argumentation, within First Order Logic, based on a notion of "acceptability of arguments", is proposed. It is shown that many logics for default reasoning can be viewed as instances of an abstract argumentation framework in which arguments and the attack relation between arguments are defined entirely abstractly, ignoring their internal structure. It is also shown that the theory captures the logic structure of many problems such as the n-persons game theory.

The acceptability/rejection of the proposition has to be evaluated together with the reasoning process involving the justification. In other words, Arguments and their constituents are judged using some rationality criteria. This distinguishes argumentation from the classical deductive reasoning viewpoint, in which proofs for propositions cannot be challenged.

Argument evaluation is essential in the study of argumentation. In AI, argument evaluation and comparison has been applied, for example, in internal agent deliberation [39] and in legal argumentation [64]. In KS context, an argument may be considered as tentative proof for some conclusion. An agent may evaluate an argument based on some objective convention such as investigating the correctness of each inference step, or by examining the validity of the underlying assumptions. For example, a classification of arguments into acceptability classes based on the strength of their construction is proposed in [23].

Arguments may also be evaluated based on their relationships with other arguments. For instance, an argument is said to be acceptable with respect to a set S of arguments if every argument attacking it is itself attacked by an argument from that set [21]. The set S is said to be admissible if it is conflict free and all its arguments are acceptable with respect to S. In [4], agents compare arguments based on preferential ordering over their constituent propositions. An agent may also choose to consider its own preferences and motivations in assessing an argument [9].

In MAS, argumentation give us means for enabling an agent to handle conflicting e information within its information state, for keeping its information state up to date with new perceptions from the environment, and for reconciling conflicting information between multiple agents through communication. It provides means for

(1) forming beliefs and decisions on the basis of incomplete, conflicting or uncertain information.
(2) structuring dialogue between participants that may have conflicting beliefs.

An inherent, almost defining, characteristic of multi-agent systems is that agents need to communicate in order to achieve their individual or collective aims. It has played an important role in the exploration of different types of dialogues in MAS [77].

Argumentation can serve both as a framework for implementing autonomous agent reasoning (e.g. about beliefs and actions) and as a means to structure communication among agents. The whole dialogue can be structured through argumentation-based protocols, based on dialogue-games. A particularly important issue on the boundary between communication and agent's internal reasoning is the specification of argumentation dialogue strategies. A strategy in an argumentation dialogue specifies what utterances to make in order to bring about some desired outcome (e.g. to persuade the counterpart to perform a particular action). It has to be said that for more complex dialogue strategies, however, only informal methodologies have been proposed [67].

An Argumentation Framework (AF) system captures and represents the constituents of arguments. These may include facts, definition, rules, regulations, theories, assumptions and defaults. They can be represented as (possibly ordered) sets of formulae. It also captures the interactions and reactions between arguments and constituents of arguments such as undercutting. Furthermore, some notion of preference over arguments may be needed in order to decide between conflicting arguments.

Definition 8.1. An argument, in T3, is a pair P = <S, A> where S is an ordered set of WFF and A is a WFF of the language of T3 such that

(1) S is consistent
(2) S ⊢$_{T3}$ A (i.e., A follows from S in T3)
(3) S is minimal. No proper sub-list of S that satisfies (1) and (2) exists.

An argument, P = <S, A>, in T3 is simply a proof of A in T3. S is called the support of P and A is its conclusion. We shall use Support(P) (resp. Conc(P)) to denote that S is a support of P (resp. A is a conclusion of P).

Since T3 contains defeasible implications (default rules) and material implication, then it would be worthwhile distinguishing between a defeasible argument and a non-defeasible/classical argument.

Definition 8.2. A defeasible argument is a proof P = <S, A> where S contains some defeasible implications. A non-defeasible/classical argument is a proof that neither contains any defeasible implication(s) nor relies on any un-discharged assumptions.

It is important to note that in T3, an agent could provide an argument for both a proposition and its negation, i.e., in default logic sense [69], the theory of the agent may have multiple extensions.

Hence, in a dialogue between two agents, a defeasible argument is a structured piece of information that might be defeated by other (defeasible) arguments. Unlike a proof, in classical monotonic logic, an argument does not establish warrant for a conclusion, in a definite way, as it may be defeated by counter-arguments which are defeasible, and so on.

Definition 8.3. A proposition B is acceptable in T3 if there is an ordered set S of propositions such that P = <S, B> is proof/argument in T3.

If P = <S, B> is attacked then B becomes non-acceptable. However, if the argument that attacked S is itself attacked, the acceptable status of B will be restored.

The idea of attack is based on conflict, and defeat, between propositions and/ sets of propositions. For instance, a non-defeasible implication defeats a defeasible implication in T3. We may define attack between arguments in many ways. One such neat way is the notion of Undercut.

Definition 8.4. Let P1 and P2 be two argument in T3. Then Undercut(P1, P2) iff (∃ B ∈ Support(P2) such that B ≡ ~Conc(P1) where " ≡" is the equivalence of classical logic as defined above.

P2 undercuts P1 if, and only if, P2 has a formula that is the negation of the conclusion of P1.

It is useful to employ some ordering or any criteria that may help inducing some preference between propositions to define a notion of defeat between conflicting arguments.

8.2 Dialogue Types for KM

We adopt the notion of a *dialogue game* in which two participants make moves to pass on *relevant* information with respect to their *goals*. Central to the model is that the agents PIS change as a result of the interpretation of the moves and that these changes trigger the production of a succeeding move. The interpretation involves some *understanding* (ability to make sense or use of) the presented information. It does involve an integration of the offered information with the PIS of the receiver. Context are represented as a consistent subset of an agent's PIS, namely those propositions which bear on the interpretation of the utterance on hand and on the propositions that are relevant to producing the goal(s). Interpretation relies on maintaining consistency in the context whilst adding the utterance's information. An agent can only interpret an utterance with respect to the knowledge it has available. Therefore, failure to complete the interpretation process/proof will point to those propositions which induce failure. Thus, part of a context is entirely local to an agent, and that agents may hold incompatible and inaccurate beliefs.

The idea is as follows: a move by an agent G is generated on the basis of some enabling conditions which G can infer from its information state and the need to satisfy some goal(s). The effect of this move after being interpreted by the other participant G1 is that G1's information state may/will undergo some change. This move may initiate the legality of other moves which G1 can employ as legal reply moves. It may also terminate the legality of some other moves and render them illegal reply moves. The initiation and termination of the legality of moves is a dynamic process. The legality of moves could partly be determined by a reply structure, i.e., a protocol. Dialogue protocols provide a lower bound on the conditions needed for dialogue coherence.

In the next turn G1 may adopt the sender's role and, subsequently, its changed information state may lead to the inference of the enabling conditions for the next move. Dialogue relevance of subsequent moves is established by the initial information states of the participants, the update rules associated with each of the primitive types of dialogue moves locutions that change a particular PIS and the rules for cooperative behavior, by the participants. Dialogue coherence relations are mainly driven by dialogue history and the dynamics of the participants' PISs with respect to the main goal of the dialogue. The coherence of a dialogue moves is tied to local interactions that are dependent on the agent's particular situation reflected in the changes in its information states and intermediary goals judged by the agent to contribute towards the main goal. Thus, the reasoning abilities and specialized knowledge available to the agents do play an important role as they do capture the agent's problem-solving and strategic reasoning ability that may affect the selection of the most appropriate legal move.

The distinction between the types of dialogue is based on collective goals, individual goals and reasons for starting the dialogue. Each type of dialogue can be formulated as a set of rules. These constitute a model, representing the ideal way by which cooperative agents participate in the type of dialogue in question. It is important to note that in the course of communication, there often occurs a shift from one type of dialogue to another. Dialogue embedding takes place when the embedded dialogue is functionally related to the first one. We now give a brief presentation of some of the types of dialogue:

1. Information seeking: When a user make a query, the system makes an attempt to extract enough information from the user as is needed to search for the required information.
2. Inquiry: The basic goal of *inquiry* is information growth so that an agreement could be reached about a conclusive answer of some question. The goal is reached by a incremental process of argumentation that employs established facts in order to prove conclusions beyond a reasonable doubt. In short, the aim is to acquire *more reliable knowledge* to the satisfaction of all involved. Inquiry is a cooperative type of dialogue and correct logic proofs are essential.
3. Negotiation dialogue: The task of negotiation dialogues is that the dialogue participants come to an agreement on an issue. Negotiation dialogues differ from many other user/system interactions because in a negotiation both parties will have their own goals and constraints.

4. Persuasion Dialogue: persuasion is an expertise that employs argumentation in order to make other agents collaborate in various activities. The goal of *persuasion dialogue* is for one participant to persuade the other participant(s) of its point of view and the method employed is to prove the adopted thesis. The initial reason for starting a persuasion dialogue is a conflict of opinion between two or more agents and the collective goal is to resolve the issue. Argument here is based on the concessions of the other participant. Proofs can be of two kinds: (i) to infer a proposition from the other participant's concessions; and (ii) by introducing new premises probably supported by evidence. Clearly, a process of learning (e.g., knowledge update/belief revision) takes place here.

5. Problem-Solving dialogue: In a problem-solving dialogue, both participants collaborate with the common goal of achieving a complex task. A problem-solving dialogue may involve all the other types of dialogue, i.e., information seeking, inquiry, negotiation and/or persuasion sub-dialogues.

8.3   Toward a Formalization of Argumentation and Dialogue

Within the framework of T3, it should be possible to formalize dialogue moves and the rules of protocols of the required types of dialogue. The rules of a protocol are nonmonotonic in the sense that the set of propositions to which an agent is committed and the validity of moves vary from one move to another. The use of PIS should allow an agent to expand consistently its viewpoint with some of the propositions to which another agent involved in a dialogue is overtly committed. In this section we present a formal framework for argumentation and dialogue. The basic structure is an explicit reply structure on moves, where each move has to be an appropriate reply to a previous (but not necessarily the last) move of the other participant. It is a important to note some types of dialogue may involve some other types of dialogue such as a problem-solving dialogue which may require information seeking, inquiry, negotiation and/or persuasion sub-dialogues.

A dialogue system [47] is a formal model that aims to represent how a dialogue should proceed. It defines the rules of the dialogue. All dialogues are assumed to be between two agents about some dialogue topic.. The topic language, L_{Topic}, is a logical language which consists of propositions that are the topics of the dialogue. The language involves defining different aspects that are necessary to describe the content of the knowledge/information to be exchanged. L_{Topic} is associated with T3 which determines the inference rules, the defeat relations between the arguments and defines the construction of proper arguments and dialogue moves.

Definition 8.5. A dialogue system is a triple $D = (L_{COM}, PROT, EFF)$ where
(1) L_{COM} is the communication language. defined below in Subsection 8.3.1.
(2) PROT is a protocol for L_{COM} and is defined in Definition 8.8 below.
(3) EFF is a set of rules that specify the effects of utterances (locutions in L_{COM}) on the participants' commitments. EFF is defined in Subsection 8.3.5.

8.3.1   The Communication Language L_{COM}
KS and agents collaboration requires a sophisticated Agent Communication Language (ACL). The main objective of an ACL is to model a suitable framework that allows

heterogeneous agents to interact and to communicate with meaningful statements that convey information about their environment or knowledge. There are two main communication languages: KQML [28] and FIPA [29, 45]. These languages have been designed to be widely applicable. This feature can be both a strength and a weakness: agents participating in conversations have too many choices of what to utter at each turn, and thus agent dialogues may endure a state-space explosion. The need for a language that allows sufficient flexibility of expression while avoiding state-space explosion had led agent communications researchers to the study of formal dialogue games. Furthermore, the semantic language for the locutions of FIPA, is an axiomatic semantics of the speech acts of the language, defined in terms of the beliefs, desires and intentions of the participating agents [29]. In this section we present LCOM that specifies the locutions which the participants are able to express or say in the dialogue. We will assume that every agent understands LCOM and that all agents have access to common argument ontology, so that the semantics of a message is the same for all agents.

8.3.2 Dialogue Move

A dialogue consists of a course of successive utterances (moves) made by the dialogue participants. Let μ be the set of possible moves which participants can make. Let $\mu_\varnothing = \mu \cup \varnothing$ where $\varnothing$ stands for the empty sequence of moves.

Definition 8.6. (Dialogue Move) A dialogue move $M \in \mu$ is a 5-tuple
$$M = <ID(M), SENDER(M), \delta(M), TOPIC(M), TARGET(M)>$$
   where
   (1) ID(M), the identifier of the move M. (i.e., ID(M) = i indicates that M is the ith element of the sequence in the dialogue).
   (2) SENDER(M) is the participant that utters $<\delta(M), TOPIC(M)>$.
   (3) $\delta(M) \in$ {Assert, Retract, Accept, Reject, Question, Challenge}
   (4) TOPIC(M) denotes the sentence which a dialogue participant wants to pass on to the other participant.
   (5) TARGET(M) is the target of the move.
   For instance, M = <3, G2, Assert, A, 2> states that M is the 3rd move in a dialogue where G2, asserts the proposition A and it is G2's reply to move M_2.
   We shall use LOC(M) = $<\delta(M), TOPIC(M)>$ to denote the locution of M.

8.3.3 Turntaking and Protocol

We shall employ the notation D_k, where $1 \leq k < \infty$, to refer to a finite sequence of moves $M_1, \ldots, M_k$. Let Finite-D = $\varnothing \cup \{D_k: 0 < k < \infty\}$ where each D_k satisfies the following conditions:

(t1) for each ith element, M, in D_k, ID(M) = i and $1 \leq i \leq k$.
(t2) TARGET(M_1) = 0, i.e., the first move in a dialogue cannot be a reply to an earlier move.
(t3) for all $1 < i \leq k$, it holds that TARGET(M_i) = j for some M_j, such that $1 \leq j < i \leq k$, in D_k.

Definition 8.7. Let G1 and G2 be two agents and P = {G1, G2}. A turntaking function, Turn, for a dialogue d ∈ Finite-D that has participants P is a function

Turn: Finite-D → P

such that Turn(∅) = G1 XOR Turn(∅) = G2 where XOR stands for exclusive OR

Turn(∅) = G1 (resp. Turn(∅) = G2) indicates that G1 (resp. G2) made the first move in the dialogue.

A turn of a dialogue is a maximal sequence of successive moves made by the same participant.

We are now in the position to define the protocol. The first part of the following definition is taken from [63].

Definition 8.8. A protocol on Finite-D is a set PROT ⊆ Finite-D satisfying the following condition (P1):

(P1) for any n, 1 ≤ n < ∞, if D_n ∈ PROT then D_{n-1} ∈ PROT

A partial function Proto: Finite-D → μ is derived from PROT as follows:

Proto(d) = Undefined if d ∉ PROT

Proto(d) = {M ∈ μ: d,M ∈ PROT} where d,M is dialogue d followed by move M.

Proto(d) is the set of moves that are allowed after d.

If d is a legal dialogue and Proto(d)=∅, then d is said to be a terminated dialogue.

Proto must satisfy the following conditions:

(C1) for all d ∈ Finite-D and M ∈ μ it holds that if d ∈ Dom(Proto) and M ∈
    Proto(d) then d,M ∈ Dom(Proto) where Dom(Proto) is the domain of Proto.

All protocols are further assumed to satisfy the following basic conditions for all moves M and all legal finite dialogues d. That is, if M ∈ Proto(d), then:

(C2) SENDER(M) ∈ Turn(d);

(C3) If d is not empty and M is not the first move. then <δ(M), TOPIC(M)> is a
    reply to <δ(M_s), TOPIC(M_s)> where TARGET(M) = s.

(C4) If M is a reply to M', then SENDER(M) ≠ SENDER(M');

(C5) If there is an M' in d such that TARGET(M) = TARGET(M') then <δ(M),
    TOPIC(M)> ≠ <δ(M'), TOPIC(M')>.

(C6) For any M' ∈ d, if M' accepts the locution uttered by SENDER(M), then M
    does not reject or challenge any of the constituents of the locution of M'.

(C7) If d = ∅ then δ(M) = Assert or δ(M) = Question.

(C1)- (C7) capture a lower bound on coherence of dialogues. (C2) says that a move is legal only if it is made by the right participant. (C3) says that a replying move must be a reply to its target. (C4) states that if a dialogue participant replies by a move M to a move M', then M' must be have been made by the other participant. (C5) states that if a participant replies in two moves to the same target, then they must have different locutions. (C6) says that "Accept" is not an appropriate reply to 'Challenge" or "Reject". (C7) says that the first move has to be an assertion or a question.

8.3.4 Basic Locutions

Every dialogue system specifies its own set of locutions. However, there are several basic types of locutions which are used in many systems.

Let G be an agent, involved in an On-Going dialogue d at stage i-1 with another agent G1. Let KB(G) (resp. KB(G1)) represent the set of propositions which G (resp. G1) accepts as true. Let $A \in L_{Topic}$, j < i, M_j a move made by participant G1, and M is a move made by G as a reply to M_j, then

(1) Assert A: An agent G can make the move "Assert A" if A is derivable from KB(G) which is not inconsistent. G can offer a proof of A.

(2) Retract A: this move is a countermove to "Assert A". G can only make the move "Retract A" as a reply to a "Assert A" move made earlier by G1. In making the move "Retract A", G in NML3, is not committed to "Assert ~A". This move deletes A from KB(G).

(3) Accept A: this move can be made by an agent G to signal that it accepts/concedes a proposition A. It has to be a reply to a previous "Assert A" made by another agent G1. G can make the move "Accept A" if A is not inconsistent with KB(G), otherwise KB(G) will be subject to revision. The impact of this move is that A will be added to KB(G). This can be possible if the locution of a previous message by another agent is   "Assert A".

(4) Reject A: a countermove to "Accept A". It is important to note that in NML3, "Reject A" by G does not commit it to "Accept ~A". This can be possible if the locution of a previous message by another agent is   "Assert A". The impact of this message could either be no change to KB(G) as it is in contradiction with A, or an update of KB(G) by retracting A.

(5) Question A: An agent G questions/asks from another, G1, for information concerning A (e.g., whether A is derivable from KB(G1)). This move does not alter either of KB(G) or KB(G1).

(6) Challenge A: This move is made by one agent G, for another G1, to explicitly state that G1 has to provide a proof for (an argument supporting) A. This move does not alter either of Context(G, D_i, G1) or Context(G1, D_i, G). In this move G is forcing G1, to explicitly state a proof (an argument supporting) A.

8.3.5 Rules of Protocols of Some Types of Dialogue

1. Information-Seeking: Assume that the information seeker is agent G and the other agent is G1. The steps in a successful information seeking dialogue are as follows:

(IS1) G makes a *Question* move such as
   M_i = <i, G, Question, A, l> where M_l is a move made earlier by G1 and l < i.

(IS2) G1 replies with the move M_k where the identifier is k and its target is the move M_j, where k > i, as follows:
   (i) M_k = <k, G1, Assert, A, i> or
   (ii) M_k = <k, G1, Assert, ~A, i> or
   (iii) M_k = <k, G1, Assert, UA, i> where UA means that for G1 the truth value of A is undefined.

(IS3) G either accepts G1's response using an *Assert* move or challenges it with a *Challenge* move. UA initiates an *inquiry* sub-dialogue between the agents or the information-seeking dialogue is terminated.

(IS4) G1 has to reply to a move "Challenge A" with a proof S using a move $M_r = <r$, G1, Assert, S, k> where S is a proof of A based on the knowledge base of G1.

(IS5) For each sentence B employed in the proof S, either G accepts B using a move such as "Assert B" or "Accept B" depending on whether it is derivable from, or it is not inconsistent with, its knowledge base. Otherwise, it may challenge B, in which case G1 has to provide a proof for B.

2. Inquiry: The following is an inquiry-protocol about a proposition A involving G and G1.

(INQ1) G seeks a support/proof for A. It begins with a move such as "Assert B $\rightarrow$A" or "Assert B $\Rightarrow$ A", for some sentence B if G believes that B $\rightarrow$A or B $\Rightarrow$ A should be derivable but it needs some confirmation and that will happen if G1 accepts the assertion. Otherwise, G will use a move such as "Assert UA".

(INQ) G1 could reply following in one of the following ways:

(I1) accepts B $\rightarrow$A or accepts B $\Rightarrow$ A as appropriate using an "Accept" move. If an accept move is made, then either the inquiry terminates successfully or G could go on asking G1 to provide a proof for B. This case is similar to one where a student asks for a clarification regarding some issue.

(I2) accepts B $\rightarrow$A or accepts B $\Rightarrow$ A as appropriate, but G1 seeks a support/proof for B, i.e., it could reply with an "Assert" move that asserts E $\rightarrow$ B or asserts E $\Rightarrow$ B, as appropriate, for some sentence E or a move that asserts UB.

(I3) challenges B $\rightarrow$ A or B $\Rightarrow$ A as appropriate with a "Challenge" move.

(INQ3) If a challenge move is made as in step (I3) by G1, Then G has to reply to the challenge with an "Assert P" move that provide a proof derived from the knowledge base of G of the last proposition challenged by G1.

(Inq4) For every sentence C in P, G1 may either accept C with an "accept" move or may challenge it with a "Challenge" move.

(Inq5) When both agents accept A, the dialogue terminates successfully.

It is important to note that the agents could switch roles and either of them could seek a support/proof for a sentence or challenge a sentence when appropriate.

3. Persuasion: The following is a persuasion protocol where agent G is trying to persuade agent G1 to accept a proposition A.

(P1) G begins with a move where it asserts A.

(P2) G1 may reply with one of the following three moves:
  (i)   "Accept A"
  (ii) "Asserts ~A"
  (iii) "Challenge A".

(P3) There are three possibilities depending on G1' reply:
  (1) If the answer of G1 in the previous step (P2) is (i), then the dialogue may successfully terminate.
  (2) If the answer of G1 in the previous step (P2) is (ii) "Asserts ~A", then go to step (P2) with the roles of the agents switched and ~A is put in place of A..
  (3) If the answer of G1 in the previous step (P2) is (iii) "Challenge A", then G should reply with a move that provide/asserts a proof P of A derived from its knowledge base.

(P4) If G has replied with a proof P of A, then for every proposition B employed in P, G1 may seek a proof/support from G1 (i.e., may invoke step P2 for B).

8.4 PIS Dialogue Control Using Theorem Proving

In KS contexts, decisions could be made, by one or more participants, dynamically, on the basis of the previous steps captured in the dialogue history, and the current PISs of the participants and their knowledge need. This approach differs form the approach adopted in some plan-based systems, where an agent G first make a complete decision regarding its solution together with a plan as to how it can it wants to proceed in order achieve its goals, before it sends it to another other agent, say G1. Nonmonotonic theorem proving of T3 can be used to determine what is accomplishable and the dialogue is used for the acquisition, if at all possible, of what is considered necessary or required to complete a specific task step. Failure to provide the needed/missing information by one agent may leave no choice but to make assumptions, i.e., by invoking nonmonotonic inference rules. In doing so, dialogue is integrated closely with KS activities and is invoked when an agent is unable to complete a task using its own knowledge and/or resources. The nonmonotonic proof method can be employed to create sub-dialogues to deal with ensuing issues needed for the overall task.

9 An Example

In this section we give an example that show that a participant, that uses T3 proof method, in a dialogue has access to every stage of the proof process, which can be used to provide cooperative and/or informative answers to questions posed by the other participants.

Assume that we have two agents G1 and G2. G1 needs a lift in a car. It notices a car parked in front of John's house and decides to ask G2 in order to decide whether to ask John for a lift.

Let Skilltodrive = john is skilled to drive
  Ownscar = John has a car
  Licensed = John has a license
  Candrive = John can drive

Scenario 1

KB(G2) ⊇ {Skilltodrive, Ownscar, License,
                Ownscar&Licensed&Skilltodrive → Candrive}
Goal(G1) ⊇ {Get-Lift}
KB(G1) ⊇ {Candrive → Get-Lift}

We employ ⇒ instead of → to express defeasibility. For instance, it is possible to
have a case where john is skilled to drive, John has a car and John has a license, but
he cannot drive because of some physical conditions. Hence, it would be more appropriate to use
  "Ownscar&Licensed&Skilltodrive ⇒ Candrive"
    instead of
  "Ownscar&Licensed&Skilltodrive → Candrive".

Dialogue 1 (Accept)
Here we present a simple example of **Information-Seeking** dialogue that terminates successfully.

First, G1 asks G2 if John can drive.
M_1: <1, G1, Question, "Candrive", 0>
M_2: <2, G2, Assert, "Candrive", 1>.

In M_2, G2 replies with a "Yes" answer to G1's question about whether John can
drive. After M_2, it is clear that G2 is committed to the "John can drive".

M_3: <3, G1, Accept, "Candrive", 2>

In M_3, G1 accepts G2's answer which was provided in M_2.
The acceptance by G1 in M_3 of the reply which G2 provided in M_2 indicates that
we have a successful termination of the dialogue.

Dialogue 1 (Challenge)
Suppose that G1 has challenged G2's answer, as in M'_3, the dialogue would have
been as follows:
M_1: <1, G1, Question, "Candrive", 0>
M_2: <2, G2, Assert, "Candrive", 1>.

M'_3: <3', G1, Challenge, "Candrive", 2>

In M'_3, G1 challenges G2's answer which was provided in M_2.

G2 has to reply with a proof that *John can drive*, "Candrive", as in M_4. Using the theorem prover, *"Candrive"* can be proved, in a proof P1, by refutation and monotonically and requires no sub-proof.

M_4: <4, G1, Assert, "P1", 3'>
G1 can either accept, as in M_5, the proof P1 presented by G2 in M_4 as in M_5 or challenge the proof.

M_5: <5, G1, Accept, "P1", 4>
In M_5, G1 accepts G2's answer which was provided in M_4.

The acceptance by G1 in M_5 of the proof P1, which G2 provided in M_4 indicates that G1 accepts its conclusion. Hence we have a successful termination of the dialogue.

Scenario 2
Suppose that the premise that John has a license is removed from KB(G2), i.e.,
KB(G2) $\supseteq$ {Skilltodrive, Ownscar,
                    Ownscar&Licensed&Skilltodrive → Candrive}
Goal(G1) $\supseteq$ {Get-Lift}
KB(G1) $\supseteq$ {Candrive → Get-Lift}

In this case, G2 may answer in two different ways as shown below in Dialogue 2.1 and Dialog 2.2.

Dialogue 2.1 (Accept)
First, G1 asks G2 if John can drive.
M_1: <1, G1, Question, "Candrive", 0>

G2 cannot prove **monotonically** "Candrive" because there is a need for the premise "Licensed" (that John has a license) to complete the proof. In this case, there are many ways along which the dialogue could proceed. One of the choices would be to assert "Licensed → Candrive", which can be interpreted as "Yes" John can drive if he has a license.

M_2: <2, G2, Assert, "Licensed → Candrive", 1>.

G1 may decide to accept G2's reply as in M_3 to indicate a successful termination of the dialogue.

M_3: <3, G1, Accept, "Licensed → Candrive", 2>

In M_3, G1 accepts G2's answer which was provided in M_2. After M_3, the situation is as follows:

Dialogue 2.2 (Challenge)
Suppose that G1 has challenged G2's answer, as M'$_3$, the dialogue would have been as follows:
\# M$_1$: <1, G1, Question, "Candrive", 0>
\# M$_2$: <2, G2, Assert, "Licensed → Candrive", 1>.
\# M'$_3$: <3', G1, Challenge, "Licensed → Candrive", 2>

In M'$_3$, G1 challenges G2's answer which was provided in M$_2$.
G2 has to reply with a proof that shows the **need to assume "Licensed"** in order to prove "Candrive" in the proof P2.

\# M$_4$: <4, G2, Assert, "P2", 3'>

G1 can either accept, as in M$_5$, the proof presented by G2 in M$_4$ or challenge the proof.

\# M$_5$: <5, G1, Accept, "P2", 4>

In M$_5$, G1 accepts G2's answer which was provided in M$_4$.
The acceptance by G1 in M$_5$ of the **nonmonotonic proof P2**, which G2 provided in M$_4$ indicates that G1 accepts its assumption and its conclusion. Hence we have a successful termination of the dialogue.

Scenario 3
Suppose that the premise that John does not have a license is added to KB(G2), i.e.,
KB(G2) ⊇ {Skilltodrive, Ownscar, ~Licensed
                  Ownscar&Licensed&Skilltodrive → Candrive}
Goal(G1) ⊇ {Get-Lift}
KB(G1) ⊇ {Candrive → Get-Lift}

Dialogue 3
G1 begins the dialogue by asking G2 if John can drive.
\# M$_1$: <1, G1, Question, "Candrive", 0>

\# M$_2$: <2, G2, Assert, "~Candrive", 1>.
In M$_2$, G2 replies with a "No" answer to G1's question about whether John can drive.

\# M$_3$: <3, G1, Challenge, "~Candrive", 2>

In M'$_3$, G1 challenges G2's answer which was provided in M$_2$.
G2 cannot prove, in P3, **monotonically** "Candrive" and thus in P3 "~Candrive" is nonmonotonically is inferred. An attempt to hold an assumption that *"John has a licence"* will, however, fail as it will contradict the premise *"~Licensed"* which states that such an assumption is false.

\# M_4: <4, G1, Assert, "P3", 3>

G1 can either accept, as in M_5, the proof presented by G2 in M_4 or challenge the proof.

\# M_5: <5, G1, Accept, "P3", 4>

In M_5, G1 accepts G2's answer which was provided in M_4.

The acceptance by G1 in M_5 of the proof P3, which G2 provided in M_4 indicates that G1 accepts its conclusion. Hence we have a successful termination of the dialogue.

10　Concluding Remark

We have in this paper proposed a multi-agent system that assists in the process of knowledge sharing between concerned knowledge worker groups. Each agent is a knowledge broker and organizer for a specialized knowledge worker group involved in the operations of a sub-system. In addition to timely access to knowledge, it should help in understanding the motivation which underlies decisions made by other groups and/or the information/knowledge bases for such decisions. Each agent is expected to learn about the activities, contexts of decisions, knowledge employed and experiences of other knowledge workers groups whose activities are considered to be relevant to the group it represents. We have employed a Partial Information State to represent the knowledge of the agents. We have employed a three-valued based nonmonotonic logic, TFONL, for representing and reasoning about PISs. We have only made use of non-temporal part, T3, of TFONL and refrained from addressing temporal issues. We have present a multi-agent based model of argumentation and dialogue for knowledge sharing. On the argumentation side, it is worthwhile investigating further the subtleties of each type of dialogue in relation to different tasks that may be accomplished by an agent. It would also be beneficial to further investigate, within the framework, strategic and tactic reasoning for rational agents.

References

1. Abdullah, M., Benest, I., Evans, A., Kimble, C.: Knowledge Modeling Techniques for Developing Knowledge Management Systems. In: 3rd European Conference on Knowledge Management, Dublin, Ireland, pp. 15–25 (2002)
2. Akman, V., Surav, M.: Steps Toward Formalizing Context. AI Magazine, 55–72 (1996)
3. Allen, T.: Managing the Flow of Technology. MIT Press, Cambridge (1977)
4. Amgoud, L., Parsons, S., Maudet, N.: Arguments, Dialogue, and Negotiation. In: Proceedings of the European Conference on Artificial Intelligence (ECAI 2000), pp. 338–342 (2000)
5. Ancona, D., Caldwell, D.: Bridging The Boundary: External Activity and Performance In Organizational Teams. Admin. Sci. Quart. 37, 634–665 (1992)
6. Argote, L., McEvily, B., Reagans: Managing Knowledge In Organizations: Creating, Retaining, and Transferring Knowledge. Management Sci. 49(4) (2003)

7. Argote, L., Ingram, P., Levine, J., Moreland, R.: Knowledge Transfer in Organizations. Organ. Behavior Human Decision Processes 82(1), 1–8 (2000)
8. Baumard, P.: Tacit Knowledge in Organizations. Sage, London (1999)
9. Bench-Capon, T.J.M.: Truth and Consequence: Complementing Logic with Value in Legal Reasoning. Information and Communications Technology Law 10(1) (2001)
10. Benerecetti, M., Bouquet, P., Ghidini, C.: Formalizing Belief Reports – the Approach and a Case Study. In: Giunchiglia, F. (ed.) AIMSA 1998. LNCS (LNAI), vol. 1480, pp. 62–75. Springer, Heidelberg (1998)
11. Benerecetti, M., Bouquet, P., Ghidini, C.: Contextual Reasoning Distilled. Journal of Experimental and Theoretical Artificial Intelligence 12(3), 279–305 (2000)
12. Booch, G.: Object-Oriented Analysis and Design with Applications. Addison-Wesley, Reading (1994)
13. Brooking, A.: Corporate Memory: Strategies for Knowledge Management. International Thomson Business Press, London (1999)
14. Bunderson, J., Sutcliffe, K.: Comparing Alternative Conceptualizations of Functional Diversity in Management Teams. Acad. Management J. 45(5), 875–893 (2002)
15. Castefranchi, C.: Modelling Social Action for AI Agents. Artificial Intelligence 103(1–2), 157–182 (1998)
16. Cohen, S., Bailey, D.: What Makes Teams Work: Group Effectiveness Research from the Shop Floor to the Executive Suite. J. Management 23(3), 239–264 (1997)
17. Crossan, M., Lane, H.W., White, R.E.: An Organizational Learning Framework: From Intuition to Institution. Academy of Management Review 24(3), 522–537 (1999)
18. Davenport, T., Prusak, L.: Working Knowledge: How Organizations Manage what they Know. Harvard Business School Press, Boston (1998)
19. De Boer, F., Van Eijk, R.M., Van der Hoek, W., Meyer, J.: A Fully Abstract Model for the Exchange of Information in Multi-Agent Systems. Theoretical Computer Science 290(3), 1753–1773 (2003)
20. Durfee, E.H.: Practically Coordinating. Artificial Intelligence Magazine 20(1), 99–116 (1999)
21. Dung, P.: The Acceptability of Arguments and its Fundamental Role in Nonmonotonic Reasoning and Logic Programming and N-Person Game. Artificial Intelligence 77, 321–357 (1995)
22. Dung, P.: An Argumentation Theoretic Foundation of Logic Programming. Journal of Logic Programming 22, 151–177 (1995)
23. Elvang-Gøransson, M., Krause, P., Fox, J.: Dialectic Reasoning with Inconsistent Information. In: Heckerman, D., Mamdani, A. (eds.) Proceedings of the 9th Conference on Uncertainty in Artificial Intelligence, pp. 114–121 (1993)
24. Faraj, S., Sproull, L.: Coordinating Expertise in Software Development Teams. Management Sci. 46(12), 1554–1568 (2000)
25. Faratin, P.: Automated Service Negotiation Between Autonomous Computational Agents. PhD thesis, University of London, Queen Mary and Westfield College, Department of Electronic Engineering (2000)
26. Fatima, S., Wooldridge, M., Jennings, N.R.: Multi-Issue Negotiation under Time Constraints. In: Proceedings of the 1st International Joint Conference on Autonomous Agents and Multiagent Systems (AAMAS-2002), pp. 143–150 (2002)
27. Ferber, J.: Multi-Agent Systems. Addison-Wesley, Reading (1999)
28. Finin, T., Labrou, Y., Mayfield, J.: KQML as an Agent Communication Language. In: Bradshaw, J.M. (ed.) Software Agent, pp. 291–315. AAAI Press, Menlo Park (1995)

29. FIPA-ACL. Communicative Act Library Specification. Technical Report XC00037H, Foundation for Intelligent Physical Agents (2001)
30. Gasser, L.: Social Conceptions of Knowledge and Action: DAI Foundations and Open System Semantics. Artificial Intelligence 47(1–3), 107–138 (1991)
31. Giunchiglia, F.: Contextual Reasoning. Epistemologia, XVI, 345–364 (1993)
32. Grice, H.: Logic and Conversation. In: Davidson, D., Harman, G. (eds.) The Logic of Grammar, Encino, California, Dickenson, pp. 64–75 (1975)
33. Guha, R.V.: Contexts: A Formalization and some Applications. PhD thesis. Stanford University (1991)
34. Hansen, M.: The Search-Transfer Problem: The Role of Weak Ties in Sharing Knowledge Across Organization Subunits. Admin. Sci. Quart. 44, 82–111 (1999)
35. Hansen, M.: Knowledge Networks: Explaining Effective Knowledge Sharing in Multiunit Companies. Organ. Sci. 13(3), 232–248 (2002)
36. Heaton, L., Taylor, J.: Knowledge Management and Professional Work. A Communication Perspective on the Knowledge-Based Organization, Management Communication Quarterly 16(2), 210–236 (2002)
37. Jennings, N.R.: Commitments and Conventions: The Foundation of Coordination in Multi-Agent Systems. Knowledge Engineering Review 8(3), 223–250 (1993)
38. Jennings, N.R., Wooldridge, M.: Agent-Oriented Software Engineering. In: Bradshaw, J. (ed.) Handbook of Agent Technology. AAAI/MIT Press (2000)
39. Kakas, A., Moraitis, P.: Argumentation Based Decision Making for Autonomous Agents. In: Proceedings of the 2nd International Joint Conference on Autonomous Agents and Multiagent Systems (AAMAS 2003), pp. 883–890 (2003)
40. Kowalczyk, R., Bui, V.: On Constraint-Based Reasoning in e-Negotiation Agents. In: Dignum, F.P.M., Cortés, U. (eds.) AMEC 2000. LNCS (LNAI), vol. 2003, pp. 31–46. Springer, Heidelberg (2001)
41. Kraus, S.: Strategic Negotiation in Multi-Agent Environments. MIT Press, Cambridge (2001)
42. Kraus, S., Sycara, K., Evenchik, A.: Reaching Agreements through Argumentation: A Logical Model and Implementation. Artificial Intelligence 104(1–2), 1–69 (1998)
43. Lenat, D.B., Guha, R.V.: Building Large Knowledge Based Systems. Addison-Wesley, Reading (1990)
44. Luo, X., Zhang, C., Jennings, N.R.: A Hybrid Model for Sharing Information between Fuzzy, Uncertain and Default Reasoning Models in Multi-Agent Systems. International Journal of Uncertainty, Fuzziness and Knowledge-Based Systems 10(4), 401–450 (2002)
45. McBurney, P., Parsons, S.: A denotational semantics for deliberation dialogues. In: Rahwan, I., Moraïtis, P., Reed, C. (eds.) ArgMAS 2004. LNCS (LNAI), vol. 3366, pp. 162–175. Springer, Heidelberg (2005)
46. Mietlewski, Z., Walkowiak, R.: Knowledge and Life Cycle of an Organization. The Electronic Journal of Knowledge Management 5(4) (2007)
47. Moubaiddin, A., Obeid, N.: Partial Information Basis for Agent-Based Collaborative Dialogue. Applied Intelligence 30(2), 142–167 (2009)
48. Moubaiddin, A., Obeid, N.: Dialogue and Argumentation in Multi-Agent Diagnosis. In: Nguyen, N.T., Katarzyniak, R. (eds.) New Challenges in Applied Intelligence Technologies. Studies in Computational Intelligence, vol. 134, pp. 13–22. Springer, Heidelberg (2008)
49. Moubaiddin, A., Obeid, N.: Towards a Formal Model of knowledge Acquisition via Cooperative Dialogue. In: Proceedings of the 9th International Conference on Enterprise Information Systems (2007)

50. Moulin, B., Chaib-Draa, B.: A Review of Distributed Artificial Intelligence. In: O'Hare, G., Jennings, N.R. (eds.) Foundations of Distributed Artificial Intelligence, ch. 1, pp. 3–55. John Wiley & Sons, Chichester (1996)

51. Nelson, R., Winter, S.: An Evolutionary Theory of Economic Change. Harvard University Press, Cambridge (1982)

52. Newell, A.: The Knowledge Level. Artificial Intelligence 18, 87–127 (1982)

53. Nonaka, I., Takeuchi, H.: The knowledge Creating Company: How Japanese Companies Create the Dynamics of Innovation. Oxford University Press, Oxford (1995)

54. Obeid, N.: A Formalism for Representing and Reasoning with Temporal Information, Event and Change. In: Applied Intelligence, Special Issue on Temporal Uncertainty, vol. 23(2), pp. 109–119. Kluwer Academic Publishers, Dordrecht (2005)

55. Obeid, N.: A Model-Theoretic Semantics for Default Logic. WSEAS Transactions on Computers 4(6), 581–590 (2005)

56. Obeid, N.: Towards a Model of Learning Through Communication, Knowledge and Information Systems, vol. 2, pp. 498–508. Springer, USA (2000)

57. Obeid, N.: Three Valued Logic and Nonmonotonic Reasoning. Computers and Artificial Intelligence 15(6), 509–530 (1996)

58. Panzarasa, P., Jennings, N.R.: Social Influence, Negotiation and Cognition. Simulation Modelling Practice and Theory 10(5–7), 417–453 (2002)

59. Panzarasa, P., Jennings, N.R., Norman, T.J.: Formalising Collaborative Decision Making and Practical Reasoning in Multi-Agent Systems. Journal of Logic and Computation 12(1), 55–117 (2002)

60. Parsons, S., Sierra, C., Jennings, N.: Agents that Reason and Negotiate by Arguing. Journal of Logic and Computation 8(3), 261–292 (1998)

61. Parsons, S., Wooldridge, M., Amgoud, L.: On the Outcomes of Formal Inter-Agent Dialogues. In: 2nd International Conference on Autonomous Agents and Multi-Agent Systems. ACM Press, New York (2003)

62. Polanyi, M.: The Tacit Dimension. Doubleday, New York (1966)

63. Prakken, H.: Coherence and Flexibility in Dialogue Games for Argumentation. Journal of Logic and Computation 15(6), 1009–1040 (2005)

64. Prakken, H., Sartor, G.: The Role of Logic in Computational Models of Legal Argument: A Criticial Survey. In: Kakas, A., Sadri, F. (eds.) Learning-Based Robot Vision. LNCS, vol. 2048, pp. 342–343. Springer, Berlin (2001)

65. Pynadath, D., Tambe, M.: Multiagent Teamwork: Analyzing Key Teamwork Theories and Models. In: Castelfranchi, C., Johnson, L. (eds.) Proceedings of the 1st International Joint Conference on Autonomous Agents and Multi-agent Systems (AAMAS 2002), pp. 873–880 (2002)

66. Rahwan, I., Ramchurn, S., Jennings, N., McBurney, P., Parsons, S., Sonenberg, L.: Argumentation-Based Negotiation. The Knowledge Engineering Review 18, 343–375 (2003)

67. Rahwan, I.: Interest-based negotiation in multi-agent systems, Ph.D. thesis, Department of Information Systems, University of Melbourne, Melbourne, Australia (2004)

68. Reed, C.A., Long, D., Fox, M., Garagnani, M.: Persuasion as a Form of Inter-Agent Negotiation. In: Lukose, D., Zhang, C. (eds.) Proceedings of the 2nd Australian Workshop on DAI. Springer, Berlin (1997)

69. Reiter, R.: A Logic for Default Reasoning. Artificial Intelligence 13, 81–132 (1980)

70. Rulke, D., Galaskiewicz, J.: Distribution of Knowledge, Group Network Structure, and Group Performance. Management Sci., 46(5), 612–625 (2000)

71. Sandholm, T.: EMediator: A Next Generation Electronic Commerce Server, Computational Intelligence. Special Issue on Agent Technology for Electronic Commerce 18(4), 656–676 (2002)
72. Sierra, C., Jennings, N.R., Noriega, P., Parsons, S.: A Framework for Argumentation-Based Negotiation. LNCS (LNAI), vol. 1365, pp. 177–192. Springer, Berlin (1998)
73. Straub, D., Karahanna, E.: Knowledge Worker Communications and Recipient Availability: Toward a Task Closure Explanation of Media Choice. Organization Science 9(2), 160–175 (1998)
74. Szulanski, G.: The Process of Knowledge Transfer: A Diachronic Analysis of Stickiness. Organizational Behaviour and Human Decision Processes 82(1), 9–27 (1999)
75. Tushman, M.: Impacts of Perceived Environmental Variability on Patterns of Work Related Communication. Acad. Management J. 22(1), 482–500 (1979)
76. van Elst, L., Dignum, V., Abecker, A.: Towards Agent-Mediated Knowledge Management. In: van Elst, L., Dignum, V., Abecker, A. (eds.) AMKM 2003. LNCS (LNAI), vol. 2926, pp. 1–30. Springer, Heidelberg (2004)
77. Walton, D.N., Krabbe, E.C.W.: Commitment in Dialogue: Basic Concepts of Interpersonal Reasoning. State University of New York Press, NY (1995)
78. Wooldridge, M., Parsons, S.: Languages for negotiation. In: Proceedings of ECAI, pp. 393–400 (2000)
79. Zack, M.: Managing Codified Knowledge. Slaon Management Review 40(4), 45–58 (1999)
80. Zander, U., Kogut, B.: Knowledge and the speed of the transfer and imitation of organizational capabilities: An empirical test. Organ. Sci. 6(1), 76–92 (1995)

Influence of the Working Strategy on A-Team Performance

Dariusz Barbucha, Ireneusz Czarnowski, Piotr Jędrzejowicz,
Ewa Ratajczak-Ropel, and Iza Wierzbowska

Department of Information Systems, Gdynia Maritime Academy,
Morska 81-87, 81-225 Gdynia, Poland
{barbucha,irek,pj,ewra,iza}@am.gdynia.pl

Abstract. An A-Team is a system of autonomous agents and the common memory. Each agent possesses some problem-solving skills and the memory contains a population of problem solutions. Cyclically solutions are being sent from the common memory to agents and from agents back to the common memory. Agents cooperate through selecting and modifying these solutions according to the user-defined strategy referred to as the working strategy. The modifications can be constructive or destructive. An attempt to improve a solution can be successful or unsuccessful. Agents can work asynchronously (each at its own speed) and in parallel. The A-Team working strategy includes a set of rules for agent communication, selection of solution to be improved and management of the population of solutions which are kept in the common memory. In this paper influence of different strategies on A-Team performance is investigated. To implement various strategies the A-Team platform called JABAT has been used. Different working strategies with respect to selecting solutions to be improved by the A-Team members and replacing the solutions stored in the common memory by the improved ones are studied. To evaluate typical working strategies the computational experiment has been carried out using several benchmark data sets. The experiment shows that designing effective working strategy can considerably improve the performance of the A-Team system.

Keywords: agents, working strategy, optimization problem, asynchronous team.

1 Introduction

The field of autonomous agents and multi-agent systems is a rapidly expanding area of research and development. One of the successful approaches to agent-based optimization is the concept of an asynchronous team (A-Team), originally introduced by Talukdar [14]. The A-Teams are open, high-performance organizations for solving difficult problems, particularly, problems from the area of planning, design, scheduling and real-time control.

An A-Team is a system of autonomous agents and the common memory. The multi agent architecture of A-Team forms a framework that enables users to easily combine disparate problem solving strategies. Autonomous agents may cooperate to evolve

E. Szczerbicki & N.T. Nguyen (Eds.): Smart Infor. & Knowledge Management, SCI 260, pp. 83–102.
springerlink.com © Springer-Verlag Berlin Heidelberg 2010

diverse and high quality solutions [11], and they may achieve it by dynamically evolving a population of solutions. Each agent possesses some problem-solving skills and the memory contains the population of problem solutions. Cyclically solutions are being sent from the common memory to agents and from agents back to the common memory. Agents cooperate through selecting and modifying these solutions according to the user-defined strategy referred to as the working strategy. The modifications can be constructive or destructive. An attempt to improve a solution can be successful or unsuccessful. Agents work asynchronously (each at its own speed) and in parallel.

The A-Team working strategy includes a set of rules for agent communication, selection of solution to be improved and management of the population of solutions which are kept in the common memory. In this chapter the influence of different strategies on A-Team performance is experimentally investigated. To implement various strategies and to carry out computational experiment the Jade-Based A-Team Platform called JABAT has been used.

In the next section of this chapter the literature reported implementations of the A-Team architectures are briefly reviewed followed by a more detailed description of the JABAT A-Team platform and its functionality. In further sections different working strategies with respect to selecting solutions to be improved by the A-Team members and replacing the solutions stored in the common memory by the improved ones, are studied. To evaluate typical working strategies the computational experiment has been carried out using several benchmark data sets. Computational experiment design and analysis of the experiment results conclude the chapter.

2   Jade-Based A-Team Platform

The reported implementations of the A-Team concept include two broad classes of systems: dedicated A-Teams and platforms, environments or shells used as tools for constructing specialized A-Team solutions. Such specialized A-Teams are usually not flexible and can be used for solving only the particular types of problems they are designed for. Among example A-Teams of such type one can mention the OPTIMA system for the general component insertion optimization problem [10] or A-Team with a collaboration protocol based on a conditional measure of agent effectiveness designed for Flow optimization of railroad traffic [4].

Among platforms and environments used to implement A-Team concept some well known include IBM A-Team written in C++ with own configuration language [11] and Bang 3 - a platform for the development of Multi-Agent Systems (MAS) [13]. Some implementations of A-Team were based on universal tools like Matlab [15]. Some other were written using algorithmic languages like, for example the parallel A-Team of [5] written in C and run under PVM operating system. The above discussed platforms and environments belong to the first generation of A-Team tools. They are either not portable or have limited portability, they also have none or limited scalability. Agents are not in conformity with the Foundation of Intelligent Psychical Agents (FIPA, http://www.fipa.org) standards and there are no interoperability nor Internet accessibility. Migration of agents is either impossible or limited to a single software platform.

In the class of A-Team platforms the middleware platform referred to as JABAT was proposed in [7]. It was intended to become the first step towards next generation A-Teams which are portable, scalable and in conformity with the FIPA standards. JABAT allows to design and implement an A-Team architecture for solving combinatorial optimization problems. JABAT can be also extended to become a fully Internet-accessible solution [2], with the ability not only to access the system in order to solve own combinatorial problems, but also to add own computer to the system with a view to increase the resources used for solving the problem at hand.

Main features of JABAT include the following:

- The system can solve instances of several different problems in parallel.
- The user, having a list of all algorithms implemented for the given problem may choose how many and which of them should be used.
- The optimization process can be carried out on many computers. The user can easily add or delete a computer from the system. In both cases JABAT will adapt to the changes, commanding the optimizing agents working within the system to migrate. JABAT may also clone some already working agents and migrate the clones, thus increasing the number of concurrently operating agents.
- The system is fed in the batch mode - consecutive problems may be stored and solved later, when the system assesses that there is enough resources to undertake new searches.

JABAT produces solutions to combinatorial optimization problems using a set of optimizing agents, each representing an improvement algorithm. The process of solving of the single task (i.e. the problem instance) consists of several steps. At first the initial population of solutions is generated. Individuals forming the initial population are, at the following computation stages, improved by independently acting agents, thus increasing chances for reaching the global optimum. Finally, when the stopping criterion is met, the best solution in the population is taken as the result. How the above steps are carried out is determined by the "working strategy". There might be different strategies defined in the system, each of them specifying:

- How the initial population of solutions is created.
- How to choose solutions which are forwarded to the optimizing agents for improvement.
- How to merge the improved solutions returned by the optimizing agents with the whole population (for example they may be added, or may replace random or worse solutions).
- When to stop searching for better solutions (for example after a given time, or after no better solution has been found within a given time).

For each user's task the system creates report files that include:

- The best solution obtained so far (that of the maximum or minimum value of fitness).
- Average value of fitness among solutions from the current population.
- Actual time of running and list of changes of the best solution (for each change of the best solution in the process of solving the time and the new value of fitness is given).

The report on the process of searching for the best solution may be later analyzed by the user. It can be easily read into a spreadsheet and converted into a summary report with the use of the pivot table.

The JABAT middleware is built with the use of JADE (Java Agent Development Framework), a software framework proposed by TILAB (http://jade.tilab.com/) for the development and run-time execution of peer-to-peer applications. JADE is based on the agents paradigm in compliance with the FIPA specifications and provides a comprehensive set of system services and agents necessary for distributed peer-to peer applications in the fixed or mobile environment. It includes both the libraries required to develop application agents and the run-time environment that provides the basic services and must be running on the device before agents can be activated [3].

JABAT itself has also been designed in such a way, that it can be easily instantiated in new environments where Java and JADE are available. To do so, JABAT with at least one set of objects (e.g. task, solutions, optimizing agent) intended to solve instances of the given problem have to be set up on a computer, and in the minimal configuration it is enough to solve an instance of the problem. Such a minimal configuration may be extended by adding more optimizing agents, more replacement strategies, new agents for solving more problems, more resources, e.g. attaching a computer (or computers) to which the optimizing agents could migrate to work on.

It is worth mentioning that JABAT has been designed in such a way, that all the above extensions that require some amount of programming (like solving new problems, solving them with new optimizing algorithms or with the use of new working strategies) can be carried-out with relatively little amount of work. JABAT makes it possible to focus only on defining these new elements, while the processes of communication and population management procedures will still work. More detailed information about extending the functionality of JABAT can be found in [1].

The JABAT environment is expected to be able to produce solutions to difficult optimization problems using a set of agents, each representing an improvement algorithm. To escape getting trapped into a local optimum an initial population of solutions called individuals is generated and stored in the common memory accessible to all agents within the A-Team. Individuals, during computations, are eventually improved by agents, thus increasing chances for reaching a good quality or, at least, a satisfactory solution.

Main functionality of the proposed tool includes generation of the initial population of solutions; application of solution improvement algorithms which draw individuals from the common memory and store them back after attempted improvement using some user defined working strategy and continuing the reading-improving-replacing cycle until a stopping criterion has been met.

The above functionality is realized by two main types of classes. The first one includes *OptiAgents*, which are implementations of improvement algorithms. The second are *SolutionManagers*, which are agents responsible for maintenance and updating of individuals in the common memory. Each *SolutionManager* is responsible for finding the best solution for a single instance of the problem. All agents act in parallel and communicate with each other exchanging solutions that are either to be improved (when solutions are sent to *OptiAgent*) or stored back (when solutions are sent to *SolutionManager*). An *OptiAgent* has two basic behaviors defined. The first behavior is sending around messages on readiness for action and indicating the required number of individuals (solutions). The second behavior is activated upon receiving a message from one of the *SolutionManagers* containing the problem instance description and the required number of individuals. This behavior involves an attempt to improve the fitness of the individual or individuals that have been supplied and resend the improved ones back to the sender.

To create a *SolutionManager* two pieces of information are needed: the name of the class that represents task and the name of the selected or proposed working strategy. The working strategy defines how the initial population is created, how solutions are selected to be sent to optimizing agents, how solutions that have been received from optimizing agents are merged with the population, and when the process of searching stops. A simple working strategy can, for example, draw a random solution and, when an improved solution is received back, replace the worst solution in the population with it. A more sophisticated strategy can introduce recursion: it allows for division of the task into smaller tasks, solving them and then merging the results to obtain solutions to the initial task. Such working strategy may be used to solve problem instances for which specific methods responsible for dividing and merging instances have been defined.

Apart from *OptiAgents* and *SolutionManagers* there are also other agents working within the system: responsible for initializing the process of solving of an instance of a problem, organizing the process of migrations and moving optimization agents among available computers that have joined the JABAT platform, or agents recording and storing computation results.

3 Computational Experiment Design

The user of the JABAT platform enjoys considerable freedom with respect to defining and implementing a working strategy understood as a procedure or a set of rules responsible for the process of maintaining the common memory and selecting individuals, which at the given computation stage should be forwarded to optimizing agents for improvement. Designing effective and useful working strategy could be supported by use of the knowledge on how a different working strategies may influence the A-Team performance. Such knowledge can be possibly obtained during the so called fine-tuning stage, which covers activities aiming at setting or identifying the "best" values of various parameters characterizing the working strategy which is to be applied within the particular A-Team implementation. Unfortunately, considering and analyzing all possible variants of the working strategy settings during the fine-tuning stage would be for obvious reasons too costly and too time consuming. On the other hand, it might be possible to extract certain pieces of knowledge that have at least some general value and are reliable enough to be used to reduce the effort needed to design an effective working strategy.

In an attempt to extract some general knowledge on how various strategies influence the A-Team performance it has been decided to design and carry-out complex computational experiment. The idea is to construct several A-Team implementations solving different and difficult combinatorial optimization problems. Next, instances of each out of the selected problems are going to be solved by the respective A-Team using different working strategies. In the next stage of the experiment effectiveness of different working strategies over the whole set of the considered problem instances for all investigated problems should be analyzed with a view to extract and validate some rules (or other pieces of knowledge).

The following four well-known combinatorial optimization problems have been selected to be a part of the experiment:

- Euclidean planar traveling salesman problem (EPTSP).
- Vehicle routing problem (VRP).
- Clustering problem (CP).
- Resource-constrained project scheduling problem (RCPSP).

The euclidean planar traveling salesman problem is a particular case of the general TSP. Given n cities (points) in the plane and their Euclidean distances, the problem is to find the shortest TSP-tour, i.e. a closed path visiting each of the n cities exactly once.

The vehicle routing problem can be stated as the problem of determining optimal routes through a given set of locations (customers) and defined on a directed graph $G=(V, E)$, where $V = \{0, 1, ..., N\}$ is the set of nodes and E is the set of edges. Node 0 is a depot with NV identical vehicles of capacity W. Each other node $i \in V\backslash\{0\}$ denotes a customer with a non-negative demand d_i. Each link $(i,j) \in E$ denotes the shortest path from customer i to j and is described by the cost c_{ij} of travel from i to j ($i,j = 1, ..., N$). It is assumed that $c_{ij} = c_{ji}$. The goal is to find vehicle routes which minimize total cost of travel (or travel distance) such that each route starts and ends at the depot, each customer is serviced exactly once by a single vehicle and the total load on any vehicle associated with a given route does not exceed vehicle capacity.

The clustering problem can be defined as follows. Given a set of N data objects, partition the data set into K clusters, such that similar objects are grouped together and objects with different features belong to different groups. Clustering arbitrary data into clusters of similar items presents the difficulty of deciding what similarity criterion should be used to obtain a good clustering. It can be shown that there is no absolute "best" criterion which would be independent of the final aim of the clustering. Euclidean distance and squared Euclidean distance are probably the most commonly chosen measures of similarity. Partition defines the clustering by giving for each data object the cluster index of the group to which it is assigned. The goal is to find such a partition that minimizes an objective function, which, for instance, is the sum of squared distances of the data objects to their cluster representatives.

The resource-constrained project scheduling problem consists of a set of n activities, where each activity has to be processed without interruption to complete the project. The dummy activities 1 and n represent the beginning and the end of the project. The duration of the activity $j, j = 1, ..., n$ is denoted by d_j where $d_1 = d_n = 0$. There are r renewable resource types. The availability of each resource type k in each time period is r_k units, $k = 1, ..., r$. Each activity j requires r_{jk} units of resource k during each period of its duration where $r_{1k} = r_{nk} = 0, k = 1, ..., r$. All parameters are non-negative integers. There are precedence relations of the finish-start type (FS) with a zero parameter value (i.e. $FS = 0$) defined between the activities. In other words, activity i precedes activity j if j cannot start until i has been completed. The structure of the project can be represented by an activity-on-node network $G = (SV, SA)$, where

SV is the set of activities and SA is the set of precedence relationships. SS_j (SP_j) is the set of successors (predecessors) of activity $j, j = 1, \ldots, n$. It is further assumed that $1 \in SP_j, j = 2, \ldots, n$, and $n \in SS_j, j = 1, \ldots, n-1$. The objective is to find a schedule S of activities starting times $[s_1, \ldots, s_n]$, where $s1 = 0$ and resource constraints are satisfied, such that the schedule duration $T(S) = s_n$ is minimized. The above formulated problem is a generalization of the classical job shop scheduling problem.

To solve instances of each of the above described problems four specialized A-Teams have been designed and implemented. A-Teams have different sets and numbers of optimization agents used with identical remaining agents (solution managers and auxiliary agents).

Instances of the euclidean planar traveling salesman problem are solved using two kinds of optimization procedures:

- *OptiDiff* randomly selects a number of points (cities) from the current solution under improvement and constructs a random sequence of these points as a part of the new (improved) solution. The rest of the original solution remains as much unchanged as feasible.
- *OptiCross* is a simple crossover operation. Out of the two solutions the offspring is produced through randomly selecting a part of the solution from the first parent and supplementing it with sequence (or sequences) of points (cities) from the second parent.

It should be mentioned that both procedures produce rather bad quality solutions when used individually. There exists a very effective heuristic known as Lin-Kerninghan algorithm [8]. It has not been used in the experiment purposefully, since in most cases it would find the optimal solution in one run, which would not give much insight into how working strategy influences A-Team performance.

Instances of the vehicle routing problem are solved using the following optimization procedures:

- *Opti3Opt* – an agent which is an implementation of the 3-opt local search algorithm, in which for all routes first three edges are removed and next remaining edges are reconnected in all possible ways.
- *Opti2Lambda* – an implementation of the local search algorithm based on λ-interchange local optimization method [9]. It operates on two selected routes and is based on the interchange/move of customers between routes. For each pair of routes from an individual a parts of routes of length less than or equal to λ are chosen and next these parts are shifted or exchanged, according to the selected operator. Possible operators are defined as pairs: (v, u), where $u, v = 1, \ldots, \lambda$. and denote the lengths of the part of routes which are moved or exchanged. For example, operator $(2,0)$ indicates shifting two customers from the first route to the second route, operator $(2,2)$ indicates an exchange of two customers between routes. Typically, $\lambda = 2$ and such value was used in the reported implementation.
- *OptiLSA* – an implementation of local search algorithm which operate on two selected routes. First, a node (customer) situated relatively far from the centroid of the first route is removed from it and next it is inserted to the second route.

Instances of the clustering problem are solved using the following optimization procedures:

- *OptiRLS* (Random Local Search) - a simple local search algorithm which finds the new solution by exchanging two randomly selected objects belonging to different clusters. The new solution replaces the current one if it is an improvement.
- *OptiHCLS* (Hill-Climbing Local Search) – a simple local search algorithm which finds the new solution by allocating the randomly selected object, from the randomly chosen cluster, to the cluster with minimal euclidean distance to the mean vector. The new solution replaces the current one if it is an improvement.
- *OptiTS* (Tabu Search) - a local search algorithm with tabu active list, which modifies the current solution by allocating the randomly selected object not on the tabu list from the randomly chosen cluster, to other randomly selected cluster. Next, the object is placed on the tabu list and remains there for a given number of iterations. The new solution replaces the current one if it is an improvement.

Instances of the resource-constrained project scheduling problem are solved using the following optimization procedures [6]:

- *OptiLSA* - a simple local search algorithm which finds the local optimum by moving each activity to all possible places within the current solution.
- *OptiCA* - a heuristic based on using the one point crossover operator. Two initial solutions are repeatedly crossed until a better solution will be found or all crossing points will be checked.
- *OptiPRA* – a procedure based on the path-relinking algorithm. For a pair of solutions from the population a path between them is constructed. Next, the best of the feasible solutions is selected.

In the computational experiment each of the four A-Team implementations has been used to solve 5 instances of the respective combinatorial optimization problem. All these instances have been taken from several well-known benchmark datasets libraries as shown in Table 1.

Table 1. Instances used in the reported experiment.

Problem	Instance names and dimensions	Source
EPTSP	pr76 (76 cities), pr107 (107), pr124 (124), pr152 (152), pr226 (226)	TSPLIB, http://www.iwr.uni-heidelberg.de/groups/comopt/software/TSPLIB95/tsp/
VRP	vrpnc1 (50 customers), vrpnc2 (75), vrpnc3 (100), vrpnc4 (150), vrpnc5 (199)	OR-Library, http://people.brunel.ac.uk/~mastjjb/jeb/orlib/vrpinfo.html
CP	Ruspini (75 objects, 2 attributes), Iris (150, 4), Cleveland heart disease (303, 13), Credit approval (690,15)	[12] UCI Machine Learning Repository, http://archive.ics.uci.edu/ml/
RCPSP	j3013_01 (30 activities), j3029_09 (30), j6009_02 (60), j9021_03 (90), j12016_01 (120)	PSPLIB, http://129.187.106.231/psplib

In the reported experiment 10 working strategies have been investigated. All strategies assume that:

- Initially the common memory of each of the respective A-Teams contains a population of solutions called individuals.
- All individuals are feasible solutions of the instance which A-Team is attempting to solve.
- All these individuals have been generated randomly.

Differences between strategies concern the way individuals are selected for improvement and the way the improved individuals are stored in the common memory. Selection of individuals for improvement can be a purely random move (RM) or can be a random move with the so called blocking (RB). Blocking means that once selected individual (or subset of individuals) can not be selected again until all other individuals or their subsets have been tried.

Storing an individual back in the common memory after an *OptiAgent* has attempted to improve it, can be based either on the replacement or on the addition. Replacement can be carried out in three ways:

- Random replacement (RR) – returning individual replaces randomly selected individual in the common memory.
- Replacement of the worse (RW) - returning individual replaces first found worse individual. Replacement does not take place if worse individual can not be found in the common memory.
- Replacement of the worse with exchange (RE) – returning individual replaces the first found worse individual. If a worse individual can not be found within a certain number of reviews (review is understood as a search for the worse individual after an improved solution is returned by the *SolutionManager*) then the worst individual in the common memory is replaced by the randomly generated one representing a feasible solution.

There are two addition procedures considered:

- Addition with the stochastic universal sampling (SUS) - returning individual is stored in the common memory increasing the number of individuals in the population by one. After the common memory grows to contain a certain number of individuals, its initial size will be restored using the stochastic universal sampling selection procedure.
- Addition with the tournament selection (TS) - returning individual is stored in the common memory increasing the number of individuals in the population by one. After the common memory grows to contain a certain number of individuals its initial size will be restored using the stochastic tournament selection procedure.

Combination of procedures for selecting individuals for improvement and storing back the improved ones in the common memory defines 10 working strategies investigated under the reported experiment as shown in Table 2.

Table 2. The investigated working strategies.

No.	Strategy	Procedures used
1.	RM-RR	Random move selection + Random replacement
2.	RM-RW	Random move selection + Replacement of the worse
3.	RM-RE	Random move selection + Replacement of the worse with exchange
4.	RM-SUS	Random move selection + Addition with the stochastic universal sampling
5.	RM-TS	Random move selection + Addition with the tournament selection
6.	RB-RR	Random move selection with blocking + Random replacement
7.	RB-RW	Random move selection with blocking + Replacement of the worse
8.	RB-RE	Random move selection with blocking + Replacement of the worse with exchange
9.	RB-SUS	Random move selection with blocking + Addition with the stochastic universal sampling
10.	RB-TS	Random move selection with blocking + Addition with the tournament selection

Pseudo codes of all of the above procedures are shown in Figures 1-7.

```
Procedure RM
Input: P - population of solutions; n - number of
solutions requested;
Output: O - solutions selected from P;
1. Choose at random n solutions from P
2. Add  chosen solutions to O
3. Return O
```

Fig. 1. Pseudo code of the RM procedure

```
Procedure RB
Input: P - population of solutions; n - number of
solutions requested; L - population of blocked
solutions;
Output: O - solutions selected from P; L -
population of blocked solutions;
1. Set m as the number of solutions in L
2. Set l as the number of solutions in P
3. If ((n + m) > l ) then L := ∅
4. Set P' := P - L
5. Choose at random n solutions from P'
6. Add chosen solutions to O and L
7. Return O, L
```

Fig. 2. Pseudo code of the RB procedure

Procedure RR
Input: *P* - population of solutions; *s* - returning solution;
Output: *P* - population of solutions;
1. If (*s*∈*P*) then goto 5
2. Select randomly individual *k* from *P*
3. Remove *k* from *P*
4. Add *s* to *P*
5. Return *P*

Fig. 3. Pseudo code of the RR procedure

Procedure RW
Input: *P* - population of solutions; *s* - returning solution;
Output: *P* - population of solution;
1. If (*s*∈*P*) then goto 5
2. Set *f* := false
3. Find *k* which is a solution worse than *s* and then set *f* := true
4. If (*f* is true) then (remove *k* from *P* and add *s* to *P*)
5. Return *P*

Fig. 4. Pseudo code of the RW procedure

Procedure RE
Input: *P* - population of solutions; *s* - returning solution; *x* - number of the allowed reviews; *Px* = maximum number of the allowed reviews;
Output: *P* - population of solution; x - number of reviews;
1. Set *f* := false
2. Find *k* which is a solution worse than *s* and then set *f*: = true
3. If (*f* is not true) goto 8
4. Remove *k* from *P*
5. Add *s* to *P*
6. Set *x*:=0
7. goto 15
8. *x*:= *x* +1
9. If (*x* < *Px*) then goto 15
10. Find *k* which is the worst solution in *P*
11. Remove *k* from *P*
12. Generate at random new solution *p*
13. Add *p* to *P*
14. Set *x*:=0
15. Return *P*, x

Fig. 5. Pseudo code of the RE procedure

Procedure SUS
Input: P – population of solutions; s – returning solution; $Pmax$ – maximum population size; $Pmin$ – minimum population size;
Output: P – population of solution;
1. If $(s \notin P)$ then add s to P
2. Set l as the number of solutions in P
3. If $(l < Pmax)$ then goto 15
4. Set $P' := \emptyset$
5. Order P by fitness
6. Calculate $total\_fitness$ as $\Sigma_{t \in P}$ fitness of t
7. Set $start$ by drawing it at random from $[0, ..., total\_fitness/l]$
8. Set $i := 0$
9. Set $ptr := start + i * total\_fitness/l$
10. Let $k =$ index of the solution from P such that: $(\Sigma_{j<k}$ fitness of the jth solution$) <= ptr$ and $(\Sigma_{j \cdot k}$ fitness of the jth solution$) > ptr$
11. Add kth solution from P to P'
12. Set $i := i+1$
13. If $(i < Pmin)$ then goto 9
14. Set $P := P'$
15. Return P

Fig. 6. Pseudo code of the SUS procedure

Procedure TS
Input: P – population of solutions; s – returning solution; $Pmax$ – maximum population size; $Pmin$ – minimum population size;
Output: P – population of solution;
1. If $(s \notin P)$ then (add s to P)
2. Set l as the number of solutions in P
3. If $(l < Pmax)$ then goto 12
4. Set $P' := \emptyset$
5. Set $m := 0$;
6. Choose randomly individuals t and r from P
7. Set q as best of the $\{t, r\}$ or worst of the $\{t, r\}$ with respective probabilities 0.75 and 0.25
8. Add q to P'
9. Set $m := m + 1$
10. If $(m < Pmin)$ then goto 6
11. Set $P := P'$
12. Return P

Fig. 7. Pseudo code of the TS procedure

In the reported experiment each of the four implemented A-Teams has been used to solve 5 instances of the respective problem using the above defined strategies. Each instance under each strategy has been solved 40 times. Hence, altogether, the experiment involved 8000 runs. The following parameter settings have been used in the computations:

- Number of individuals in the initial population (l) – 50.
- Maximum number of the allowed reviews in the RE Procedure (Px) – 5.
- Maximum population size ($Pmax$) – 100.
  Minimum population size ($Pmin$) – 50.
- Termination condition – 120 sec. without a solution improvement.

Experiment has been carried out on the cluster Holk of the Tricity Academic Computer Network built of 256 Intel Itanium 2 Dual Core with 12 MB L3 cache processors with Mellanox InfiniBand interconnections with 10Gb/s bandwidth.

4   Computational Experiment Results

Tables 3-6 shows mean values and standard percentage deviations of the respective fitness functions calculated for all computational experiment runs as specified in the experiment design described in the previous section.

The main question addressed by the reported experiment is to decide whether the choice of the working strategy has influence on the quality of solutions obtained by A-Teams? To answer this question the one-way analysis of variance (ANOVA) for each of the considered problems has been carried out. For the purpose of the analysis the following hypotheses have been formulated:

- H_0 - zero hypothesis: the choice of strategy does not influence the quality of solutions (mean values of the fitness function are not statistically different).
- H_1 - alternative hypothesis: the quality of solutions is not independent from the working strategy used (mean values of the fitness function are statistically different).

The analysis has been carried out at the significance level of 0.05. In Table 7 values of F-test statistics ratios calculated for each of the considered problem instances are shown. The value of F-distribution calculated for 9 degrees of freedom between groups and 190 degrees of freedom within groups is equal to 1.92692532432991.

The one-way analysis of variance (ANOVA) which result are shown in Table 7 allows to observe the following:

- Zero hypothesis stipulating that choice of strategy does not influence the quality of solutions (mean values of the fitness function are not statistically different) is accepted in 8 out of 20 considered cases.
- In 12 out of 20 cases zero hypothesis is rejected in favor of the alternative hypothesis suggesting that the quality of solutions is not independent from working strategy used (mean values of the fitness function are statistically different).

Table 3. Mean values and standard deviations of the fitness function calculated for EPTSP instances and working strategies.

Instance	RM-RR	RM-RW	RM-RE	RM-SUS	RM-TS	RB-RR	RB-RW	RB-RE	RB-SUS	RB-TS
pr226	370433.11	134131.47	130604.33	607786.97	230627.32	385272.63	143745.57	187894.67	719787.33	144604.51
	+/- 8.2%	+/- 10.1%	+/- 12.1%	+/- 12.4%	+/- 8.9%	+/- 9.3%	+/- 11.2%	+/- 10.2%	+/- 6.3%	+/- 10.8%
pr152	219594.53	91802.20	91054.06	313883.69	130833.75	221802.27	90334.50	104501.48	406450.26	234647.65
	+/- 9.5%	+/- 5.9%	+/- 4.1%	+/- 9.2%	+/- 8.7%	+/- 7.9%	+/- 3.0%	+/- 5.7%	+/- 7.5%	+/- 10.5%
pr124	153862.72	65416.75	65030.01	202245.54	96026.38	147691.59	65631.03	68580.27	251220.26	149472.16
	+/- 6.1%	+/- 2.6%	+/- 2.9%	+/- 8.6%	+/- 6.1%	+/- 8.4%	+/- 3.4%	+/- 4.0%	+/- 7.8%	+/- 12.5%
pr107	144604.51	50302.34	50334.45	141219.92	63358.44	95846.15	50471.94	51517.01	173062.54	91369.22
	+/- 10.8%	+/- 3.9%	+/- 3.1%	+/- 10.9%	+/- 4.6%	+/- 9.5%	+/- 2.8%	+/- 4.1%	+/- 9.4%	+/- 10.3%
pr76	198386.06	111362.53	111397.66	197078.51	127565.39	171968.02	111164.74	112566.87	220678.21	155398.92
	+/- 5.7%	+/- 1.2%	+/- 1.1%	+/- 6.3%	+/- 3.6%	+/- 3.7%	+/- 1.2%	+/- 1.6%	+/- 3.9%	+/- 5.0%

Table 4. Mean values and standard deviations of the fitness function calculated for VRP instances and working strategies.

Instance	RM-RR	RM-RW	RM-RE	RM-SUS	RM-TS	RB-RR	RB-RW	RB-RE	RB-SUS	RB-TS
vrpnc1	524.61	524.96	524.61	524.72	524.72	524.67	525.03	524.61	525.15	524.61
	+/- 0.0%	+/- 0.1%	+/- 0.0%	+/- 0.1%	+/- 0.1%	+/- 0.0%	+/- 0.2%	+/- 0.0%	+/- 0.2%	+/- 0.0%
vrpnc2	855.98	850.41	852.79	854.40	852.76	852.37	854.37	852.05	854.36	854.03
	+/- 0.6%	+/- 0.8%	+/- 0.6%	+/- 0.6%	+/- 0.7%	+/- 0.7%	+/- 0.6%	+/- 0.6%	+/- 0.5%	+/- 0.5%
vrpnc3	841.07	839.32	839.76	840.59	838.26	839.11	840.64	838.40	840.07	841.35
	+/- 0.3%	+/- 0.4%	+/- 0.4%	+/- 0.4%	+/- 0.5%	+/- 0.4%	+/- 0.4%	+/- 0.5%	+/- 0.4%	+/- 0.5%
vrpnc4	1067.17	1068.92	1068.82	1065.27	1064.36	1066.96	1068.72	1064.45	1067.15	1067.61
	+/- 0.5%	+/- 0.8%	+/- 0.5%	+/- 0.6%	+/- 0.6%	+/- 0.6%	+/- 0.4%	+/- 0.7%	+/- 0.6%	+/- 0.6%
vrpnc5	1365.43	1368.23	1369.59	1367.51	1366.29	1366.17	1366.20	1363.18	1367.97	1366.65
	+/- 0.7%	+/- 0.6%	+/- 0.5%	+/- 0.5%	+/- 0.6%	+/- 0.7%	+/- 0.5%	+/- 0.6%	+/- 0.6%	+/- 0.5%

Table 5. Mean values and standard deviations of the fitness function calculated for CP instances and working strategies.

Instance	RM-RR	RM-RW	RM-RE	RM-SUS	RM-TS	RB-RR	RB-RW	RB-RE	RB-SUS	RB-TS
Ruspini (k=4)	1759.65 +/- 14.4%	1641.78 +/- 9.3%	1520.54 +/- 13.5%	1644.00 +/- 13.1%	1623.90 +/- 13.4%	1576.84 +/- 12.8%	1693.61 +/- 13.2%	1629.18 +/- 14.2%	1721.62 +/- 12.9%	1694.31 +/- 14.9%
Ruspini (k=10)	30.61 +/- 9.5%	36.66 +/- 6.7%	38.53 +/- 9.0%	36.86 +/- 10.6%	26.63 +/- 7.1%	38.54 +/- 9.2%	38.83 +/- 10.2%	35.23 +/- 2.5%	39.24 +/- 12.2%	38.32 +/- 8.9%
Iris	178.56 +/- 5.3%	175.15 +/- 6.2%	174.57 +/- 3.6%	179.82 +/- 5.9%	178.66 +/- 4.5%	177.47 +/- 3.9%	171.78 +/- 3.3%	173.69 +/- 6.4%	181.56 +/- 6.4%	176.79 +/- 5.2%
Credit	790.60 +/- 8.1%	964.09 +/- 6.1%	799.86 +/- 5.1%	822.89 +/- 4.9%	801.37 +/- 6.3%	807.35 +/- 5.3%	846.22 +/- 4.9%	787.10 +/- 5.4%	914.80 +/- 6.4%	842.39 +/- 6.2%
Hart	353.33 +/- 6.7%	408.98 +/- 7.3%	345.62 +/- 10.5%	398.35 +/- 4.9%	339.61 +/- 5.4%	369.83 +/- 6.7%	385.47 +/- 7.2%	377.02 +/- 5.2%	373.95 +/- 5.8%	351.83 +/- 7.9%

Table 6. Mean values and standard deviations of the fitness function calculated for RCPSP instances and working strategies.

Instance	RM-RR	RM-RW	RM-RE	RM-SUS	RM-TS	RB-RR	RB-RW	RB-RE	RB-SUS	RB-TS
j3013_01	60.35 +/- 1.3%	60.35 +/- 1.2%	60.15 +/- 1.2%	60.25 +/- 1.1%	60.30 +/- 1.4%	60.40 +/- 1.1%	59.95 +/- 1.5%	60.35 +/- 1.1%	60.70 +/- 0.8%	60.45 +/- 1.0%
j3029_09	98.80 +/- 1.1%	99.40 +/- 1.3%	98.75 +/- 0.9%	99.25 +/- 1.2%	98.55 +/- 1.1%	98.75 +/- 1.0%	99.10 +/- 1.1%	98.45 +/- 1.0%	99.20 +/- 1.0%	99.25 +/- 1.2%
j6009_02	85.20 +/- 0.6%	85.20 +/- 0.8%	85.40 +/- 0.7%	85.20 +/- 0.8%	85.30 +/- 0.8%	85.25 +/- 0.7%	85.35 +/- 0.6%	85.40 +/- 0.6%	85.40 +/- 0.7%	85.20 +/- 0.7%
j9021_03	129.45 +/- 0.8%	129.05 +/- 0.9%	129.75 +/- 0.9%	129.25 +/- 0.8%	129.40 +/- 0.8%	129.30 +/- 0.8%	129.20 +/- 0.9%	129.25 +/- 0.7%	129.25 +/- 1.0%	129.00 +/- 1.1%
j12016_01	214.50 +/- 0.6%	215.60 +/- 1.0%	217.35 +/- 1.6%	212.55 +/- 0.6%	215.10 +/- 2.5%	213.25 +/- 0.8%	213.30 +/- 0.8%	213.55 +/- 0.7%	213.15 +/- 0.7%	213.70 +/- 0.5%

Table 7. F-test statistics for the reported experiment.

Problem		Value of F-test	H_0 Accepted
EPTSP	pr226	743.866557	-
	pr152	730.45726	-
	pr124	661.101448	-
	pr107	480.721529	-
	pr76	791.200912	-
VRP	vrpnc1	3.00992674	-
	vrpnc2	1.77428553	YES
	vrpnc3	1.93417536	-
	vrpnc4	1.37679857	YES
	vrpnc5	0.91781823	YES
CP	Ruspini (k=4)	1.44484895	YES
	Ruspini (k=10)	3.02387219	-
	Iris	2.24965623	-
	Credit	3.25508912	-
	Heart	2.94237222	-
RCPSP	j3013_01	1.45465253	YES
	j3029_09	1.8316856	YES
	j6009_02	0.44827037	YES
	j9021_03	0.69595177	YES
	j12016_01	6.94431695	-

Taking into account that the problems under consideration are not homogenous it has not been possible to carry out   analysis of variance with additional factor with reference to suitability of particular working strategies to particular problems. Instead it has been decided to use the non-parametric Friedman test to obtain the answer to the question weather particular working strategies are equally effective independently of the kind of problem being solved.

The above test has been based on weights (points) assigned to working strategies used in the experiment. To assign weights the 10 point scale has been used with 10 points for the best and 1 point for the worst strategy. The test aimed at deciding among the following hypotheses:

- H_0 – zero hypothesis: working strategies are statistically equally effective regardless the kind of problem which instances are being solved.
- H_1 - alternative hypothesis: not all working strategies are equally effective.

Again the analysis has been carried out at the significance level of 0.05. The respective value of the χ^2 statistics with 10 working strategies and 20 instances of the

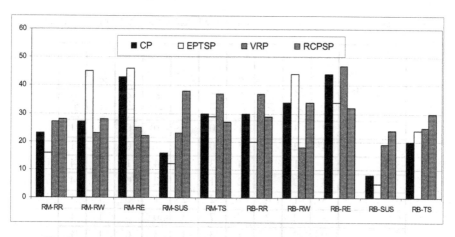

Fig. 8. The Friedman test weights (in points) for each problem and each strategy.

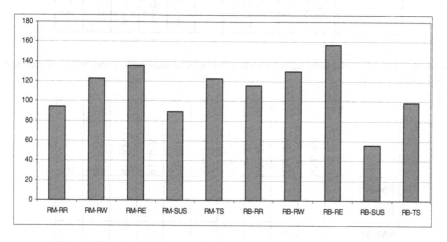

Fig. 9. The total of the Friedman weights for the investigated strategies.

considered problems is 349.68 and the value of χ^2 distribution is equal to 16.919 for 9 degrees of freedom. Thus it can be observed that not all working strategies are equally effective regardless of the kind of problem which instances are being solved.

In Fig. 8 and 9 sums of weights obtained by the working strategies for each of the considered problems as well as overall total of weights for each working strategy are, respectively, compared. Besides, in Table 8 distributions of points obtained from the Friedman test by working strategies are shown.

Friedman test results as well as totals from Fig. 8 and 9 show that strategies RM-RW, RM-RE, RM-TS, RB-RR, RB-RW, RB-RE performed above the average. The highest score in total and with respect to each of the analyzed problem has been obtained by the working strategy based on selection with blocking and replacement

Table 8. Distribution of points obtained from the Friedman test by working strategies.

Scale (in points)	RM-RR	RM-RW	RM-RE	RM-SUS	RM-TS	RB-RR	RB-RW	RB-RE	RB-SUS	RB-TS
1	2	3	3	1	0	0	0	0	9	1
2	4	2	1	5	0	1	1	0	3	1
3	1	1	1	4	3	2	4	1	2	4
4	4	0	0	2	1	4	2	0	2	4
5	3	1	0	3	2	2	1	0	1	5
6	2	2	2	0	7	3	0	3	1	1
7	0	2	1	1	3	1	3	6	1	1
8	0	2	4	2	1	6	4	0	0	0
9	2	4	3	0	1	1	0	6	1	0
10	2	3	5	2	2	0	5	4	0	3

Table 9. Pairs of the working strategies showing no statistical differences in the Tukey test (pairs marked by X).

	RM-RR	RM-RW	RM-RE	RM-SUS	RM-TS	RB-RR	RB-RW	RB-RE	RB-SUS	RB-TS
RM-RR										
RM-RW										
RM-RE		X								
RM-SUS										
RM-TS										
RB-RR										
RB-RW		X	X							
RB-RE		X					X			
RB-SUS										
RB-TS										

with exchange (RB-RE). This strategy as a rule obtained scores between 5 and 10. This leads to the conclusion that in majority of cases RB-RE working strategy assures obtaining solutions of the above average quality.

To identify a significant differences between working strategies the Tukey test, based on the ANOVA results has been carried out. This test aimed at comparing pairs of the working strategies with a view to evaluate which pairs are statistically different and which are not. Strategies which do not differ in either of the pairs are shown in Table 9. The data from Table 9 together with ranges shown earlier in Fig. 8 and 9 allow to draw the following observation: working strategies with "Replacement of the worse" and "Replacement of the worse with exchange" procedures produce results of the statistically comparable quality. This remains true regardless of the procedure of selecting individuals to be sent for improvement. The above conclusion does not change the earlier observation that working strategy combining selection with blocking and replacement with exchange should be recommended as the one with the highest probability of producing the above average quality result.

5 Conclusions

The presented research confirms the importance of choosing an effective working strategy when employing A-Teams searching for solutions of computationally difficult combinatorial problems. The following identified pieces of knowledge can be useful in the process of designing an implementation of the A-Team dedicated to solving a particular optimization problem:

- The quality of solutions produced by the A-Team will not be independent from the working strategy used.
- Not all working strategies are equally effective, moreover the choice of strategy might be problem dependent.
- In majority of cases working strategy based on selection with blocking (RB) and replacement of the worse with exchange (RE) assures obtaining solutions of the above average quality.
- Good performance of the working strategy based on selection with blocking (RB) and replacement of the worse with exchange (RE) can be attributed to the fact that such selection assures that all individuals from the common memory have a chance to contribute to finding an improved solution. On the other hand replacement with exchange (RE) provides a useful mechanism of escaping from the local optimum.

Acknowledgments. Calculations have been performed in the Academic Computer Centre TASK in Gdańsk.

References

[1] Barbucha, D., Czarnowski, I., Jedrzejowicz, P., Ratajczak, E., Wierzbowska, I.: JADE-Based A-Team as a Tool for Implementing Population-Based Algorithms. In: Chen, Y., Abraham, A. (eds.) Intelligent Systems Design and Applications, ISDA, Jinan Shandong China, pp. 144–149. IEEE, Los Alamitos (2006)

[2] Barbucha, D., Czarnowski, I., Jędrzejowicz, P., Ratajczak-Ropel, E., Wierzbowska, I.: Web Accessible A-Team Middleware. In: Bubak, M., van Albada, G.D., Dongarra, J., Sloot, P.M.A. (eds.) ICCS 2008, Part III. LNCS, vol. 5103, pp. 624–633. Springer, Heidelberg (2008)

[3] Bellifemine, F., Caire, G., Poggi, A., Rimassa, G.: JADE. A White Paper, Exp. 3(3), 6–20 (2003)

[4] Blum, J., Eskandarian, A.: Enhancing intelligent agent collaboration for flow optimization of railroad traffic. Transportation Research Part A 36, 919–930 (2002)

[5] Correa, R., Gomes, F.C., Oliveira, C., Pardalos, P.M.: A parallel implementation of an asynchronous team to the point-to-point connection problem. Parallel Computing 29, 447–466 (2003)

[6] Jedrzejowicz, P., Ratajczak, E.: Population Learning Algorithm for Resource-Constrained Project Scheduling. In: Pearson, D.W., Steele, N.C., Albrecht, R.F. (eds.) Artificial Neural Nets and Genetic Algorithms, pp. 223–228. Springer Computer Science, Wien (2003)

[7] Jedrzejowicz, P., Wierzbowska, I.: JADE-Based A-Team Environment. In: Alexandrov, V.N., et al. (eds.) ICCS 2006. LNCS, vol. 3993, pp. 719–726. Springer, Heidelberg (2006)

[8] Lin, S., Kerningham, B.W.: An Effective Heuristic Algorithm for the Travelling Salesman Problem. Operations Research 21, 498–516 (1973)

[9] Osman, I.H.: Metastrategy Simulated Annealing and Tabu Search Algorithms for Vehicle Routing Problem. Annals of Operations Research 41, 421–451 (1993)

[10] Rabak, C.S., Sichman, J.S.: Using A-Teams to optimize automatic insertion of electronic components. Advanced Engineering Informatics 17, 95–106 (2003)

[11] Rachlin, J., Goodwin, R., Murthy, S., Akkiraju, R., Wu, F., Kumaran, S., Das, R.: A-Teams: An Agent Architecture for Optimization and Decision-Support. In: Rao, A.S., Singh, M.P., Müller, J.P. (eds.) ATAL 1998. LNCS (LNAI), vol. 1555, pp. 261–276. Springer, Heidelberg (1999)

[12] Ruspini, E.H.: Numerical method for fuzzy clustering. Inform. Sci. 2(3), 19–150 (1970)

[13] Neruda, R., Krusina, P., Kudova, P., Rydvan, P., Beuster, G.: Bang 3: A Computational Multi-Agent System. In: Proc. of the IEEE/WIC/ACM International Conference on Intelligent Agent Technology, IAT 2004 (2004)

[14] Talukdar, S.N., Souza, P., Murthy, S.: Organizations for Computer-Based Agents. Engineering Intelligent Systems 1(2) (1993)

[15] Talukdar, S., Baerentzen, L., Gove, A., Souza, P.: Asynchronous teams: cooperation schemes for autonomous agents. Journal of Heuristics 4, 295–332 (1998)

Incremental Declarative Process Mining

Massimiliano Cattafi, Evelina Lamma, Fabrizio Riguzzi, and Sergio Storari

ENDIF – Università di Ferrara – Via Saragat, 1 – 44100 Ferrara, Italy
{massimiliano.cattafi,evelina.lamma}@unife.it
{fabrizio.riguzzi,sergio.storari@unife.it}

Abstract. Business organizations achieve their mission by performing a number of processes. These span from simple sequences of actions to complex structured sets of activities with complex interrelation among them. The field of Business Processes Management studies how to describe, analyze, preserve and improve processes. In particular the subfield of Process Mining aims at inferring a model of the processes from logs (i.e. the collected records of performed activities). Moreover, processes can change over time to reflect mutated conditions, therefore it is often necessary to update the model. We call this activity Incremental Process Mining. To solve this problem, we modify the process mining system DPML to obtain IPM (Incremental Process Miner), which employs a subset of the $\mathcal{S}$CIFF language to represent models and adopts techniques developed in Inductive Logic Programming to perform theory revision. The experimental results show that is more convenient to revise a theory rather than learning a new one from scratch.

Keywords: Business Processes, Process Mining, Theory Revision.

1 Introduction

In the current knowledge society, the set of business processes an organization performs in order to achieve its mission often represents one of the most important assets of the organization. It is thus necessary to be able to describe them in details, so that they can be stored and analyzed. In this way we can preserve and/or improve them. These problems are studied in the field of Business Processes Management (BPM) (see e.g. [1]).

Often organizations do not have a formal or precise description of the processes they perform. The knowledge necessary to execute the processes is owned by the individual workers but not by the organization as a whole, thus exposing it to possible malfunctions if a worker leaves.

However, modern information systems store all the actions performed by individual workers during the execution of a process. These action sequences are called traces and the set of all the traces recorded in a period of time is called a log. The Process Mining research area [2] proposes techniques for inferring a model of the process from a log.

Very often processes change over time to reflect mutated external or internal conditions. In this case, it is necessary to update their models. In particular,

E. Szczerbicki & N.T. Nguyen (Eds.): Smart Infor. & Knowledge Management, SCI 260, pp. 103–127.
springerlink.com   © Springer-Verlag Berlin Heidelberg 2010

given a process model and a new log, we want to modify the model so that it conforms also with the new log. We call this activity Incremental Process Mining. In this paper we show that revising an existing model may be more effective than learning a new model ex novo from the previous and new log. Moreover, in some cases the previous log may not be available, thus making model updating necessary.

We choose Logic Programming for the representation of traces and process models in order to exploit its expressiveness for the description of traces together with the wide variety of learning techniques available for it. An activity can be represented as a logical atom in which the predicate indicates the type of action and the arguments indicate the attributes of the action. One of the attributes is the time at which the action has been performed. Thus a trace can be represented as a set of instantiated atoms, i.e., a logical interpretation.

In order to represent process models, we use a subset of the $\mathcal{S}$CIFF language [3,4]. A model in this language is a set of logical integrity constraints in the form of implications. Given a $\mathcal{S}$CIFF model and a trace, there exists an interpreter that checks whether the trace satisfies or not the model. Such a representation of traces and models is declarative in the sense that we do not explicitly state the allowed execution flows but we only impose high level constraints on them.

DPML is able to infer a $\mathcal{S}$CIFF theory from a set of positive and negative traces. Positive traces represent correct executions of the business process, while negative traces represent process executions that have been judged incorrect or undesirable.

Given that these traces are represented as logical interpretations and that $\mathcal{S}$CIFF integrity constraints are similar to logical clauses, DPML (Declarative Process Model Learner) [5] employs Inductive Logic Programming techniques [6] for learning from interpretations [7]. In particular, it modifies the ICL system [8] that learns sets of logical clauses from positive and negative interpretations.

In this paper, we present the system IPM (Incremental Process Miner) that faces the problem of revising an existing theory in the light of new evidence. This system is an adaptation of DPML and adopts techniques developed in Inductive Logic Programming (such as [9]) to perform theory revision.

IPM generalizes theories that do not satisfy new positive traces, as well as specializes theories that do not exclude new negative examples. To this purpose, we exploit the generalization operator presented in [5] for $\mathcal{S}$CIFF theories. Moreover, we define a specialization operator.

IPM is experimentally evaluated on processes regarding the management of a hotel and an electronic auction protocol. In the "hotel management" case the available traces are divided into two sets: one containing "old" traces and one containing "new" traces. Then two experiments are performed: in the first we learn a theory with DPML using the "old" traces and we revise the theory with IPM using the "new" traces, while in the latter we learn a theory with DPML from "old" and "new" traces. Then we compare the accuracy of the final theories obtained and the running time. A similar comparison is performed on the auction protocol, except that in this case the initial theory is not learned using some "old" traces but it is a modified version of the actual model to simulate the revision of an imperfect theory written down by a user. Results associated to these experiments show that revising a theory is more efficient that inducing it from scratch. Moreover, the models obtained are more accurate on unseen data.

The paper is organized as follows. In Section 2 we recall the basic notions of Logic Programming, Inductive Logic Programming and Business Process Management. In Section 3 we discuss the representation of traces and models using Logic Programming. Section 4 illustrates the IPM algorithm. In Section 5 we report on the experiments performed. In Section 6 we discuss related works and we conclude with Section 7.

2   Preliminaries

We start by briefly recalling the basic concepts of Logic Programming, Inductive Logic Programming and Business Process Management.

2.1   Logic Programming

A *first order alphabet* Σ is a set of predicate symbols and function symbols (or functors) together with their arity. A *term* is either a variable or a functor applied to a tuple of terms of length equal to the arity of the functor. If the functor has arity 0 it is called a *constant*. An *atom* is a predicate symbol applied to a tuple of terms of length equal to the arity of the predicate. A *literal* is either an atom a or its negation $\neg a$. In the latter case it is called a *negative literal*. In logic programming, predicate and function symbols are indicated with alphanumeric strings starting with a lowercase character while variables are indicated with alphanumeric strings starting with an uppercase character.

A *clause* is a formula C of the form

$$h_1 \vee \ldots \vee h_n \leftarrow b_1, \ldots, b_m$$

where $h_1, \ldots, h_n$ are atoms and $b_1, \ldots, b_m$ are literals. A clause can be seen as a set of literals, e.g., C can be seen as

$$\{h_1, \ldots, h_n, \neg b_1, \ldots, \neg b_m\}.$$

In this representation, the disjunctions among the elements of the set are left implicit.

The form of a clause that is used in the following will be clear from the context. $h_1 \vee \ldots \vee h_n$ is called the *head* of the clause and $b_1, \ldots, b_m$ is called the *body*. We will use $head(C)$ to indicate either $h_1 \vee \ldots \vee h_n$ or $\{h_1, \ldots, h_n\}$, and $body(C)$ to indicate either $b_1, \ldots, b_m$ or $\{b_1, \ldots, b_m\}$, the exact meaning will be clear from the context. When $m = 0$, C is called a *fact*. When $n = 1$, C is called a *program clause*. When $n = 0$, C is called a *goal*. The conjunction of a set of literals is called a *query*. A clause is *range restricted* if all the variables that appear in the head appear as well in positive literals in the body.

A *theory* P is a set of clauses. A *normal logic program* P is a set of program clauses.

A term, atom, literal, goal, query or clause is *ground* if it does not contain variables. A *substitution* θ is an assignment of variables to terms: $\theta = \{V_1/t_1, \ldots, V_n/t_n\}$. The *application of a substitution to a term, atom, literal, goal, query or clause* C, indicated with $C\theta$, is the replacement of the variables appearing in C and in θ with the terms specified in θ.

The *Herbrand universe* $H_U(P)$ is the set of all the terms that can be built with function symbols appearing in P. The *Herbrand base* $H_B(P)$ of a theory P is the set of all the ground atoms that can be built with predicate and function symbols appearing in P. A *grounding* of a clause C is obtained by replacing the variables of C with terms from $H_U(P)$. The grounding $g(P)$ of a theory P is the program obtained by replacing each clause with the set of all of its groundings. A *Herbrand interpretation* is a set of ground atoms, i.e. a subset of $H_B(P)$. In the following, we will omit the word Herbrand.

Let us now define the truth of a formula in an interpretation. Let I be an interpretation and ϕ a formula, ϕ is true in I, written $I \models \phi$ if

- $a \in I$, if ϕ is a ground atom a;
- $a \notin I$, if ϕ is a ground negative literal $\neg a$;
- $I \models a$ and $I \models b$, if ϕ is a conjunction $a \wedge b$;
- $I \models a$ or $I \models b$, if ϕ is a disjunction $a \vee b$;
- $I \models \psi\theta$ for all θ that assign a value to all the variables of $\mathbf{X}$ if $\phi = \forall\mathbf{X}\psi$;
- $I \models \psi\theta$ for a θ that assigns a value to all the variables of $\mathbf{X}$ if $\phi = \exists\mathbf{X}\psi$.

A clause C of the form

$$h_1 \vee \ldots \vee h_n \leftarrow b_1, \ldots, b_m$$

is a shorthand for the formula

$$\forall\mathbf{X}h_1 \vee \ldots \vee h_n \leftarrow b_1, \ldots, b_m$$

where $\mathbf{X}$ is a vector of all the variables appearing in C. Therefore, C is true in an interpretation I iff, for all the substitutions θ grounding C, if $I \models body(C)\theta$ then $I \models head(C)\theta$, i.e., if $(I \models body(C)\theta) \rightarrow (head(C)\theta \cap I \neq \emptyset)$. Otherwise, it is false. In particular, a program rule is true in an interpretation I iff, for all the substitutions θ grounding C, $(I \models body(C)\theta) \rightarrow h \in I$.

A theory P is true in an interpretation I iff all of its clauses are true in I and we write

$$I \models P.$$

If P is true in an interpretation I we say that I is a *model* of P. It is sufficient for a single clause of a theory P to be false in an interpretation I for P to be false in I.

For normal logic programs, we are interested in deciding whether a query Q is a logical consequence of a theory P, expressed as

$$P \models Q.$$

This means that Q must be true in the model $M(P)$ of P that is assigned to P as its meaning by one of the semantics that have been proposed for normal logic programs (e.g. [10,11,12]).

For theories, we are interested in deciding whether a given theory or a given clause is true in an interpretation I. This can be achieved with the following procedure [13]. The truth of a range restricted clause C on a finite interpretation I can be tested by asking the goal ?-$body(C), \neg head(C)$ against a database containing the atoms of I as facts. By $\neg head(C)$ we mean $\neg h_1, \ldots, \neg h_m$. If the query fails, C is true in I, otherwise C is false in I.

In some cases, we are not given an interpretation I completely but we are given a set of atoms J and a normal program B as a compact way of indicating the interpretation $M(B \cup J)$. In this case, if B is composed only of range restricted rules, we can test the truth of a clause C on $M(B \cup J)$ by running the query ?-$body(C), \neg head(C)$ against a Prolog database containing the atoms of J as facts together with the rules of B. If the query fails C is true in $M(B \cup J)$, otherwise C is false in $M(B \cup J)$.

2.2   Inductive Logic Programming

Inductive Logic Programming (ILP) [6] is a research field at the intersection of Machine Learning and Logic Programming. It is concerned with the development of learning algorithms that adopt logic for representing input data and induced models. Recently, many techniques have been proposed in the field that were successfully applied to a variety of domains. Logic proved to be a powerful tool for representing the complexity that is typical of the real world. In particular, logic can represent in a compact way domains in which the entities of interest are composed of subparts connected by a network of relationships. Traditional Machine Learning is often not effective in these cases because it requires input data in the flat representation of a single table.

The problem that is faced by ILP can be expressed as follows:

Given:

- a space of possible theories $\mathcal{H}$;
- a set E^+ of positive example;
- a set E^- of negative examples;
- a background theory B.

Find a theory $H \in \mathcal{H}$ such that;

- all the positive examples are covered by H
- no negative example is covered by H

If a theory does not cover an example we say that it rules the example out so the last condition can be expressed by saying the "all the negative examples are ruled out by H".

The general form of the problem can be instantiated in different ways by choosing appropriate forms for the theories in input and output, for the examples and for the covering relation.

In the *learning from entailment* setting, the theories are normal logic programs, the examples are (most often) ground facts and the coverage relation is entailment, i.e., a theory H covers an example e iff

$$H \models e.$$

In the learning from interpretations setting, the theories are composed of clauses, the examples are interpretations and the coverage relation is truth in an interpretation, i.e., a theory H covers an example interpretation I iff

$$I \models H.$$

Similarly, we say that a clause C covers an example I iff $I \models C$.

In this paper, we concentrate on learning from interpretation so we report here the detailed definition:

Given:

- a space of possible theories $\mathcal{H}$;
- a set E^+ of positive interpretations;
- a set E^- of negative interpretations;
- a background normal logic program B.

Find a theory $H \in \mathcal{H}$ such that;

- for all $P \in E^+$, H is true in the interpretation $M(B \cup P)$;
- for all $N \in E^-$, H is false in the interpretation $M(B \cup N)$.

The background knowledge B is used to encode each interpretation parsimoniously, by storing separately the rules that are not specific to a single interpretation but are true for every interpretation.

The algorithm ICL [8] solves the above problem. It performs a covering loop (function ICL in Figure 1) in which negative interpretations are progressively ruled out and removed from the set E^-. At each iteration of the loop a new clause is added to the theory. Each clause rules out some negative interpretations. The loop ends when E^- is empty or when no clause is found.

The clause to be added in every iteration of the covering loop is returned by the procedure FindBestClause (Figure 1). It looks for a clause by using beam search with $p(\ominus|\overline{C})$ as a heuristic function, where $p(\ominus|\overline{C})$ is the probability that an example interpretation is classified as negative given that it is ruled out by the clause C. This heuristic is computed as the number of ruled out negative interpretations over the total number of ruled out interpretations (positive and negative). Thus we look for clauses that cover as many positive interpretations as possible and rule out as many negative interpretations as possible. The search starts from the clause $false \leftarrow true$ that rules out all the negative interpretations but also all the positive ones and gradually refines that clause.

The refinements of a clause are obtained by generalization. A clause C is *more general* than a clause D if the set of interpretations covered by C is a superset of those covered by D. This is true if $D \models C$. However, using logical implication

```
function ICL(E⁺, E⁻, B)
initialize H := ∅
do
      C := FindBestClause(E⁺, E⁻, B)
      if best clause C ≠ ∅ then
            add C to H
            remove from E⁻ all interpretations that are false for C
while C ≠ ∅ and E⁻ is not empty
return H

function FindBestClause(E⁺, E⁻, B)
initialize Beam := {false ← true}
initialize BestClause := ∅
while Beam is not empty do
      initialize NewBeam := ∅
      for each clause C in Beam do
            for each refinement Ref ∈ δ(C) do
                  if Ref is better than BestClause then BestClause := Ref
                  if Ref is not to be pruned then
                        add Ref to NewBeam
                        if size of NewBeam > MaxBeamSize then
                              remove worst clause from NewBeam
      Beam := NewBeam
return BestClause
```

Fig. 1. ICL learning algorithm

as a generality relation is impractical because of its high computational cost. Therefore, the syntactic relation of θ-*subsumption* is used in place of implication: D θ-*subsumes* C (written $D \geq C$) if there exist a substitution θ such that $D\theta \subseteq C$. If $D \geq C$ then $D \models C$ and thus C is more general than D. The opposite, however, is not true, so θ-subsumption is only an approximation of the generality relation. For example, let us consider the following clauses:

$$C_1 = accept(X) \leftarrow true$$
$$C_2 = accept(X) \vee refusal(X) \leftarrow true$$
$$C_3 = accept(X) \leftarrow invitation(X)$$
$$C_4 = accept(alice) \leftarrow invitation(alice)$$

Then $C_1 \geq C_2, C_1 \geq C_3$ but $C_2 \not\geq C_3$, $C_3 \not\geq C_2$ so C_2 and C_3 are more general than C_1, while C_2 and C_3 are not comparable. Moreover $C_1 \geq C_4$, $C_3 \geq C_4$ but $C_2 \not\geq C_4$ and $C_4 \not\geq C_2$ so C_4 is more general than C_1 and C_3 while C_2 and C_4 are not comparable.

From the definition of θ-subsumption, it is clear that a clause can be refined (i.e. generalized) by applying one of the following two operations on a clause

- adding a literal to the (head or body of the) clause
- applying a substitution to the clause

FindBestClause computes the refinements of a clause by applying one of the above two operations. Let us call $\delta(C)$ the set of refinements so computed for a clause C. The clauses are gradually generalized until a clause is found that covers all (or most of) the positive interpretations while still ruling out some negative interpretations.

The literals that can possibly be added to a clause are specified in the *language bias*, a collection of statements in an ad hoc language that prescribe which refinements have to be considered. Two languages are possible for ICL: $\mathcal{D}$LAB and rmode (see [14] for details). Given a language bias which prescribes that the body literals must be chosen among $\{invitation(X)\}$ and that the head disjuncts must be chosen among $\{accept(X), refusal(X)\}$, an example of refinements sequence performed by FindBestClause is the following:

$false \leftarrow true$
$accept(X) \leftarrow true$
$accept(X) \leftarrow invitation(X)$
$accept(X) \vee refusal(X) \leftarrow invitation(X)$

The refinements of clauses in the beam can also be pruned: a refinement is pruned if it is not statistical significant and if it cannot produce a value of the heuristic function larger than that of the best clause. As regards the first type of pruning, a statistical test is used, while as regards the second type, the best refinement that can be obtained is a clause that covers all the positive examples and rules out the same negative examples as the original clause.

When a new clause is returned by FindBestClause, it is added to the current theory. The negative interpretations that are ruled out by the clause are ruled out as well by the updated theory, so they can be removed from E^-.

2.3   Incremental Inductive Logic Programming

The learning framework presented in Section 2.2 assumes that all the examples are provided to the learner at the same time and that no previous model exists for the concepts to be learned. In some cases, however, the examples are not all known at the same time and an initial theory may be available. When a new example is obtained, one approach consists of adding the example to the previous training set and learning a new theory from scratch. This approach may turn out to be too inefficient, especially if the amount of previous examples is very high. An alternative approach consists in revising the existing theory to take into account the new example, in order to exploit as much as possible the computations already done. The latter approach is called Theory Revision and can be described by the following definition:

Given:

- a space of possible theories $\mathcal{H}$;
- a background theory B
- a set E^+ of previous positive example;
- a set E^- of previous negative examples;

- a theory H that is consistent with E^+ and E^-
- a new example e.

Find a theory $H' \in \mathcal{H}$ such that;

- H' is obtained by applying a number of transformations to H
- H' covers e if e is a positive example, or
- H' does not cover e if e is a negative example.

Theory Revision has been extensively studied in the learning from entailment setting of Inductive Logic Programming. In this case, examples are ground facts, the background theory is a normal logic program and the coverage relation is logical entailment. Among the systems that have been proposed for solving such a problem are: RUTH [15], FORTE [16] and Inthelex [9]

All these systems perform the following operations:

- given an uncovered positive example e, they generalize the theory T so that it covers it
- given a covered negative example e, they specialize the theory T so that it does not cover it

As an example of an ILP Theory Revision system, let us consider the algorithm of Inthelex that is shown in Figure 2.

Function Generalize is used to revise the theory when the new example is positive. Each clause of the theory is considered in turn and is generalized. The resulting theory is tested to see whether it covers the positive example. Moreover, it is tested on all the previous negative examples to ensure that the clause is not generalized too much. As soon as a good refinement is found it is returned by the function.

Function Specialize is used to revise the theory when the new example is negative. Each clause of the theory involved in the derivation of the negative example is considered in turn and is specialized. The resulting theory is tested to see whether it rules out the negative example. Moreover, it is tested on all the previous positive examples to ensure that the clause is not specialized too much. As soon as a good refinement is found it is returned by the function.

While various systems exist for Theory Revision in the learning from entailment setting, to the best of our knowledge no algorithm has been proposed for Theory Revision in the learning from interpretation setting.

2.4 Business Process Management

The performances of an organization depend on how accurately and efficiently it enacts its business processes. Formal ways of representing business processes have been studied in the area of Business Processes Management (see e.g. [17]), so that the actual enactment of a process can be checked for compliance with a model.

Recently, the problem of automatically inferring such a model from data has been studied by many authors (see e.g. [18,2,19]). This problem has been called

---

function Generalize(E^-, e, B, H)
repeat
    pick a clause C from H
    obtain a set of generalizations $\delta(C)$
    for each clause $C' \in \delta(C)$
        let $H' := H \setminus \{C\} \cup \{C'\}$
        test H' over e and over all the examples in E^-
        if H' cover e and does not cover any negative example then
            return H'
until all the clauses of H have been considered
// no generalization found
add a new clause to H that covers e and is consistent with E^-
let H' be the new theory
return H'

function Specialize(E^+, e, B, H)
repeat
    pick a clause C used in the derivation of e in H
    obtain a set of specializations $\rho(C)$
    for each clause $C' \in \delta(C)$
        let $H' := H \setminus \{C\} \cup \{C'\}$
        test H' over e and over all the examples in E^-
        if H' does not cover e and covers all positive examples then
            return H'
until all the clauses of H used in the derivation of e have been considered
// no specialization found
add e to H as an exception
let H' be the new theory
return H'

Fig. 2. Inthelex Theory Revision algorithm

Process Mining or Workflow Mining. The data in this case consists of execution traces (or histories) of the business process. The collection of such data is made possible by the facility offered by many information systems of logging the activities performed by users.

Let us now describe in detail the problem that is solved by Process Mining. A *process trace* T is a sequence of events. Each event is described by a number of attributes. The only requirement is that one of the attributes describes the event type. Other attributes may be the executor of the event or event specific information.

An example of a trace is

$\langle a, b, c \rangle$

that means that activity a was performed first, then b and finally c.

A *process model PM* is a description of the process in a language that expresses the conditions a trace must satisfy in order to be compliant with the process, i.e., to be a correct enactment of the process. An interpreter of the language must exists that, when applied to a model PM and a trace T, returns yes if the trace is compliant with the description and false otherwise. In the first case we write $T \models PM$, in the second case $T \not\models PM$. A bag of process traces L is called a *log*. Often, in Process Mining, only compliant traces are used as input of the learning algorithm, see e.g. [18,2,19]. We consider instead the case in which we are given both compliant and non compliant traces. This is the case when we want distinguish successful process executions from unsuccessful ones.

The approaches presented in [18,2,19] aim at discovering complex and procedural process models, and differ by the structural patterns they are able to mine. While recognizing the extreme importance of such approaches, recently [20] pointed out the necessity of discovering declarative logic-based knowledge, in the form of process fragments or business rules/policies, from execution traces. Declarative languages seem to fit better complex, unpredictable processes, where a good balance between support and flexibility is of key importance.

[20] presents a graphical language for specifying process flows in a declarative manner. The language, called ConDec, leaves the control flow among activities partially unspecified by defining a set of constraints expressing policies/business rules for specifying either what is forbidden as well as mandatory in the process. Therefore, the approach is inherently open and flexible, because workers can perform actions if those are not explicitly forbidden. ConDec adopts an underlying semantics by means of Linear Temporal Logics (LTL), and can also be mapped onto the logic programming-based framework $\mathcal{S}$CIFF [3,21] that provides a declarative language based on Computational Logic. In $\mathcal{S}$CIFF constraints are imposed on activities in terms of reactive rules (namely Integrity Constraints). Such reactive rules mention in their body occurring activities, i.e., *events*, and additional constraints on their variables. $\mathcal{S}$CIFF rules contain in their head expectations over the course of events. Such expectations can be positive, when a certain activity is required to happen, or negative, when a certain activity is forbidden to happen.

Most works in Process Mining deal with the discovery of procedural process models (such as Petri Nets or Event-driven Process Chains [22,23]) from data. Recently, some works have started to appear on the discovery of logic-based declarative models: [24,5,25] study the possibility of inferring essential process constraints, easily understandable by business analysts and not affected by procedural details.

3   Representing Process Traces and Models with Logic

A process trace can be represented as a logical interpretation: each event is modeled with an atom whose predicate is the event type and whose arguments store the attributes of the action. Moreover, an extra argument is added to the atom indicating the position in the sequence. For example, the trace:

$\langle a, b, c \rangle$

can be represented with the interpretation

$\{a(1), b(2), c(3)\}$.

If the execution time is an attribute of the event, then the position in the sequence can be omitted.

Besides traces, we may have some general knowledge that is valid for all traces. We assume that this background information can be represented as a normal logic program B. The rules of B allow to complete the information present in a trace I: rather than simply I, we now consider $M(B \cup I)$, the model of the program $B \cup I$ according to one of the semantics for normal logic programs.

For example, consider the trace

$I = \{ask\_price(bike, 1), tell\_price(500, 2), buy(bike, 3)\}$

of a bike retail store and the background theory

$B = \{high\_price(T) \leftarrow tell\_price(P, T), P \geq 400\}$.

that expresses information regarding price perceptions by clients. Then $M(B \cup I)$ is

$\{ask\_price(bike, 1), tell\_price(500, 2), high\_price(2), buy(bike, 3)\}$

in which the information available in the trace has been enlarged by using the background information.

The process language we consider was proposed in [5] and is a subset of the $\mathcal{SCIFF}$ language, originally defined in [3,4], for specifying and verifying interactions in open agent societies.

A process model in our language is a set of *integrity constraints* (ICs for short). An IC, C, is a logical formula of the form

$$Body \rightarrow \exists(ConjP_1) \vee \ldots \vee \exists(ConjP_n) \vee \forall\neg(ConjN_1) \vee \ldots \vee \forall\neg(ConjN_m) \quad (1)$$

where $Body$, $ConjP_i$ $i = 1, \ldots, n$ and $ConjN_j$ $j = 1, \ldots, m$ are conjunctions of literals built over event atoms, over predicates defined in the background or over built-in predicates such as $\leq, \geq, \ldots$. The variables appearing in the body are implicitly universally quantified with scope the entire formula. The quantifiers in the head apply to all the variables appearing in the conjunctions and not appearing in the body.

We will use $Body(C)$ to indicate $Body$ and $Head(C)$ to indicate the formula $\exists(ConjP_1) \vee \ldots \vee \exists(ConjP_n) \vee \forall\neg(ConjN_1) \vee \ldots \vee \forall\neg(ConjN_m)$ and call them respectively the *body* and the *head* of C. We will use $HeadSet(C)$ to indicate the set $\{ConjP_1, \ldots, ConjP_n, ConjN_1, \ldots, ConjN_m\}$.

$Body(C)$, $ConjP_i$ $i = 1, \ldots, n$ and $ConjN_j$ $j = 1, \ldots, m$ will be sometimes interpreted as sets of literals, the intended meaning will be clear from the context. We will call P *conjunction* each $ConjP_i$ for $i = 1, \ldots, n$ and N *conjunction* each $ConjN_j$ for $j = 1, \ldots, m$. We will call P *disjunct* each $\exists(ConjP_i)$ for $i = 1, \ldots, n$ and N *disjunct* each $\forall\neg(ConjN_j)$ for $j = 1, \ldots, m$.

An example of an IC is

$$a(bob, T), T < 10$$
$$\rightarrow \exists(b(alice, T1), T < T1)$$
$$\vee \qquad\qquad\qquad (2)$$
$$\forall\neg(c(mary, T1), T < T1, T1 < T + 10)$$

The meaning of IC (2) is the following: if *bob* has executed action a at a time $T < 10$, then *alice* must execute action b at a time $T1$ later than T or *mary* must not execute action c for 9 time units after T. The disjunct $\exists(b(alice, T1), T < T1)$ stands for $\exists T1(b(alice, T1), T < T1)$ and the disjunct $\forall\neg(c(mary, T1), T < T1, T1 < T + 10)$ stands for $\forall T1\neg(c(mary, T1), T < T1, T1 < T + 10)$.

An *IC C is true in an interpretation* $M(B \cup I)$, written $M(B \cup I) \models C$, if, for every substitution θ for which *Body* is true in $M(B \cup I)$, there exists a disjunct $\exists(ConjP_i)$ or $\forall\neg(ConjN_j)$ that is true in $M(B \cup I)$. If $M(B \cup I) \models C$ we say that the trace I is *compliant* with C. [5] showed that the truth of an IC in an interpretation $M(B \cup I)$ can be tested by running the query:

?-*Body*, $\neg(ConjP_1), \ldots \neg(ConjP_n), ConjN_1, \ldots, ConjN_m$

against a Prolog database containing the clauses of B and atoms of I as facts. If the N conjunctions in the head share some variables, then the following query must be issued

?-*Body*, $\neg(ConjP_1), \ldots \neg(ConjP_n), \neg(\neg(ConjN_1)), \ldots, \neg(\neg(ConjN_m))$

that ensures that the N conjunctions are tested separately without instantiating the variables.

Thus, for IC 2, the query is

?-$a(bob, T), T < 10, \neg(b(alice, T1), T < T1), \neg(\neg(c(mary, T1), T < T1, T1 < T + 10))$

If the query finitely fails, the IC is true in the interpretation. If the query succeeds, the IC is false in the interpretation. Otherwise nothing can be said. It is the user's responsibility to write the background B in such a way that no query generates an infinite loop. For example, if B is acyclic then a large class of queries will be terminating [26].

A *process model H is true in an interpretation* $M(B \cup I)$ if every IC of H is true in it and we write $M(B \cup I) \models H$. We also say that trace I is *compliant with H*.

The ICs we consider are more expressive than clauses, as can be seen from the query used to test them: for ICs, we have the negation of conjunctions, while for clauses we have only the negation of atoms. This added expressiveness is necessary for dealing with processes because it allows us to represent relations between the execution times of two or more activities.

4   Incremental Learning of ICs Theories

In order to induce a theory that describes a process, we must search the space of ICs. To this purpose, we need to define a generality order in such a space.

IC C is *more general than* IC D if C is true in a superset of the traces where D is true. If $D \models C$, then C is more general than D.

Similarly to the case of clauses, [5] defined the notion of θ-subsumption also for ICs.

Definition 1 (Subsumption). *An IC D θ-subsumes an IC C, written $D \geq C$, iff it exists a substitution θ for the variables in the body of D or in the N conjunctions of D such that*

- $Body(D)\theta \subseteq Body(C)$ *and*
- $\forall ConjP(D) \in HeadSet(D),\ \exists ConjP(C) \in HeadSet(C) : ConjP(C) \subseteq ConjP(D)\theta$ *and*
- $\forall ConjN(D) \in HeadSet(D),\ \exists ConjN(C) \in HeadSet(C) : ConjN(D)\theta \subseteq ConjN(C)$

For example, IC 2 is subsumed by the IC

$$
\begin{aligned}
& a(bob, 4) \\
& \rightarrow \exists(b(alice, T1), 4 < T1, T1 < 4 + 10) \\
& \vee \\
& \forall \neg (c(mary, 5), 4 < 5)
\end{aligned}
\tag{3}
$$

with the substitution $\{T/4, T1/5\}$.

It was proved in [5] that implication and θ-subsumption for ICs share the same relation as in the case of clauses.

Theorem 1 ([5]). $D \geq C \Rightarrow D \models C$.

Thus, θ-subsumption can be used for defining a notion of generality among ICs, which can be used in learning algorithms.

In order to define a refinement operator, we must first define the language bias. We use a language bias that consists of a set of IC templates. Each template specifies

- a set of literals BS allowed in the body,
- a set of disjuncts HS allowed in the head. Each disjunct is represented as a couple $(Sign, DS)$ where
  - $Sign$ is either $+$ or $-$ and specifies where it is a P or an N disjunct,
  - DS is the set of literals allowed in the disjunct.

[5] defined a refinement operator from specific to general (upward operator) in the following way: given an IC D, the set of *upward refinements* $\delta(D)$ of D is obtained by performing one of the following operations

- adding a literal from BS to the body;
- removing a literal from a P disjunct in the head;
- adding a literal to an N disjunct in the head where the literal must be allowed by the language bias;
- adding a disjunct from HS to the head: the disjunct can be

- a formula $\exists(d_1 \wedge \ldots \wedge d_k)$ where $DS = \{d_1, \ldots, d_k\}$ is the set of literals allowed by the IC template for D for a P disjunct,
- a formula $\forall\neg(d)$ where d is allowed by the IC template for D for a N disjunct.

In order to perform theory revision, we also define a refinement operator from general to specific (downward operator). The operator inverts the operations performed in the upward operator, i.e., given an IC D, the set of *downward refinements* $\rho(D)$ of D is obtained by performing one of the following operations

- removing a literal from the body of D;
- adding a literal to a P disjunct in the head, the literal must be allowed by the language bias;
- removing a literal from an N disjunct in the head;
- removing a disjunct from the head when
  - it is a P disjunct $\exists(d_1 \wedge \ldots \wedge d_k)$ where $\{d_1, \ldots, d_k\}$ is the set of literals allowed by the IC template for D for the P disjunct,
  - it is an N disjunct containing a single literal $\forall\neg(d)$.

We define the algorithm for performing theory revision starting from the algorithm DPML (Declarative Process Model Learner) that is an adaptation of ICL [8]. DPML solves the following learning problem
Given

- a space of possible process models $\mathcal{H}$
- a set E^+ of positive traces;
- a set E^- of negative traces;
- a background normal logic program B.

Find: a process model $H \in \mathcal{H}$ such that

- for all $T^+ \in E^+$, $M(B \cup T^+) \models H$;
- for all $T^- \in E^-$, $M(B \cup T^-) \not\models H$;

If $M(B \cup T) \models C$ we say that IC C *covers* the trace T and if $M(B \cup T) \not\models C$ we say that C *rules out* the trace T.

DPML is obtained from ICL by using the testing procedure and the refinement operator defined for $\mathcal{SCIFF}$ ICs in place of those for logical clauses.

The system IPM (Incremental Process Mines) modifies DPML in order to deal with theory revision. As in Section 2.3, we call $\mathcal{H}$ the space of possible theories, B the background theory, E^+ the set of previous positive examples, E^- the set of previous negative ones and T the theory (obtained by DPML or expressed by a human expert) we would like to refine to make it consistent with the new examples: $Enew^-$ and $Enew^+$. Figure 3 shows the IDPML algorithm.

The initial theory, together with old and new positive examples and old negative ones, is given as input to RevisePositive whose aim is to revise the theory in order to cover as many positive examples as possible. The output of RevisePositive is then given as input, together with all sets of examples, to ReviseNegative,

```
function IPM(T, E⁺, E⁻, Enew⁺, Enew⁻, B)
    H:= RevisePositive(T, E⁺, E⁻, Enew⁺, B)
    H:= ReviseNegative(H, E⁺, E⁻, Enew⁺, Enew⁻, B)
    H:=H∪ DPML(E⁺ ∪ Enew⁺, Covered(Enew⁻, H), B)
    return H

function RevisePositive(T, E⁺, E⁻, Enew⁺, B)
    foreach e⁺ ∈ Enew⁺
        VC:=FindViolatedConstraints(T, e⁺)
        T:= T − VC
        E⁺:=E⁺ ∪ {e⁺}
        foreach vc ∈ VC
            c:= Generalize(vc, E⁺, E⁻, B)
            T:= T ∪ {Best(c, vc)}
    return T

function Generalize(vc, E⁺, E⁻, B)
    Beam:={vc}
    BestClause:= ∅
    while Beam ≠ ∅
        foreach c ∈ Beam
            foreach ref of c
                BestClause:= Best(ref, c)
                Beam:= Beam ∪ {ref}
                if size(Beam) > MaxBeamSize
                    Beam:= Beam − {Worst(Beam)}
    return Beam

function ReviseNegative(T, E⁺, E⁻, Enew⁺, Enew⁻, B)
    Enew⁻:=TestNegative(T, E⁻, Enew⁻)
    E⁺:=E⁺ ∪ Enew⁺
    H:=∅
    while T ≠ ∅ ∧ Enew⁻ ≠ ∅
        pick randomly an IC c from T
        T:=T − {c}
        nc:= Specialize(c, E⁺, Enew⁻, B)
        H:=H ∪ {Best(c, nc)}
        Enew⁻:= Enew⁻ − RuledOut(Enew⁻, Best(c, nc))
    return H
```

Fig. 3. IPM algorithm

whose revision tries to rule out the negative examples, and whose output is the overall revised theory.

RevisePositive cycles on new positive examples and finds out which constraints (if any) of the previous theory are violated for each example. An inner cycle generalizes all such constraints in order to make the theory cover the ruled out positive example. The generalization function performs a beam search with $p(\ominus|\overline{C})$ as the heuristic (see Section 2.2) and δ as the refinements operator (see Section 4). For theory revision, however, the beam is not initialized with the most specific constraint (i.e. $\{false \leftarrow true\}$) but with the violated constraint.

Since some of the previously ruled out negative examples may be again covered after the generalization process, ReviseNegative checks at first which negative examples, either old or new, are not ruled out. Then it selects randomly an IC from the theory and it performs a specialization cycle until no negative example is covered. The Specialize function is similar to the Generalize one with δ replaced with ρ as the refinement operator (see Section 4).

It is also possible that some negative examples can't be ruled out just by specializing existing constraints, so after ReviseNegative a covering loop (as the one of DPML) has to be performed on all positive examples and on the negative ones which are still to be ruled out.

5   Experiments

In this section we present some experiments that have been performed for investigating the effectiveness of IPM. In particular, we want to demonstrate that, given an initial theory H and a new set of examples $Enew$, it can be more beneficial to revise H in the light of $Enew$ than to learn a theory from $E \cup Enew$. Another use case consists in the revision of an (imperfect) theory written down by a user and its comparison with the theory learned from scratch using the same set of examples.

5.1   Hotel Management

Let's first consider a process model regarding the management of a hotel and inspired by [27]. We generated randomly a number of traces for this process, we classified them with the model and then we applied both DMPL and IDMPL.

The model describes a simple process of renting rooms and services in a hotel. Every process instance starts with the registration of the client name and her preferred way of payment (e.g., credit card). Data can also be altered at a later time (e.g the client may decide to use another credit card). During her stay, the client can require one or more room and laundry services. Each service, identified by a code, is followed by the respective registration of the service costs into the client bill. The cost of each service must be registered only if the service has been effectively provided to the client and it must be registered only once. The cost related to the nights spent in the hotel must be billed. It is possible for the total bill to be charged at several stages during the stay.

This process was modeled by using eight activities and eight constraints. Activities *register_client_data*, *check_out* and *charge* are about the check-in/check-out of the client and expense charging. Activities *room_service* and *laundry_service* log which services have been used by the client, while billings for each service are represented by separate activities. For each activity, a unique identifier is introduced to correctly charge the clients with the price for the services they effectively made use of.

Business related aspects of our example are represented as follows:

- (C.1) every process instance starts with activity *register_client_data*. No limits on the repetition of this activity are expressed, hence allowing alteration of data;
- (C.2) *bill_room_service* must be executed after each *room_service* activity, and *bill_room_service* can be executed only if the *room_service* activity has been executed before;
- (C.3) *bill_laundry_service* must be executed after each *laundry_service* activity, and *bill_laundry_service* can be executed only if the *laundry_service* activity has been executed before;
- (C.4) *check_out* must be performed in every process instance;
- (C.5) *charge* must be performed in every process instance;
- (C.6) *bill_nights* must be performed in every process instance.
- (C.7) *bill_room_service* must be executed only one time for each service identifier;
- (C.8) *bill_laundry_service* must be executed only one time for each service identifier;

The process model is composed by the following ICs:

$(C.1)$ *true*
$\quad \rightarrow \exists (register\_client\_data(Trcd) \wedge Trcd = 1).$

$(C.2)$ $room\_service(rs\_id(IDrs), Trs)$
$\quad \rightarrow \exists (bill\_room\_service(rs\_id(IDbrs), Tbrs) \wedge$
$IDrs = IDbrs \wedge Tbrs > Trs).$
$bill\_room\_service(rs\_id(IDbrs), Tbrs)$
$\quad \rightarrow \exists (room\_service(rs\_id(IDrs), Trs) \wedge$
$IDbrs = IDrs \wedge Trs < Tbrs).$

$(C.3)$ $laundry\_service(la\_id(IDls), Tls)$
$\quad \rightarrow \exists (bill\_laundry\_service(la\_id(IDbls), Tbls) \wedge$
$IDls = IDbls \wedge Tbls > Tls).$
$bill\_laundry\_service(la\_id(IDbls), Tbls)$
$\quad \rightarrow \exists (laundry\_service(la\_id(IDls), Tls) \wedge$
$IDbls = IDls \wedge Tls < Tbls).$

$(C.4)$ *true*
$\quad \rightarrow \exists (check\_out(Tco)).$

$(C.5)$ *true*
$\quad \rightarrow \exists (charge(Tch)).$

$[t](C.6)$ $true$

$\rightarrow \exists(bill\_nights(Tbn))$.

$(C.7)$ $bill\_room\_service(rs\_id(IDbrs1), Tbrs1)$

$\rightarrow \forall \neg (bill\_room\_service(rs\_id(IDbrs2), Tbrs2) \wedge$
$IDbrs1 = IDbrs2 \wedge Tbrs2 > Tbrs1)$.

$(C.8)$ $bill\_laundry\_service(la\_id(IDbls1)), Tbls1)$

$\rightarrow \forall \neg (bill\_laundry\_service(la\_id(IDbls2), Tbls2) \wedge$
$IDbls1 = IDbls2 \wedge Tbls2 > Tbls1)$.

For this process, we randomly generated execution traces and we classified them with the above model. This was repeated until we obtained four training sets each composed of 2000 positive examples and 2000 negative examples. Each training set was randomly split into two subset, one containing 1500 positive and 1500 negative examples, and the other containing 500 positive and 500 negatives examples. The first subset is used for getting an initial theory, while the second is used for the revision process.

DPML was applied to each training sets with 3000 examples. The theories that were obtained were given as input to IPM together with the corresponding 1000 examples training set. Finally, DPML was applied to each of the complete 4000 examples training sets.

The models obtained by IPM and by DPML on a complete training set were then applied to each example of the other three training set. Accuracy is then computed as the number of compliant traces that are correctly classified as compliant plus the number of non-compliant traces that are correctly classified as non-compliant divided by the total number of traces.

In table 1 we show a comparison of time spent (in seconds) and resulting accuracies from the theory revision process and from learning based on the full dataset. The μ sub-columns for accuracy present means of results from tests on the three datasets not used for training, while in the σ one standard deviations can be found. The last row shows aggregated data for the correspondent columns.

As it can be noticed, in this case revising the theory to make it compliant with the new logs is faster than learning it again from scratch, and the accuracy of the results is higher.

5.2   Auction Protocol

Let us now consider an interaction protocol among agents participating in an electronic auction [28].

The auction is sealed bid: the auctioneer communicates the bidders the opening of the auction, the bidders answer with bids over the good and then the auctioneer communicates the bidders whether they have won or lost the auction.

Table 1. Revision compared to learning from full dataset for the hotel scenario

	revision				full dataset			
dataset	time		accuracy		time		accuracy	
			μ	σ			μ	σ
1	4123		0.732539	0.0058	18963		0.702367	0.0068
2	4405		0.757939	0.0223	17348		0.686754	0.0269
3	6918		0.825067	0.0087	13480		0.662302	0.0180
4	3507		0.724764	0.0257	17786		0.679003	0.0248
	μ	σ	μ	σ	μ	σ	μ	σ
global	4738	1299	0.760	0.0433	16894	2057	0.682	0.0255

The protocol is described by the following ICs [29].

$$\begin{aligned} &bid(B, A, Quote, TBid) \\ \rightarrow &\exists(openauction(A, B, TEnd, TDL, TOpen), \\ &TOpen < TBid, TBid < TEnd) \end{aligned} \tag{4}$$

This IC states that if a bidder sends the auctioneer a *bid*, then there must have been an *openauction* message sent before by the auctioneer and such that the bid has arrived in time (before $TEnd$).

$$\begin{aligned} &openauction(A, B, TEnd, TDL, TOpen), \\ &bid(B, A, Quote, TBid), \\ &TOpen < TBid \\ \rightarrow &\exists(answer(A, B, lose, Quote, TLose), \\ &TLose < TDL, TEnd < TLose) \\ \vee &\exists(answer(A, B, win, Quote, TWin), \\ &TWin < TDL, TEnd < TWin) \end{aligned} \tag{5}$$

This IC states that if there is an *openauction* and a valid *bid*, then the auctioneer must answer with either *win* or *lose* after the end of the bidding time ($TEnd$) and before the deadline (TDL).

$$\begin{aligned} &answer(A, B, win, Quote, TWin) \\ \rightarrow &\forall\neg(answer(A, B, lose, Quote, TLose), TWin < TLose) \end{aligned} \tag{6}$$

$$\begin{aligned} &answer(A, B, lose, Quote, TLose) \\ \rightarrow &\forall\neg(answer(A, B, win, Quote, TWin), TLose < TWin) \end{aligned} \tag{7}$$

These two ICs state that the auctioneer can not answer both *win* and *lose* to the same bidder.

A graphical representation of the protocol is shown in Figure 4.

The traces have been generated in the following way: the first message is always *openauction*, the following messages are generated randomly between

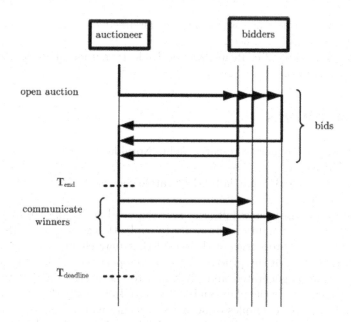

Fig. 4. Sealed bid auction protocol.

Table 2. Revision compared to learning from dataset for the auction scenario

dataset	revision			full dataset		
	time	accuracy		time	accuracy	
		μ	σ		μ	σ
1	820	0.962812	0.0043	1373	0.921687	0.0043
2	1222	0.962937	0.0043	1403	0.939625	0.0041
3	806	0.96375	0.0039	1368	0.923312	0.0044
4	698	0.961125	0.0018	1618	0.937375	0.0020
5	743	0.963875	0.0038	1369	0.92350	0.0042
	μ / σ	μ	σ	μ / σ	μ	σ
global	857 / 187	0.962	0.0039	1426 / 96	0.929	0.0086

bid and *answer*. For *answer*, *win* and *lose* are selected randomly with equal probability. The bidders and auctioneer are always the same. The times are selected randomly from 2 to 10. Once a trace is generated, it is tested with the above ICs. If the trace satisfies all the ICs it is added to the set of positive traces, otherwise it is added to the set of negative traces. This process is repeated until 500 positive and 500 negative traces are generated for length 3, 4, 5 and 6. Five datasets are obtained in this way, containing each 2000 positive and 2000 negative traces.

We then considered 500 randomly selected traces (half positive and half negative). We applied both DPML and IPM to this dataset, the latter starting with a version of the model that was manually modified to simulate an imperfect theory written down by an user.

The results in Table 2 confirm those of Table 1: revision offers benefits both in time and accuracy.

6   Related Works

Process mining is an active research field. Notable works in such a field are [18,22,23,19,30,31].

In particular in [30,31] (partially) declarative specifications (thus closer to our work) are adopted.

In [31] activities in business process are seen as planning operators with pre--conditions and post-conditions. In order to explicitly express them, *fluents* besides activities (i.e., properties of the world that may change their truth value during the execution of the process) have to be specified. A plan for achieving the business goal is generated and presented to the user which has to specify whether each activity of the plan can be executed. In this way the system collects positive and negative examples of activities executions that are then used in a learning phase. Our work remains in the traditional domain of BPM in which the pre-conditions and post-conditions of activities are left implicit.

In [30] sets of process traces, represented by Petri nets, are described by high level process *runs*. Mining then performs a merge of runs regarding the same process and the outcome is a model which may contain sets of activities that must be executed, but for which no specific order is required. However, runs are are already very informative of the process model; in our work, instead, mining starts from traces, which are simply a sequence of events representing activity executions.

In [32] events used as negative examples are automatically generated in order to partially take away from the user the burden of having to classify activities. We are interested in the future to investigate automatic generation of negative traces.

A useful survey about theory revision methods, including a brief description of common refinement operators and search strategies, can be found in [33]. With regard to the taxonomy proposed there, we deal with proper revision (not restructuring) both under the specializing and generalizing points of view.

In [34], the authors address structural process changes at run-time, once a process is implemented, in the context of adaptive process-aware information systems. Basically, adaptive systems are based on loosely specified models able to deal with uncertainty, i.e. able to be revised to cover unseen positive examples. The implemented process must be able to react to exceptions, i.e. it must be revised to rule-out unseen negative examples. Both kinds of revision must guarantee that compliant traces with the previous model are still compliant with the revised one. In [34], the authors consider process models expressed as Petri nets, where structural adaptation is based on high-level change patterns, previously defined in [35]. They review structural and behavioral correctness criteria needed to ensure the compliance of process instances to the changed schema. Then, they show how and under which criteria it is possible to support dynamic changes in the ADEPT2 system, also guaranteeing compliance. Similarly to them, we address the problem of updating a (declarative and rule-based, in our case) process model while preserving the compliance of process instances to the changed model. This is guaranteed, however, not by identifying correctness criteria, but rather by the theory revision algorithm itself. We think that their approach is promising, and subject for future work, in order to identify both change patterns to be considered (up to now we consider one generalization and one refinement operator only) and correctness criteria under which the revision algorithm might be improved.

7   Conclusions

In previous work we have presented the system DPML that is able to infer a process model composed of a set of logical integrity constraints starting from a log containing positive and negative traces.

In this paper we introduce the system IPM that modifies DMPL in order to be able to revise a theory in the light of new evidence. This allows to deal with the case in which the process changes over time and new traces are periodically collected. Moreover, it does not need to store all previous traces. IPM revises the current theory by generalizing it if it does not cover some positive traces and by specializing it if it covers some negative traces.

IPM has been tested on artificial data regarding a hotel management process and on an electronic auction protocol. The result shows that, when new evidence becomes available, revising the current theory in the light of the new evidence is faster than learning a theory from scratch, and the accuracy of the theories obtained in the first way is higher. This supports our claim that updating can be better that inducing a new theory.

In the future, we plan to perform more experiments in order to further analyze the performance difference between learning from scratch and revision. Moreover, we plan to investigate techniques for learning from positive only traces and for taking into account change patterns.

References

1. Dumas, M., Reichert, M., Shan, M.-C. (eds.): BPM 2008. LNCS, vol. 5240. Springer, Heidelberg (2008)
2. van der Aalst, W.M.P., van Dongen, B.F., Herbst, J., Maruster, L., Schimm, G., Weijters, A.J.M.M.: Workflow mining: A survey of issues and approaches. Data Knowl. Eng. 47(2), 237–267 (2003)
3. Alberti, M., Chesani, F., Gavanelli, M., Lamma, E., Mello, P., Torroni, P.: Verifiable agent interaction in abductive logic programming: The sciff framework. ACM Trans. Comput. Log. 9(4) (2008)

4. Alberti, M., Gavanelli, M., Lamma, E., Mello, P., Torroni, P.: An abductive interpretation for open societies. In: Cappelli, A., Turini, F. (eds.) AI*IA 2003. LNCS (LNAI), vol. 2829. Springer, Heidelberg (2003)
5. Lamma, E., Mello, P., Riguzzi, F., Storari, S.: Applying inductive logic programming to process mining. In: Blockeel, H., Ramon, J., Shavlik, J., Tadepalli, P. (eds.) ILP 2007. LNCS (LNAI), vol. 4894, pp. 132–146. Springer, Heidelberg (2008)
6. Muggleton, S., De Raedt, L.: Inductive logic programming: Theory and methods. Journal of Logic Programming 19 20, 629–679 (1994)
7. Raedt, L.D., Dzeroski, S.: First-order jk-clausal theories are pac-learnable. Artif. Intell. 70(1-2), 375–392 (1994)
8. De Raedt, L., Van Laer, W.: Inductive constraint logic. In: Zeugmann, T., Shinohara, T., Jantke, K.P. (eds.) ALT 1995. LNCS (LNAI), vol. 997, Springer, Heidelberg (1995)
9. Esposito, F., Semeraro, G., Fanizzi, N., Ferilli, S.: Multistrategy theory revision: Induction and abduction in inthelex. Machine Learning 38(1-2), 133–156 (2000)
10. Clark, K.L.: Negation as failure. In: Logic and Databases. Plenum Press, New York (1978)
11. Gelfond, M., Lifschitz, V.: The stable model semantics for logic programming. In: Proceedings of the Fifth International Conference and Symposium on Logic Programming, pp. 1070–1080 (1988)
12. Van Gelder, A., Ross, K.A., Schlipf, J.S.: The well-founded semantics for general logic programs. Journal of the ACM 38(3), 620–650 (1991)
13. De Raedt, L., Dehaspe, L.: Clausal discovery. Machine Learning 26(2-3), 99–146 (1997)
14. Van Laer, W.: ICL manual, http://www.cs.kuleuven.be/~ml/ACE/DocACEuser.pdf
15. Adé, H., Malfait, B., Raedt, L.D.: Ruth: an ilp theory revision system. In: Ras, Z.W., Zemankova, M. (eds.) ISMIS 1994. LNCS, vol. 869, pp. 336–345. Springer, Heidelberg (1994)
16. Richards, B.L., Mooney, R.J.: Automated refinement of first-order horn-clause domain theories. Machine Learning 19(2), 95–131 (1995)
17. Georgakopoulos, D., Hornick, M.F., Sheth, A.P.: An overview of workflow management: From process modeling to workflow automation infrastructure. Distributed and Parallel Databases 3(2), 119–153 (1995)
18. Agrawal, R., Gunopulos, D., Leymann, F.: Mining process models from workflow logs. In: Schek, H.-J., Saltor, F., Ramos, I., Alonso, G. (eds.) EDBT 1998. LNCS, vol. 1377, pp. 469–483. Springer, Heidelberg (1998)
19. Greco, G., Guzzo, A., Pontieri, L., Saccà, D.: Discovering expressive process models by clustering log traces. IEEE Trans. Knowl. Data Eng. 18(8), 1010–1027 (2006)
20. Pesic, M., van der Aalst, W.M.P.: A declarative approach for flexible business processes management. In: Eder, J., Dustdar, S. (eds.) BPM Workshops 2006. LNCS, vol. 4103, pp. 169–180. Springer, Heidelberg (2006)
21. Montali, M., Pesic, M., van der Aalst, W.M.P., Chesani, F., Mello, P., Storari, S.: Declarative specification and verification of service choreographies. ACM Transactions on The Web (2009)
22. van der Aalst, W.M.P., Weijters, T., Maruster, L.: Workflow mining: Discovering process models from event logs. IEEE Trans. Knowl. Data Eng. 16(9), 1128–1142 (2004)
23. van Dongen, B.F., van der Aalst, W.M.P.: Multi-phase process mining: Building instance graphs. In: Atzeni, P., Chu, W., Lu, H., Zhou, S., Ling, T.-W. (eds.) ER 2004. LNCS, vol. 3288, pp. 362–376. Springer, Heidelberg (2004)

24. Lamma, E., Mello, P., Montali, M., Riguzzi, F., Storari, S.: Inducing declarative logic-based models from labeled traces. In: Alonso, G., Dadam, P., Rosemann, M. (eds.) BPM 2007. LNCS, vol. 4714, pp. 344–359. Springer, Heidelberg (2007)
25. Chesani, F., Lamma, E., Mello, P., Montali, M., Riguzzi, F., Storari, S.: Exploiting inductive logic programming techniques for declarative process mining. In: ToPNoC II 2009. LNCS, vol. 5460, pp. 278–295. Springer, Heidelberg (2009)
26. Apt, K.R., Bezem, M.: Acyclic programs. New Generation Comput. 9(3/4), 335–364 (1991)
27. Pesic, M., Schonenberg, H., van der Aalst, W.M.P.: Declare: Full support for loosely-structured processes. In: 11th IEEE International Enterprise Distributed Object Computing Conference (EDOC 2007), pp. 287–300. IEEE Computer Society Press, Los Alamitos (2007)
28. Chavez, A., Maes, P.: Kasbah: An agent marketplace for buying and selling goods. In: Proceedings of the First International Conference on the Practical Application of Intelligent Agents and Multi-Agent Technology (PAAM 1996), London, April 1996, pp. 75–90 (1996)
29. Chesani, F.: Socs protocol repository, http://edu59.deis.unibo.it:8079/SOCSProtocolsRepository/jsp/index.jsp
30. Desel, J., Erwin, T.: Hybrid specifications: looking at workflows from a run-time perspective. Int. J. Computer System Science & Engineering 15(5), 291–302 (2000)
31. Ferreira, H.M., Ferreira, D.R.: An integrated life cycle for workflow management based on learning and planning. Int. J. Cooperative Inf. Syst. 15(4), 485–505 (2006)
32. Goedertier, S.: Declarative techniques for modeling and mining business processes. PhD thesis, Katholieke Universiteit Leuven, Faculteit Economie en Bedrijfsweten-schappen (2008)
33. Wrobel, S.: First order theory refinement. In: Raedt, L.D. (ed.) Advances in Inductive Logic Programming, pp. 14–33. IOS Press, Amsterdam (1996)
34. Reichert, M., Rinderle-Ma, S., Dadam, P.: Flexibility in process-aware information systems. T. Petri Nets and Other Models of Concurrency 2, 115–135 (2009)
35. Mutschler, B., Reichert, M., Rinderle, S.: Analyzing the dynamic cost factors of process-aware information systems: A model-based approach. In: Krogstie, J., Opdahl, A.L., Sindre, G. (eds.) CAiSE 2007 and WES 2007. LNCS, vol. 4495, pp. 589–603. Springer, Heidelberg (2007)

A Survey on Recommender Systems
for News Data

Hugo L. Borges and Ana C. Lorena

Centro de Matemática, Computação e Cognição, Universidade Federal do ABC
Rua Santa Adélia, 166 - Bairro Bangu, 09.210-170, Santo André, SP, Brazil
{hugo.borges,ana.lorena}@ufabc.edu.br

Abstract. The advent of online newspapers broadened the diversity of available news' sources. As the volume of news grows, so does the need for tools which act as filters, delivering only information that can be considered relevant to the reader. Recommender systems can be used in the organization of news, easing reading and navigation through newspapers. Employing the users' history on items consumption, user profiles or other source of knowledge, these systems can personalize the user experience, reducing the information overload we currently face. This chapter presents these recommender filters, explaining their particularities and applications in the news' domain.

Keywords: News data, recommender systems.

1 Introduction

Recommender systems (RS) arose as a research area in the nineties, when researchers began to focus on the structure of evaluations. The origins of RS are linked to cognitive science, approximation theory, information retrieval, prediction theories and have also links to management science and to the process of modelling options made by consumers in marketing [3].

Before presenting a formal definition of RS, it is important to explain the concept of information filtering, since RS act as information filters. Belvin [5] characterizes the expression "information filtering" for processes where there is a delivery of relevant information for certain persons. This type of system normally works with huge quantities of unstructured data: text, images, voice and video. Filtering may be made based on the description of the preference of individuals or groups of individuals.

RS have also been employed in the organization of news data, easing reading and navigation through online newspapers. As the volume of news grows in this type of media, so does the need for tools which act as filters, delivering only information that can be considered relevant to the reader.

This chapter will summarize the topic of Recommender Systems, explaining the main decisions involved in the design of this in. It will also present some works of RS related to news data, highlighting their peculiarities and showing how RS were employed in these systems.

E. Szczerbicki & N.T. Nguyen (Eds.): Smart Infor. & Knowledge Management, SCI 260, pp. 129–151.
springerlink.com   © Springer-Verlag Berlin Heidelberg 2010

Initially, Section 2 defines RS formally and discusses the way users evaluations are usually carried out. Next, Section 3 presents different approaches employed in the literature for recommendation. Next sections discuss, respectively, the main similarity measures (Section 4), evaluation measures (Section 5) and the applications (Section 6) of RS in commercial systems and in the literature. Section 7 presents work related to RS for news data. Finally, Section 8 presents the main conclusions of this chapter.

2   Definition and Therminology

The problem involved in RS is that of estimating the evaluations of items unknown to an user and using these evaluations to recommend to the user a list of items better evaluated, that is, those items which will more probably be of the user's interest. To make such estimates, one may use the evaluations of other items made by the same user or the evaluations made by other users with similar interests to a particular user.

Formalizing the problem, given a set of users U and a set of items I, let s be an utility function which defines the punctuation (evaluation or note) of an item i for an user u. That is: $s : U \times I \to P$, in which P is a completely ordered set, formed by non-negative values within an interval, 0 to 10, for example. The system must recommend an item i' which maximizes the utility function for an user:

$$i' = \arg \max_{i \in I} s(u, i) \tag{1}$$

An element in the set U may be defined by several characteristics, which corresponds to the user's profile. Equally, elements from set I may be also defined by several characteristics, these related to the domain of the items. A film, for instance, may have as characteristics its title, genders, year of release and the names of artists, directors and writers involved in the film production.

Since function s is not defined in all space $U \times I$, it must be extrapolated, allowing presenting to the users items unevaluated by them and which will probably be of their interest. This is the central problem in RS. This extrapolation may be carried out through the use of heuristics defining the utility function, which are empirically validated, or it may be carried out through an estimate of the utility function by optimizing a certain performance criterion, as the mean squared error. More specifically, the estimate of the evaluations may be obtained using methods from approximation theory, heuristic formulas as cosine similarity and Machine Learning (ML) techniques as Bayesian classifiers, Support Vector Machines (SVMs), Artificial Neural Networks (ANNs), and clustering techniques [3].

Concerning the evaluations performed by users, a RS may have either an implicit or an explicit feedback. In the first case, the system must capture the user preferences without making direct questions to him regarding his opinion about the item. One approach to obtain implicit feedback is the analysis of the user's consuming profile. For news, one may analyze which news were accessed by the user and the time spent in their reading. In the case of explicit feedback, the RS must present the user an input interface where he can evaluate the item considering some evaluation parameter.

Evaluation may be performed in several ways: integer numbers varying from 0 to 10; "stars" (one star indicating awful, two bad, three medium, four good and five outstanding); continuous bars where the right side indicates a negative evaluation, the center indicates a neutral evaluation and the left side indicates a positive evaluation; or by a binary system: upwards arrow (or positive sign) indicating positive evaluation and downwards arrow (or negative sign) indicating negative evaluation. It is also possible to have a one class (or unary) evaluation: the user says only if he liked the item or not. Other possibility is the insertion of a textual description of the item. The RS must then convert the evaluation into a punctuation, which will be employed in the recommendation of new items. Although this is a topic studied less frequently, Cosley et Al. [15] concluded that the choice of a correct interface for the recommendation can improve significantly the user's experience with the system.

3 Recommendation Approaches

RS are usually classified according to the sources of information and how they are employed. Generally, RS have: background data, information the system has before initiating the recommending process; input data, information communicated to the system by the user and which is necessary to generate a recommendation; an algorithm which combines the first and second sources in order to suggest new items for the users [11]. Using these three components, two main recommendation approaches may be identified: content-based and collaborative, described in sections 3.1 and 3.2, respectively. Other classifications presented by Burke [11] are: demographic RS, knowledge-based RS and utility-based RS.

Both collaborative and content-based use the user's (implicit or explicit) ratings as the input to the classification. The main difference between these techniques is in the background data. While collaborative filters employ previous ratings given by different users to an item as background data, content-based filters employ the item's features.

Demographic RS use information from the user as sex, age and scholarity level to trace an user profile. Herewith, the recommendation is made according to the demographic class to which the user belongs and there is no need of user ratings. This technique has been under-explored, since it needs information from the users usually harder to obtain than the items' evaluations.

Knowledge-based recommendations use knowledge about the items or users to perform the recommendation. In this case the users need to provide a description of their needs or interests regarding the item. Burke [11], for example, developed a knowledge-based RS for restaurants. The system suggests restaurants based on the choices made by users: starting from a known restaurant, step by step the user indicates some preference that will be considered for the restaurant recommendation.

Utility-based RS, similarly to content-based RS, uses characteristics of the items to be recommended. The difference between these filters is that utility-based RS uses as input an utility function about the items which describe an user's preferences, while the inputs from the content-based RS are the evaluations of items chosen by the user.

In an earlier work, Malone et al. [29] propose three types of information filters based on the filtering decisions made by users in different scenarios of information filtering within an organization. They classified the filters as cognitive, that are similar to the content based filters and social, in which the value of a message is mainly based on the characteristics of its sender. They also suggest the economic approach, in which the selection is made considering the cost-benefit brought by the consumption of the item. One example given by the authors is the size of the message as a cost estimate. A person may be not willing to read a long text unless it is of extreme importance. Other example would be to analyze the cost of a message to estimate the information gain it brings. Mass e-mails, for example, aggregate low information, while having a low production cost.

When two or more RS are combined, by using the same recommending approach with different characteristics or using distinct recommending approaches, hybrid RS are identified, which will be presented in Section 3.3. Next, the content-based and collaborative approaches, which are the most used in hybrid combinations, are presented in more details.

3.1   Collaborative Approach

In collaborative RS (CF - Collaborative Filtering), the subjective opinion of other users is employed. The system will recommend to a new user items that were well evaluated in the past by people with similar preferences to those of the user. Systems employing CF may be memory-based, comparing the users among themselves using some similarity measure (see Section 4), or may derive a model using the history of evaluations made by the users. CF can be seen as an automation of the *word-of-mouth* recommendation, where a person receives or asks for recommendations of friends about some product or service.

CF presents four main problems that may influence in the accuracy of a RS employing this recommending approach:

– First evaluations problem (*early rater, ramp up, cold start*) - a collaborative filter cannot make predictions for items that were not evaluated by any user yet. Predictions made from the evaluations of few users tend to be of low accuracy. New users will also face a similar problem because of having evaluated few or no items, needing to evaluate a minimum quantity of items to begin receiving recommendations.
– Sparsity problem - the number of items in many systems is much higher than the number of users. Evaluation matrices are therefore very sparse (reaching 99% of sparsity), which difficult finding items evaluated by a sufficient number of persons.

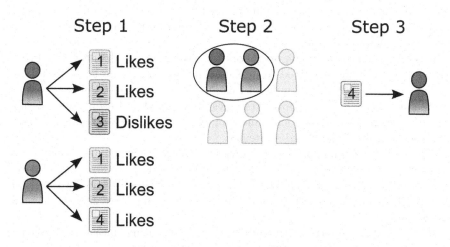

Fig. 1. User-based Collaborative Recommender System - In Step one, the system obtains user ratings for some items; In Step 2, the system finds users with similar interests; In Step 3, the system delivers to the user a new item that will probably interest the user based on his similarity with other users.

- Gray sheep problem - in small and medium size communities, some individuals may not benefit by collaborative filters because of having an opinion that is not different or similar to that of other users, making the system predictions usually bad for this particular user.
- Portfolio problem - the RS may recommend items too similar to those already consumed by the user.

Related to content-based RS, the main advantage of the CF approach is that it is independent of the items' representation, and may be directly employed for different types of items not easily interpretable by computers, as images or videos. Therefore, prior knowledge about the domain is not necessary.

Furthermore, CF allows the identification of niches of multiple genres, since it is independent of content and it is adaptive, that is, the quality of recommendation tends to increase with time. Nevertheless, a disadvantage is the need for a relatively large amount of historical data to get a good recommendation. A last advantage of CF relies in the fact that it is able to obtain good results even using only implicit feedback.

Candillier et. al [13] identified three different strategies for CF for the prediction of new items: user-based, item-based and model-based. In Figure 1 the three basic steps for a user-based collaborative recommendation are presented.

In the user-based approach, employed most frequently, the evaluation of a new item is based on the evaluations made by neighbor users. First a similarity measure between users must be defined and then a method to combine the evaluations performed by the neighbors must be chosen. Commonly a weighted sum of the evaluations made by the nearest neighbor users that already evaluated that item is employed.

In the item-based strategy, neighbors for each item are defined. The evaluation of a new item is derived from the evaluations made by the user for the item's neighborhood. An example of this approach is presented in the work by Deshpande & Karypis [18]. Both strategies, item-based and user-based, employ similarity measures for the determination of the neighborhoods, and are also known as neighborhood-based strategies.

The quadratic complexity of the previous approaches is not desirable for systems where the evaluations must be quickly obtained. Therefore, model-based strategy aims to reach a modelling between users and items. Some of the possible techniques employed in this modelling are: clustering, Bayesian models and association rules.

In experiments reported in Candillier et. al [13] using data sets *Movielens* and *Netflix*, the item-based strategy obtained the lowest mean quadratic error in the evaluation of new items, followed by the user-based strategy and by the model-based strategy. An important parameter of item-based and user-based approaches is the number of neighbors to be considered. As advantage, the model-based approach had learning and prediction times largely reduced when used in a problem with a large amount of users and items.

3.2   Content-Based Approach

In content-based or cognitive RS (CBF - Content Based Filtering), the user preference for certain items is used to implicitly build an user profile, by searching common points in the description (content) of the items positively or negatively evaluated by this user. A model of the user may be also obtained by providing descriptions of these items for a supervised Machine Learning (ML) [32] algorithm. Using the CBF approach, the system will recommend items similar to those for which the user has shown interest in the past. When an explicit profile of the user is built, in which some sort of questionnaire is presented to him, a knowledge-based RS is characterized. In Figure 2 the three basic steps for a content-based recommender system are presented.

As disadvantage, CBF can not distinguish between content of high and low quality, which is much easier for a human being. There is also the fact that no 'surprise' or novelty may be present in the new recommendations, an effect known as serendipity. Therefore, the problem of portfolio can be even more pronounced in the CBF when compared to CF.

This approach is subject to specialization, in which only items of the categories assessed by the user in the past are recommended. Moreover, for certain types of media such as video, audio and images, the analysis and evaluation of the content is more difficult, making the use of CBF complex or making necessary the use of some sort of textual description of the item.

Similar to CF, in CBF the user needs to perform a sufficient amount of evaluations so that the system can begin to make recommendations with a good

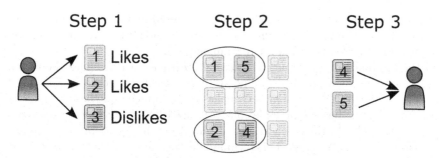

Fig. 2. Content-based Recommender System - In Step one, the system obtains user ratings for some items; In Step 2, the system finds new items that have a similar content to those the user likes; In Step 3, the system delivers to the user the most relevant items.

accuracy and the quality of assessments tends to improve with time. As an advantage over CF, CBF may recommend a new item even if it has not yet been rated by any user.

In general, the items used in the CBF approach are semi-structured texts, which allows the use of Information Retrieval techniques and representations.

3.3 Hybrid Approach

Many RS currently proposed are hybrid, trying to aggregate the benefits and minimize the disadvantages of different approaches. The most common combination is of collaborative (CF) and content-based (CBF) filters. Burke [11] identifies seven possible types of hybrid RS.

In a weighted hybrid RS, the score of a recommended item is measured by the combination of all techniques present in the system. A simple example would be a linear combination of two scores as in *P-Tango* [14]. This technique assumes that the relative value of the different techniques combined is approximately uniform in the space of possible items, which is not always true, since collaborative classifiers have lower performance for items with few evaluators.

An alternate hybrid system alternates between the different recommendation techniques according to some given criterion. RS *Daily Learner* [8], for example, uses a CF or a CBF if the confidence of the recommendation made by CBF is not good. This technique adds complexity, since it is necessary to determine a criterion for switching, including a new level of parametrization.

There is also a mixed approach for recommendations. In this case, recommendations from different recommenders are presented at the same time. This technique is interesting when the performance of multiple simultaneous recommendations is practicable. Examples of works using this approach are: the PTV system [16] for the recommendation of TV programs, *ProfBuilder* [43] and *Pick-AFlick* [10], which have multiple side by side recommendations. It is common that, when presenting the recommendation, they are properly ranked. In this case it is necessary to employ some technique for combining the different RS.

It is also possible to obtain a hybrid RS by combining the attributes of different RS. The attributes of the different sources of knowledge are combined and employed in a single algorithm for recommendation.

In the approach of recommender cascade (*cascade*), a RS refines the recommendation made by a second RS. The system *EntreeC* proposed by Burke [11] uses this approach. Initially, *EntreeC* makes use of a technique based on knowledge. Afterwards, the recommendations produced by this technique are sorted according to a CF. In the cascade system, the RS with highest priority (applied first) should be the strongest. This technique is therefore tolerant to noise in the second RS applied, but can not readjust the first RS applied.

Finally, a hybrid RS with increased attributes uses the output of a recommending technique as input attribute for a second recommending technique. A similar approach to this is the *meta-level*. Nevertheless, in this case, instead of using the output produced by a recommender, a model generated by such recommender is employed as input for the second technique applied.

4   Similarity Measures

Both in content-based approach as in the neighborhood-based collaborative approach (of items or users), the definition of the similarity between some sort of entities is important for the quality of the recommendations made.

Manouselis & Costopoulou [36] discuss different options for projecting a neighborhood-based collaborative RS, featuring some possible similarity measures. Three simple measures are the square difference of the averages used by Shardanand & Maes [42] to calculate the similarity of two users in one of the first works on collaborative filtering, the Euclidean distance (L_2) employed by Kim & Yang [27] in the same context and the Manhattan distance (L_1).

Pearson's correlation is probably most popular for the user-based collaborative approach, being employed in *Grouplens* [38], for example. This measure is used to obtain the similarity between an user and all related neighbors available to then compute a final prediction based on the weighted average of the standard deviation of the neighborhood. It corresponds to the cosine of the standard deviations from the average.

Given two users and a and u, the Pearson's correlation between them can be calculated by Equation 2.

$$Pearson(a, u) = \frac{\sum_{\{i \in S_a \cap S_u\}} (r_{ai} - \bar{r}_a) \times (r_{ui} - \bar{r}_u)}{\sqrt{\sum_{\{i \in S_a \cap S_u\}} (r_{ai} - \bar{r}_a)^2 \sum_{\{i \in S_a \cap S_u\}} (r_{ui} - \bar{r}_u)^2}} \qquad (2)$$

Where S_a is the set of items evaluated by user a, S_u is the set of items evaluated by user u, r_{ai} represents the evaluation of item i made by user a, r_{ui} represents the evaluation made to item i by user u, $\bar{r}_a$ and $\bar{r}_u$ are the averages of the recommendations made by users a and u.

Shardanand and Maes propose the *constrained* Pearson [42] in which only positive ratings are taken into account. They observed that, in their system, negative ratings are very few so they represent little difference for this calculation.

Another popular measure is based on cosine, and is also known as vectorial or L2-norm. It can be used in the user-based or item-based collaborative approaches. The similarity between the vector of evaluations of a user a and the vector of a user u is measured for the set of items rated by both users (Equation 3).

$$cosine(a, u) = \frac{\sum_{\{i \in S_a \cap S_u\}} r_{ai} \times r_{ui}}{\sqrt{\sum_{\{i \in S_a \cap S_u\}} r_{ai}^2 \sum_{\{i \in S_a \cap S_u\}} r_{ui}^2}} \qquad (3)$$

It is a simpler measure compared to Pearson's correlation as it does not take in account the averages of user ratings. This measure can also be employed in the content-based approach. In this case (Equation 4), vectors V_i and V_j represent a set of attributes from items i and j, as keywords.

$$cosine = \frac{V_i \times V_j}{|V_i||V_j|} \qquad (4)$$

Pearson and cosine measures consider only the common set of attributes between two vectors. So, for these measures, if two users have only one rated item in common and they gave the same rating for this item, they will have a similarity of one, no matter if they have rated other items differently. On the other hand, the Jaccard similarity measure (Equation 5) considers the difference between two sets of items but does not take in account the difference of ratings values, so it is more adequate to binary or one class-ratings.

$$Jaccard(a, u) = \frac{|\{S_a \cap S_u\}|}{|\{S_a \cup S_u\}|} \qquad (5)$$

The work of Candillier et al. [12] indicates that the combination of similarity measures used for collaborative filtering can bring substantial improvements to the recommendations. These authors conclude that the use of two measures with different properties allows to consider a smaller number of neighbors, therefore reducing the time spent on the recommender training and prediction.

Other measures presented by Manouselis & Costopoulou [36] are: Spearman correlation [21], adjusted cosine, Clark's distance [33] and variance weighting [21].

5 Evaluation of Recommender Systems

Generally, the performance of a RS is assessed by measuring its accuracy to find new relevant items. However, there is no consensus on which measures are more appropriate and most of the work in the area use their own methodology for the evaluation of the proposed recommender systems.

Candillier et. al [13] highlight three measures as the most applied to measure the performance of a RS: the mean absolute error (MAE), the root mean squared error (RMSE) and precision measures.

Equations 6 and 7 present MAE and RMSE for a set T, composed by users u, items i and rates r. p_{ui} is the prediction of an item i obtained for an user u.

$$MAE = \frac{1}{|T|} \sum_{(u,i,r) \in T} |p_{ui} - r| \tag{6}$$

$$RMSE = \sqrt{\frac{1}{|T|} \sum_{(u,i,r) \in T} (p_{ui} - r)^2} \tag{7}$$

MAE measures the average absolute difference between a predicted and a true rating. Compared to MAE, RMSE gives more emphasis on large errors, as the errors are squared before they are summed.

In RS with graded evaluation, as the ratings using one to five stars, the predicted values are usually rounded to integers (3;5) or multiple of means (0.5;2.5), which makes easy the interpretation of the rating value. These values must then be employed in the calculations of the measures presented.

For systems that present a ranked list of recommended items, MAE and RMSE are not reliable measures, given that wrong items in the top of the list have a different weight from those on the bottom. For these systems, it is better to employ *Rank Accuracy Metrics* like precision and *half-life utility* [41]. Precision is given as the proportion of relevant items selected relative to the total number of items selected. The *half-life utility* measures the difference between the user's rating for an item and a default rating that is usually set to a neutral or slightly negative rating.

Herlocker et al. [22] point out two main reasons that make the evaluation of RS difficult. The first one is related to the fact that the performance of a RS may vary for different data sets. Some characteristics of the data sets, like number of items, number of users, sparsity of ratings and ratings scale can make one algorithm inappropriate for a given domain. The second reason is that the evaluation of an RS may have different goals. In systems that are used to support decisions, for example, it is more desirable to measure how often the systems leads to right or wrong decisions. Some authors also believe that the user satisfaction is the most important measure for a recommender system.

It is also important to consider other factors beyond accuracy in the choice of an appropriate RS. Factors such as coverage, which measures the percentage of items for which is possible to make predictions, computational complexity, scalability, novelty detection and reliability. An additional factor to be considered, especially in CF, is the robustness of the RS, which measures the power of the algorithm to achieve good predictions in the presence of noisy data that usually comes from attacks to the system.

6   Applications

In the case of RS developed in the academy, there is a variety of applications using different approaches. One of the first systems was *Ringo* [42], which recommends music albums and artists using collaborative filtering. *Movielens* [20] and *Video Recommender* [23] are examples of systems geared to recommend movies and also make use of collaborative filters. *MovieLens* dataset is one of the most used sources for evaluation and comparison of RS.

Fab system [4] uses a content-based RS for the recommendation of web pages. *Jester* [19] (available at [26]) is a system for the recommendation of jokes employing a collaborative filtering algorithm known as *Eigentaste*. Jester's users must, first, rate eight jokes so the system can start to recommend new items.

An example of hybrid RS that incorporates the content of the data in a collaborative filter is proposed in [31], obtaining superior results to usual collaborative and content-based RS.

The most common examples of the use of RS in commercial web applications are *Amazon* virtual store and *Netflix*, a service for movie rentals. Both companies use hybrid approaches in their recommendations.

Amazon is seen as one of the pioneers in the use of RS for e-commerce. When accessing the a book's page, for example, there is the recommendation of other items purchased by other customers with the same item. Users can evaluate the items with stars and make their comments on the items. With this assessment, the site can rate the best items in each category. The purchase history of user searches is also used to recommend new items.

Netflix is a company that rents movies in DVD and *Blu-ray* formats by mail and streams of movies and TV series. A major difference from their service is in its recommendation system called *Cinematch*. The creators of the system estimate that between 2010 and 2012 it will have over 10 billion ratings, receiving more than 10 million of them per day and doing about 5 billion predictions every day [6].

Recognizing the strategic importance of their system, the company launched the *Netflix* Prize in 2006, offering a U.S.$ 1 million prize for the team who can develop a system of recommendation to present a solution with 10% better accuracy compared to their current system. In 2008 "Belkor in BigChaos" team obtained an improvement of 9.44% over *Cinematch*. The proposed solution [1,2] combines 22 different models and a search algorithm for automatically tuning the parameters of the different models in order to minimize the error.

Two other examples of applications that use RS are *Last.fm* and *Pandora* that, similarly to *Ringo*, suggest songs and artists to users using collaborative filtering techniques.

7   RS Applied to the Domain of News

News recommendation is one of the main from the various applications in which the recommender systems have been applied. In a study by Montaner et al. [34], 37 Internet systems which employed recommender agents were analyzed. Of these, 11 were used for the recommendation of stories, both coming from *Newsgroups* and from online newspapers. In this section some of these systems are analyzed, presenting the particularities in this domain and the techniques for recommendation adopted in the different scenarios proposed by the authors.

Table 1. RS applied to the domain of news stories

Author(s)	System	Recommendation Approach
Resnick et al. [38]	*GroupLens*	Collaborative
Lang [28]	*Newsweeder*	Hybrid
Claypool et al. [14]	*P-Tango*	Hybrid
Billsus and Pazzani [8]	*Daily Learner*	Content-based
Das et al. [17]	*Google News*	Collaborative
Bogers and Bosch [9]	-	Content-based

Table 1 gives a summary of some of the most important work involving RS for news, presenting the approaches and techniques employed for the recommendation. Afterwards, these works are presented in greater detail.

7.1 *GroupLens*: An Open Architecture for Collaborative Filtering of Netnews

Resnick et al. [38] define *GroupLens* as a collaborative filtering system for *News-Groups*. The architecture proposed by the authors is composed by a client where the user views the news's ratings prediction and, based on this information, chooses which of them he will read. This client also allows the user to evaluate news. Besides the client, there are servers responsible for collecting the ratings given by users, estimating the scores of stories and then disseminating them to the clients. The objectives of this design are the openness, ease of use, compatibility, scalability and privacy. The system is based on the heuristic that users who had similar opinions in the past will probably have similar opinions in the future.

The rating system correlates rates of items already evaluated to determine weights to be assigned to each person when making predictions for them. Initially it calculates the correlation coefficients that indicate how a user tended to agree with other users in articles that they both evaluated.

Since users may have similar interests in some subjects but not in others, the matrices of ratings are built for each group of stories. Clients used to capture the news then have the task of deciding how to use the predicted ratings. The authors propose three different implementations for the client. First, a column was added to the side of stories showing the predicted rating numerically. The second implementation used the ratings to change the order in which items were displayed. The ratings were also visible in an additional column. To ease the visualization of the ratings, the numbering scheme was changed to letters ranging from A to E. The third client showed the predicted ratings in bar graphs, making it even faster to visually scan the items with higher predicted ratings.

Resnick et al. [38] comment that few experiments were performed. Users of the pilot test stated that, after some time they started to rate the news, the predictions became relevant. The authors also comment that it is likely that there are better mechanisms for evaluation and that it would be interesting to use the contents of the article and take into account the time that people take to read the news in estimating predictions.

7.2   *News Weeder*: Learning to Filter Netnews

The filtering system for *News Weeder*, proposed by Lang [28], uses collaborative filtering to discover the interests of the user. Lang used this approach to ensure that the system is not dependent on creating and maintaining a profile in which the user has to describe his interests. The system also employs a content-based filtering, therefore a hybrid approach. The final prediction is obtained by the weighted average of content-based predictions for the user, predictions made by other users and content-based predictions for all the users. However, in this work, Lang examines only the content-based prediction made for each user.

In *News Weeder*, the user selects the news he wants to read and evaluates it after reading with ratings ranging from 1 (essential reading) to 5 (no interest) and 'skip' for articles that the user, only by the summary , decided not to read. Each of the six possible rates represents a category. The author argues that the use of explicit feedback is better because it facilitates the measurement of performance and also the training of the recommender system. On the other hand, it requires the user to take the extra effort of having to choose a rate for each story read.

Lang gathered information from evaluations of two users who used the system for almost a year. With this information, he conducted experiments related to the generation of models of user's interests. Text preprocessing was performed using bag-of-words. Beyond words, the punctuation marks were kept. Lang then compared both approaches for the prediction of ratings.

The first approach uses a technique known as TF-IDF (Term Frequency - Inverse Document Frequency) of IR. To classify a document, the vector that represents it is compared with the prototype vectors of each category. The prototype vectors correspond to the average of all vectors belonging to the category. The classification results are then converted to a continuous value by linear regression with the least squares method. The chosen similarity measure for the comparison between the vectors is the cosine of the angle formed between the vectors of both documents.

The second technique chosen is based on the Minimum Description Length (MDL) principle [39]. Similarly to the TF-IDF approach, the MDL approach initially classifies the items in the various degrees of evaluation (categories) and the measures of similarity are converted into a continuous space.

Lang stressed that the use of a stopwords list could lead to loss of important data. The author removed the most frequent words and obtained better accuracy in predicting the ratings for the case where 100 to 400 words with the highest value of TF-IDF weight were removed. Lang chose to measure the accuracy of evaluation, in this case defined as the percentage of items evaluated as interesting found in 10% of the files with the highest predicted ratings. This measure was evaluated for the two approaches, varying the amount of training examples for all the data from two users.

The MDL method achieved an accuracy of 44% and 58% for users A and B, respectively, 21% higher than the precision provided by the TF-IDF technique for the user A and 16% higher for the user B.

7.3   *P-Tango*: Combining Content-Based and Collaborative Filters in an Online Newspaper

Claypool et al. [14] publication presents the design of a filtering system for the *Worcester Telegram and Gazette on-line* newspaper which, according to the authors, tries to combine the speed and coverage of content-based filters with the depth of collaborative filters. In this work the authors present preliminary results and suggest the benefits of the chosen methodology.

In the researchers approach, the rating of each story is obtained by the weighted average between the content-based prediction and the collaborative prediction. The weight given to collaborative prediction is varied and generally increases as an item is rated by more people, avoiding the problem of the first ratings. The weighting of the predictions is defined for each user of the system, which minimizes the 'gray sheep' problem. The system is also hierarchical, as components and subcomponents may receive different weights. In the content-based filter, for example, the components keyword and section receive the same weight.

The collaborative filter system employed by *P-Tango* system uses *Pearson* correlation coefficient to calculate the similarity between users and the content-based filter uses keywords. The user can, optionally, set the preference in his profile for categories of the newspaper or specify keywords. A profile also has a list of implicit keywords. These words are obtained from 25% of the articles that received the best ratings from the user. The generation of keywords for each article is obtained as follows: stopwords are removed, stemming is applied, and finally the words that occur more frequently are selected. The predictions for the three indicators (categories, implicit and explicit keywords) initially have the same weight. Over time the errors are calculated for each prediction and the weights are adjusted for each user.

The architecture of the system is composed of a front-end (web interface with the user), a back-end and the database (DB). The back-end imports the articles from *Tango* web site, calculates the similarities between users, generates the keywords and calculates the predictions. The DB is responsible for the storage of articles, users' profiles, evaluations and calculated predictions.

The experiments were conducted with 18 users who used the system for three weeks. These have made 0.5% of the possible evaluations. The mean absolute error (called by the authors as inaccuracy) between predictions made by the system and the evaluations made by the users was evaluated. This measure was calculated for the collaborative filter, the content-based filter and the combination of both. The inaccuracy showed up next in the first few weeks for the three filters. However, in the last week, the inaccuracy of the collaborative filter was considerably lower than that of the content-based filter. The hybrid approach was generally more accurate than the other two. The authors said that more testing time would be necessary to better evaluate the system. Despite the fact that content-based filter did not bring better accuracies, it is valid in situations which there are no ratings yet. However, it is difficult to measure this benefit numerically.

Claypool et al. finalize their work saying that there are many issues involved in the personalization of newspapers online. It can be obtained in relation to content, layout, types of media (text only, text with pictures, video), advertising, among others. It is possible, for example, to customize the number of articles per page, rearrange the articles and even change the hierarchical structure of newspaper categories.

7.4 *Daily Learner*: User Modelling for Adaptive News Access

The work of Billsus and Pazzani [8] concerns the project, implementation and evaluation of a framework for adaptive access to news. This system employs a client-server architecture. The Adaptive Information Server (AIS) uses ML algorithms to obtain models from users based on their implicit and explicit feedback. In this work, two versions of the *Daily Learner* agent are presented. The first version shares the news via a web interface, while the second is geared for mobile devices like cell phones and PDAs. The authors use a *Palm VII* device as an example.

The AIS uses a multi-strategy ML algorithm in order to capture the interests of individual users for daily news. The news source service used was *Yahoo News*. The authors consider various aspects to create their solution: the use of a central repository of data, which eases data collection and incorporation of preferences of users in the recommendation algorithm; scalability; and the system must be able to produce recommendations for a large number of users and, therefore, the recommendation algorithm should be efficient. Other requirements are: platform independence of devices, the system should be not limited to a single platform so that it can attract a large number of users; ubiquitous access, where access to individual profiles can be done by any device that has access to Internet; and response time, in which waiting time should be reduced at maximum.

The newspaper created for the web interface contains nine different categories and uses explicit feedback, while the mobile version uses only implicit feedback due to bandwidth limitations of the devices and for usability reasons.

To create the model of users, some points were considered. The model should represent the different interests of users in various categories. It needs to be flexible to adapt quickly to the user's changes in interest. Furthermore, it should consider the fact that the user needs for information changes due to his own interaction with the information.

Therefore, Billsus and Pazzani [8] have chosen a model composed of two parts: one representing the long-term interests and other the short-term interests of the user. According to the authors, this choice has several advantages for areas with temporal characteristics, such as news.

The model considers the 100 short-term latest news read by a user. In this model the news are represented by a vector of TF-IDF terms. The prediction of the rating for a new history is obtained as follows. Calculating the distance of the news concerned with all the news already read by the user by using the k-NN algorithm. The stories that reach a threshold t-minimum are chosen. The rate is then predicted as the average between the weighted evaluation of all the stories chosen, where the weight is given by the similarity between a news story chosen and a new story. If the threshold value of the news is over a t-max, it is marked as "known" because it is very similar to a story already read. If there are no news between the thresholds t-minimum and t-max, the news are passed to the long-term model. The advantage of the nearest neighbors algorithm is that it can provide a similar story using only a single training example, which does not happens for other algorithms that require larger amounts of training examples.

The long-term model aims to capture the user's general preferences. It also uses the representation of the news by a TF-IDF vector, selecting the ten attributes with greater weight. It is also selects the 150 words that appear most in each category. This selection of different attributes was used to ensure better system scalability and also better rates of correct classification. Furthermore, the words that appear most in each category are used to build explanations for the recommendations made. The approach used for the construction of explanations is detailed in [7]. The Multinomial Naïve Bayes (MNB) algorithm [30] is used and the news are classified as interesting or non-interesting. The authors also included two restrictions to prevent the system of long-term rating items with little evidence of belonging to a class. An attribute is only considered informative if it appears at least ten times in all training. An item is classified if it contains at least two attributes informative.

Billsus and Pazzani [8] evaluated the F-measure and the percentage of correct predictions for the four stories with higher predicted evaluation. Two experimental approaches were used: the first measures the performance of the system daily. Systems' predictions are compared with the actual evaluations made by the user and then the mean of the results is taken. The second quantifies the performance of the prediction system as a function of the number of training data available. The performance is measured regarding the training sessions, when the user enters the system and evaluates at least four stories. Finally, the mean of the performance for each user of the system is taken. The web client had about 150 users and the wireless client had 185 users.

The web client performance showed significantly better results compared to the mobile client. This is explained by the fact that it first receives the explicit ratings. The F-measure of the hybrid approach against the F-measure of short and long-term approaches was statistically higher in both the web client and the wireless client. The value of F-measure increased over three sessions of the web client. In the wireless client, 10 sessions were evaluated. After climbing a few days, the values of F-measure began to fluctuate as a result of the different distribution of the daily news. The results indicate the superiority of the hybrid approach proposed by the authors.

7.5   *Google News* Personalization: Scalable Collaborative Online Filter

Das et al. [17] present the approach used by *Google News* for the recommendation of news to its users. The authors highlight the fact that it is a system with millions of users and items and that is very dynamic, i.e., has high turnover of items, that is, the news. The approach presented is independent of content, can be used for other areas. The predictions are obtained by a weighted combination of three different approaches: the grouping *Min-Hashing*, the probability of semantic latent indexing (PLSI - *Probabilistic Latent Semantic Indexing*) and counting covisitations.

Google News is a system that aggregates headlines from more than 4,500 sources of information, grouping similar stories and showing them to the user according to their personal interests. Recommendations are made for users logged into the system. They are based on implicit user feedback, i.e., the system stores the news that were accessed by the user. Therefore, there is a single class configuration, that is, a visit counts as a positive vote, and therefore there is no negative class. Das et al. [17] stress that this setup is more subject to noise than cases in which there are two classes (positive or negative evaluation) or a range of scores. As advantage, the implicit feedback does not require any user intervention in the system. There is also the restriction of time: the recommendations should be generated in a few hundred milliseconds to keep the application response time down.

Min-Hashing is a probabilistic clustering method which links a pair of users to the same cluster according to a probability proportional to the overlap between the items which received positive feedback (were visited) by both users. Therefore, the similarity between two users is given by the Jaccard coefficient between both users.

PLSI technique was developed by Hofmann [24,25] for performing collaborative filtering in a constant prediction time. Users and items are modeled as random variables. The relationship between items s and users u is modeled by a combined distribution. A variable Z is then introduced to capture this relationship, which can be seen as the identification of communities of users, i.e. those users with similar preferences.

The prediction of a recommendation using Min-Hashing and PLSI for a news i and a user u is calculated by obtaining the clusters of which the user is part of and, for each of these clusters, check how many times, discounted by 'age', the members of this cluster visited the news i. These values are summed and normalized in a range from 0 to 1.

The third technique used by the recommendation of *Google News* is covisitation. Unlike the first two techniques, it is based on memory and not on models. It aims to shape the preferences of the user in a short period of time. Covisitation is defined as the event in which a user visits two stories within a specified period of time. This can be represented in a graph where the nodes are the items (in this case, the news) and the edges that connect the nodes have weights proportional to the number of covisitations decayed by time, that is usually set to a few hours. It is also possible to include the direction of the edges where the order of access to items is important. In the implementation of the authors, this graph is stored as an adjacency list.

When a news s_k is accessed by the user u_i, the system iterates through the list C_{u_i} of items recently visited by the user in the last hours or days. All these items have their adjacency lists updated and the system also updates the list for item K. If there is an edge between the pair of news, the weight is decreased by the counting time. The nearest neighbors of a particular item are given for items that were covisited with this item, weighted by the counts decayed by the time of how often these items were visited. Finally, the predicted ratings by covisitation are normalized.

They have evaluated these three techniques in three different data sets, one using data from *MovieLens* [35] and two using stories from *Google News* itself. For the three data sets, they evaluated the precision and sensitivity curves. The PLSI obtained the best performance followed by *Min-Hashing* and then the method of cosine correlation, used for comparison. The second test was performed on live traffic to evaluate the three proposed algorithms combined with different weights. They were compared to the algorithm Popular, where the predictions are made by the popularity of the news, that is, by their age discounted count. Ranked lists were generated for each of the algorithms to perform this evaluation. These lists were interlaced and presented to the user. The order of interlacing was varied to avoid a bias in positioning, i.e., the user choosing a story only because it is in a higher position in the list. Then they evaluate which algorithms have most visited stories. The experiment took place over six months with millions of users. The algorithm with a higher weight to covisitation (2, 1, 1) had the best performance, in general followed by the algorithm with more weight to the Min-Hashing and PLSI (1, 2, 2). Both performances were on average approximately 40% higher than that of the algorithm Popular. The authors emphasize that the last algorithm is superior to the other two when the stories are very popular. There was also a final experiment comparing only PLSI and Min-Hashing, individually and combined over 20 days. Individually, it was not possible to say which one was better. When combined, however, the obtained performance was always lower.

7.6 Comparing and Evaluating Information Retrieval Algorithms for News Recommendation

Bogers and Bosch [9] claim that improving the performance of content-based news recommender systems has been slow due to the use of matching algorithms relatively old, as a simple comparison of keywords. Therefore, in their study, these authors employ recent probabilistic algorithms, with the objective of achieving significant gains of performance.

The recommendation algorithms described by the authors are based only on the content of the items, and independent ratings of users. Bogers and Bosch [9] examine three areas where they believe it is possible to bring benefits to content-based RS. Regarding retrieval models, most content-based systems uses keywords or TF-IDF vector representation. The representative size of stories is a factor considered in few experiments. Regarding evaluation, it has been difficult to compare systems, because the different approaches use different data sets and the efforts in evaluation have been divided between binary (relevant or non-relevant) and graded (multiple rates) evaluation.

For the experiments, Bogers and Bosch [9] created their own collection with full texts on RCV1 (Reuters Corpora Volume 1). The similarity between the articles was measured by people who gave a score of 0 to 4 (0 for non-related and 4 to highly related).

The authors conducted the experiments with different variations in the size of the articles to show how this variable influences the task of recommendation. The first including only the title. The second using the title and first sentence and the third using the title and the first two sentences. And so on, until it included the title and the first ten sentences. For the binary evaluation, related articles are considered those with assessment of 3 or 4. The measure used for evaluation was the Mean Uninterpolated Average Precision (MAP). For the evaluation of multiple ratings (graduated), they measured *Kendall's* tau correlation, a statistical measure used to assess the degree of correlation between two rankings and the significance of that relationship.

Three recommendation algorithms were compared. The first is *Okapi* [40] function, which represents the classical probabilistic model for Information Retrieval. The second is a framework of language modeling (LM) [37], which constructs a probabilistic model of language for each document. As a comparative basis, they employed TF-IDF weight considering the size of stories for calculating the weights. A text retrieval based on 50 articles using the title and body was performed. Table 2 presents the results for the different algorithms employing only the article's title, title plus the first five lines, title plus the first ten lines and all the text. Finally, Bogers and Bosch assessed the correlation between MAP and *Kendall's* tau and found a reasonably correlation of 0.4865, showing that binary and graded evaluation do not always share equivalent results.

When using the entire text story, the MAP for *Okapi* and LM were about 0.7 and there was no significant difference between them. The MAP of the TF-IDF

Table 2. MAP and *Kendall's* tau scores for LM, *Okapi* and TF-IDF [9]

	LM		Okapi		TF-IDF	
	MAP	K's tau	MAP	K's tau	MAP	K's tau
Title Only	0.1598	0.1431	0.2680	0.2667	0.2122	0.2212
Title + 5 lines	0.6032	0.2668	0.5978	0.2856	0.5849	0.2560
Title + 10 lines	0.6665	0.2822	0.6683	0.2871	0.6342	0.2630
Full text	0.6973	0.2932	0.7016	0.2968	0.6136	0.2293

system was lower, 0.61. The authors also note that the implementation of LM is in average about five times faster than that of *Okapi*. When considering bigger articles the MAP always rose, although in the case of TF-IDF it stabilized at a certain point and even got worse, confirming that TF-IDF approach has problems with long documents. When evaluating the values of *Kendall's* tau, there is a similar behavior. Only in the case of TF-IDF it is clear that the use of all the text reduces the value of this metric.

8 Conclusions

In this chapter the main issues related to recommender systems were presented by pointing the main approaches employed, their advantages and disadvantages and some of the possible applications of RS. Then, a special attention was given to the domain of news stories, summarizing the main works surrounding the topic.

Throughout the points mentioned in this chapter, one can say that RS are one of the major tools that help to reduce information overload, acting as filters and customizing content for the final user. The collaborative approach has been the most studied and used because of its good results and by the fact that it can be applied to any type of item.

The hybrid approach, in turn, has increasingly been studied because it allows to tackle some of the deficiencies found when only one approach is employed.

The choice of an appropriate RS for an application depends on the analysis of various factors such as the number of users and items, and domain characteristics of the items. In the case of systems geared to recommend news, it is possible to observe some characteristics through the study of literature presented. First, there is the pronounced problem of the sparsity due to the high churn of items. News stories appear in minutes and become obsolete in a matter of hours. Moreover, the interests of the user can change suddenly, making approaches that consider short and long term interests, as proposed by Billsus and Pazzani [8], very interesting. Another important point is the fact that an RS for news that employ different sources, like *Google News*, must avoid presenting redundant news to the user, given that is common that several sources report the same event just using different words, which can be solved with the use of an RS with a content-based approach. Regarding the collaborative approach to the domain of news, it is important to choose a technique based on models or to employ clustering techniques such as those proposed by [17]. The use of techniques based on the neighborhood can become impractical from the viewpoint of computational time for large problems, since the news recommendation should be performed, preferably, within few minutes or even instantly.

Acknowledgments

To the financial support of Universidade Federal do ABC and of the Brazilian research agency CNPq.

References

1. Tscher, M.J.A.: The BigChaos Solution to the Netflix Prize (2008)
2. Tscher, R.L.A., Jahrer, M.: Improved neighborhood-based algorithms for large-scale recommender systems. In: SIGKDD Workshop on Large-Scale Recommender Systems and the Netflix Prize Competition, ACM Press, New York (2008)
3. Adomavicius, G., Tuzhilin, A.: Toward the Next Generation of Recommender Systems: A Survey of the State-of-the-Art and Possible Extensions. IEEE Transactions on Knowledge and Data Engineering, 734–749 (2005)
4. Balabanović, M., Shoham, Y.: Fab: content-based, collaborative recommendation. Communications of the ACM 40(3), 66–72 (1997)
5. Belkin, N.J., Croft, W.B.: Information filtering and information retrieval: two sides of the same coin?. Communications of the ACM 35(12), 29–38 (1992)
6. Bennet, J.: The cinematch system: Operation, scale, coverage, accuracy impact (2006)
7. Billsus, D., Pazzani, M.: A personal news agent that talks, learns and explains. In: Proceedings of the third annual conference on Autonomous Agents, pp. 268–275. ACM, New York (1999)
8. Billsus, D., Pazzani, M.J.: User Modeling for Adaptive News Access. User Modeling and User-Adapted Interaction 10(2), 147–180 (2000)
9. Bogers, T., van den Bosch, A.: Comparing and evaluating information retrieval algorithms for news recommendation. In: Proceedings of the 2007 ACM conference on Recommender systems, pp. 141–144. ACM Press, New York (2007)
10. Burke, R.: Knowledge-based recommender systems. Encyclopedia of Library and Information Systems 69(suppl. 32), 175–186 (2000)
11. Burke, R.: Hybrid Recommender Systems: Survey and Experiments. User Modeling and User-Adapted Interaction 12(4), 331–370 (2002)
12. Candillier, L., Meyer, F., Fessant, F.: Designing specific weighted similarity measures to improve collaborative filtering systems. In: Perner, P. (ed.) ICDM 2008. LNCS (LNAI), vol. 5077, pp. 242–255. Springer, Heidelberg (2008)
13. Candillier, L., Meyer, F., Fessant, F.: Collaborative and Social Information Retrieval and Access: Techniques for Improved User Modeling. In: Candillier, L., Meyer, F., Fessant, F. (eds.) State-of-the-Art Recomender Systems, ch. 1, pp. 1–22. Idea Group Inc., IGI (February 2009)
14. Claypool, M., Gokhale, A., Miranda, T., Murnikov, P., Netes, D., Sartin, M.: Combining content-based and collaborative filters in an online newspaper. In: ACM SIGIR Workshop on Recommender Systems (1999)
15. Cosley, D., Lam, S., Albert, I., Konstan, J., Riedl, J.: Is seeing believing?: how recommender system interfaces affect users' opinions. In: Proceedings of the SIGCHI conference on Human factors in computing systems, pp. 585–592. ACM, New York (2003)
16. Cotter, P., Smyth, B.: Ptv: Intelligent personalised tv guides. In: Proceedings of the Seventeenth National Conference on Artificial Intelligence and Twelfth Conference on Innovative Applications of Artificial Intelligence, pp. 957–964. AAAI Press/The MIT Press (2000)
17. Das, A., Datar, M., Garg, A., Rajaram, S.: Google news personalization: scalable online collaborative filtering. In: Proceedings of the 16th international conference on World Wide Web, pp. 271–280. ACM Press, New York (2007)
18. Deshpande, M., Karypis, G.: Item-based top-N recommendation algorithms. ACM Transactions on Information Systems (TOIS) 22(1), 143–177 (2004)

19. Goldberg, K., Roeder, T., Gupta, D., Perkins, C.: Eigentaste: A Constant Time Collaborative Filtering Algorithm. Information Retrieval 4(2), 133–151 (2001)
20. Good, N., Schafer, J., Konstan, J., Borchers, A., Sarwar, B., Herlocker, J., Riedl, J.: Combining Collaborative Filtering with Personal Agents for Better Recommendations. In: Procedings of The National Conference on Aritificial Inteligence, pp. 439–446. John Wiley & Sons Ltd, Chichester (1999)
21. Herlocker, J., Konstan, J.A., Riedl, J.: An Empirical Analysis of Design Choices in Neighborhood-Based Collaborative Filtering Algorithms. Information Retrieval 5(4), 287–310 (2002)
22. Herlocker, J.L., Konstan, J.A., Terveen, L.G., Riedl, J.T.: Evaluating Collaborative Filtering Recommender Systems. ACM Transactions on Information Systems 22(1), 5–53 (2004)
23. Hill, W., Stead, L., Rosenstein, M., Furnas, G.: Recommending and evaluating choices in a virtual community of use. In: Proceedings of the SIGCHI conference on Human factors in computing systems, pp. 194–201. ACM Press/Addison-Wesley Publishing Co., New York (1995)
24. Hofmann, T.: Probabilistic latent semantic indexing. In: SIGIR 1999: Proceedings of the 22nd annual international ACM SIGIR conference on Research and development in information retrieval, pp. 50–57. ACM Press, New York (1999)
25. Hofmann, T.: Latent Semantic Models for Collaborative Filtering. ACM Transactions on Information Systems 22(1), 89–115 (2004)
26. Ken Goldberg, E.B.: Tavi Nathanson. Jester 4.0
27. Kim, T.H., Yang, S.B.: Using Attributes to Improve Prediction Quality in Collaborative Filtering. LNCS, 1–10 (2004)
28. Lang, K.: Newsweeder: Learning to filter netnews. In: Proceedings of the 12th International Machine Learning Conference (ML 1995), pp. 331–339 (1995)
29. Malone, T.W., Grant, K.R., Turbak, F.A., Brobst, S.A., Cohen, M.D.: Intelligent information-sharing systems. Communications of the ACM 30(5), 390–402 (1987)
30. Mccallum, A., Nigam, K.: A comparison of event models for naive bayes text classification. In: AAAI 1998 Workshop on Learning for Text Categorization, pp. 41–48. AAAI Press, Menlo Park (1998)
31. Melville, P., Mooney, R.J., Nagarajan, R.: Content-boosted collaborative filtering. In: Proceedings of the SIGIR-2001 Workshop on Recommender Systems, vol. 9, pp. 187–192 (2001)
32. Mitchel, T.M.: Machine Learning. McGraw-Hill, New York (1997)
33. Montaner, M., Lopez, B., de la Rosa, J.: Opinion-Based Filtering through Trust. LNCS, pp. 164–178. Springer, Heidelberg (2002)
34. Montaner, M., López, B., de la Rosa, J.: A Taxonomy of Recommender Agents on the Internet. Artificial Intelligence Review 19(4), 285–330 (2003)
35. Movielens data sets
36. Nikos Manouselis, C.C.: Personalized Information Retrieval and Access. In: Overview of Design Options for Neighbourhood-Based Collaborative Filtering Systems, ch. 2, pp. 30–54. Idea Group Inc., IGI (2008)
37. Ponte, J., Croft, W.: A language modeling approach to information retrieval. In: Proceedings of the 21st annual international ACM SIGIR conference on Research and development in information retrieval, pp. 275–281. ACM, New York (1998)
38. Resnick, P., Iacovou, N., Suchak, M., Bergstrom, P., Riedl, J.: Grouplens: an open architecture for collaborative filtering of netnews. In: CSCW 1994: Proceedings of the 1994 ACM conference on Computer supported cooperative work, pp. 175–186. ACM Press, New York (1994)

39. Rissanen, J.: Modeling by shortest data description. Automatica 14(5), 465–471 (1978)
40. Robertson, S., Walker, S.: Some simple effective approximations to the 2-Poisson model for probabilistic weighted retrieval. In: Proceedings of the 17th annual international ACM SIGIR conference on Research and development in information retrieval, pp. 232–241. Springer, New York (1994)
41. Schafer, J.B., Frankowski, D., Herlocker, J., Sen, S.: Collaborative filtering recommender systems. In: Brusilovsky, P., Kobsa, A., Nejdl, W. (eds.) Adaptive Web 2007. LNCS, vol. 4321, pp. 291–324. Springer, Heidelberg (2007)
42. Shardanand, U., Maes, P.: Social information filtering: algorithms for automating "word of mouth". In: CHI 1995: Proceedings of the SIGCHI conference on Human factors in computing systems, pp. 210–217. ACM Press/Addison-Wesley Publishing Co, New York (1995)
43. Wasfi, A.: Collecting user access patterns for building user profiles and collaborative filtering. In: Proceedings of the 4th international conference on Intelligent user interfaces, pp. 57–64. ACM Press, New York (1998)

Negotiation Strategies with Incomplete Information and Social and Cognitive System for Intelligent Human-Agent Interaction

Amine Chohra, Arash Bahrammirzaee, and Kurosh Madani

Images, Signals, and Intelligent Systems Laboratory (LISSI / EA 3956), Paris-XII University, Senart Institute of Technology, Avenue Pierre Point, 77127 Lieusaint, France
{chohra,bahrammirzaee,madani}@univ-paris12.fr

Abstract. Finding the adequate (*win-win* solutions for both parties) negotiation strategy with *incomplete* information for autonomous agents, even in one-to-one negotiation, is a complex problem. Elsewhere, negotiation behaviors, in which the characters such as conciliatory, neutral, or aggressive define a '*psychological*' aspect of the negotiator personality, play an important role. More, *learning* in negotiation is fundamental for understanding human behaviors as well as for developing new solution concepts of teaching methodologies of negotiation strategies (skills). First part of this Chapter aims to develop negotiation strategies for autonomous agents with incomplete information, where negotiation behaviors, based on time-dependent behaviors, are suggested to be used in combination (inspired from empirical human negotiation research). The suggested combination of behaviors allows agents to improve the negotiation process in terms of agent utilities, round number to reach an agreement, and percentage of agreements. Second part of this Chapter aims to develop a SOcial and COgnitive SYStem (SOCOSYS) for learning negotiation strategies from interaction (human-agent or agent-agent), where the characters conciliatory, neutral, or aggressive, are suggested to be integrated in negotiation behaviors (inspired from research works aiming to analyze human behavior and those on social negotiation psychology). The suggested strategy displays the ability to provide agents, through a basic buying strategy, with a first intelligence level, with SOCOSYS to learn from interaction (human-agent or agent-agent).

1 Introduction

Negotiations have received wide attention from the distributed Artificial Intelligence (AI) community as a pervasive mechanism for distributed conflict resolution between intelligent computational agents [1]. In a context where agents must reach agreements (deals) on matters of mutual interest, *negotiation* techniques for reaching agreements (deals) are required. In general, any negotiation settings will have four different components [2]: - a negotiation set, the space of possible proposals that agents can make ; - a protocol, the legal proposals that agents can make ; - a collection of strategies, one for each agent, which determine what proposals agents will make ; and - an agreement

E. Szczerbicki & N.T. Nguyen (Eds.): Smart Infor. & Knowledge Management, SCI 260, pp. 153–175.
springerlink.com

rule that determines the reach agreements (deals) stopping the negotiation. Negotiation usually proceeds in a series of rounds, with every agent making a proposal at every round. The proposals that agents make are defined by their strategy (a way to use the protocol), must be drawn from the negotiation set, and must be legal, as defined by the protocol (which defines possible proposals at different rounds). If agreement (deal) is reached, as defined by the agreement rule, then negotiation terminates with the agreement deal. These four parameters lead to an extremely rich and complex environment for analysis. Another source of complexity in negotiation is the number of agents involved in the process, and the way in which these agents interact [2]. First possibility is one-to-one negotiation, second possibility is many-to-one negotiation (which can be treated as a number of concurrent one-to-one negotiations), and third possibility is many-to-many negotiation (which is hard to handle).

An interesting survey on negotiation models in the AI field is given in [3], [4], [5]. Elsewhere, Lomuscio *et al.* [6] identified the main parameters on which any automated negotiation depends and provided a classification scheme for negotiation models. The environment that a negotiator is situated in greatly impacts the course of negotiation actions. Instead of focusing on analyzing the strategy equilibrium as a function of (the distribution of) valuations and historical information as in game theory, AI researchers are interested in designing flexible and sophisticated negotiation agents in complex environments with incomplete information. Agents have incomplete and uncertain information about each other, and each agent's information (e.g., deadline, utility function, strategy, ...) is its private knowledge.

Faratin *et al.* [7] devised a negotiation model that defines a range of strategies and behaviors for generating proposals based on time, resource, and behaviors of negotiators. Moreover, a type of combinations was earlier suggested in [7], it concerns linear combinations of different behaviors using a matrix of weights. Then, several research works suggested negotiation strategies based on such type of combinations [8], [9]. In the work developed in [8], experiments (one-to-one single issue), a linear combination (50-50% weighting) of behaviors is suggested for (time-dependent behavior, time-dependent behavior), for (time-dependent behavior, resource-dependent behavior), and for (time-dependent behavior, behavior-dependent behavior). In the work developed in [9], experiments (one-to-one four issues), a linear combination (10-90% weighting) of behaviors is suggested for (time-dependent behavior, behavior-dependent behavior).

Elsewhere, another type of combination consists of using different behaviors during a thread, i.e., a certain behavior during the first part of the thread and another behavior during the second part of the thread. Such type of combinations is based on empirical research on human negotiation. In fact, much of the research on human negotiation concerns the effect of demand level and concession rate on the outcome of negotiation. A negotiator's demand level is the level of benefit to the self associated with the current offer. Concession rate is the speed at which demand level declines over time [10]. Most studies consist of laboratory experiments on two-party, single issue negotiation. These studies support the following two conclusions [11], [10], [12]:

- higher initial demands and slower concessions make agreement less likely and less rapidly reached,
- lower initial demands and faster concessions produce smaller outcomes for the party employing them and larger outcomes for the other party, if agreement is reached.

These two conclusions imply a third, that there is an inverted U-shaped relationship between level of demand and the negotiation outcome: negotiators who start with high demands and concede slowly often fail to reach agreement, which usually leads to inferior outcomes ; those who start with low demands and concede rapidly usually reach agreement on the other party's terms, also yielding inferior outcomes ; those between these extremes ordinarily achieve better outcomes. The work developed in [13] suggested negotiation strategies based on this second type of combination.

By another way, in the research works developed aiming to analyze and describe the human behavior in [14], twelve categories representing three major parts of the behavior have been defined: the positive socio-emotional part, a neutral task part, and the negative socio-emotional part. In another side, in research works on the social psychology of the negotiation of Rubin and Brown developed in [15], the interpersonal orientation of a person has an influence on his negotiating behavior. More, it is possible to manipulate the interpersonal orientation, at least indirectly, in various ways. For instance, the communication can be manipulated for example by varying the type and amount of interpersonal information made available to a party. This will influence the interpersonal orientation, which in turn will influence the quality of the negotiation. According to Rubin and Brown, the interpersonal orientation is predominantly concerned with the degree of a person's responsiveness. If he is not responsive, he stands to gain much in the negotiating situation due to the deliberateness of his behaviour. Responsive people are more co-operative and therefore expect positive results. The personality type should therefore be determined first to obtain the best results in negotiation. Elsewhere, a great number of successes in AI can be attributed to a straightforward strategy [16]: linear evaluation of several simple features, trained by temporal-difference learning, and combined to an appropriate search algorithm. Games provide interesting case studies for this approach. In games as varied as Chess, Checkers, Othello, Backgammon and Scrabble, computers have exceeded human levels of performance.

Finding a shared solution to a problem within a group requires *negotiation*, a potentially exhausting and time-consuming process. To negotiate successfully, members have to involve the whole group, explain their position clearly and do their best to understand those of others [17]. However, in reality, groups often fail to negotiate, even when negotiation would be useful for each part of the group. Sometimes the problem lies in sheer size of the group, or in hierarchical organizational structures or in impediments to communication deriving from language, culture or history. In other cases, the main barriers lie in the individual psychology of specific group members. Typical problems include weak communications skills, lack of empathy with others, and poor control over the emotions arising during prolonged discussion. Such problems can be directly or indirectly related to the personality of each group member participating to the negotiation. Thus, negotiation behaviors, in which the characters such as Conciliatory (Con), Neutral (Neu), or Aggressive (Agg) define a 'psychological' aspect of the personality of a negotiation member (negotiator), play an important role [14], [15].

Elsewhere, learning from interaction in negotiation is fundamental, from embodied cognitive science and understanding natural intelligence perspectives [18], [19], for understanding human behaviors and developing new solution concepts [20]. Learning from interaction is a foundational idea underlying nearly all theories of learning. Indeed, whether a human is learning to drive a car or to hold a conversation (during a negotiation), he is acutely aware of how his environment responds to what he does, and he seeks to influence what happens through his behavior. Elsewhere, reinforcement learning is much more focused on goal-directed learning from interaction than other approaches to machine learning [21], [22], [23]. More, reinforcement learning approaches offer two important advantages over classical dynamic programming [24]. First, they are on-line having capability to take into account dynamics nature of real environments. Second, they can employ function approximation techniques as neural networks [25], [26], [27], to represent their knowledge, and to generalize so that learning time scales much better.

Humans have developed advanced skills in the intentions and the bodily expressions of the other human being, particularly important in high level communication which is at the basis of any "successful" negotiation (interaction process). Dealing with this, the SOcial and COgnitive SYStem, SOCOSYS, is developed for an intelligent human-agent interaction in our research laboratory.

This Chapter deals with the one-to-one negotiation (single issue settings) with incomplete information, in order to automate the negotiation process. Indeed, finding the adequate (*win-win* solutions for both parties) negotiation strategy with incomplete information is a complex problem, even in one-to-one negotiation.

First part of this Chapter aims to examine the performance of different strategies, based on time-dependent behaviors developed by Faratin *et al.* [7], and intends to suggest combinations inspired from empirical research on human negotiation in order to improve the performance (in terms of agent utilities, round number to reach an agreement, and percentage of agreements) of the negotiation process, with incomplete information, for autonomous agents. For this purpose, one-to-one bargaining process, in which a buyer agent and a seller agent negotiate over single issue (price), is developed in Section 2. In this work, this process is mainly based on time-dependent behaviors developed by Faratin *et al.* [7]. These behaviors, which can be used individually (a kind of strategies), are suggested to be used in *combination* (another kind of strategies inspired from empirical human negotiation research). Afterwards, experimental environments and measures, allowing a set of experiments carried out for short term deadlines, are detailed for these two kinds of strategies in Section 3. Also, experimental results of different strategies (using behaviors individually or in combination), are presented and analyzed in Section 3.

Second part of this Chapter aims to suggest a social and cognitive system for learning negotiation strategies from interaction (human-agent or agent-agent), where the characters conciliatory, neutral, or aggressive (which define the 'psychological aspect of the human personality) are integrated in negotiation behaviors (inspired from research works aiming to analyze human behavior and those on social negotiation psychology). For this purpose, a SOcial and COgnitive SYStem (SOCOSYS) is developed, in Section 4, to interact with an integrated system of simulation for negotiation allowing to learn, from interaction, a negotiation strategy essentially based on such human personality behaviors. Then, reinforcement learning (Q-learning and Sarsa-Learning) approaches are developed, analyzed, and compared in order to acquire the strategy negotiation behaviors in Section 4. Afterwards, a Fuzzy ArtMap Neural Network (FAMNN) is developed to acquire this strategy in Section 4.

2 One to One Negotiation

In this Section, one-to-one bargaining process, in which a buyer agent and a seller agent negotiate, based on time-dependent behaviors (negotiation decision functions), over single issue (price), is developed.

2.1 Negotiation Set

A negotiation set is the space of possible proposals that agents can make. The negotiation set (objects): the range of issues over which an agreement must be reached.

Let i represents the negotiating agents, in bargaining bilateral negotiation i ∈ {buyer(b), seller(s)}, and j the issues under negotiation, in bargaining single issue negotiation j = *price*. The value for issue *price* acceptable by each agent i is $x^i \in [min^i, max^i]$.

2.2 Negotiation Protocol

A protocol consists of the legal proposals that agents can make. The process of negotiation can be represented by rounds, where each round consists of an offer from agent b (buyer) at time t_1 and a counter-offer from an agent s (seller) at time t_2. Then, a negotiation consists in a sequence of rounds: round1 (t_1, t_2), round2 (t_3, t_4), ... Thus, for a negotiation between agents b and s, and if agent b starts first, then agent b should offer in times (t_1, t_3, t_5, ..., t_{max}^b), and agent s provides counter-offers in (t_2, t_4, t_6, ..., t_{max}^s), where t_{max}^b and t_{max}^s denote the negotiation deadline for agent b and agent s, respectively.

Note that the different following deadline cases are allowed:

- $t_{max}^b > t_{max}^s$, in this case the considered deadline is $T_{max} = t_{max}^s$;
- $t_{max}^b = t_{max}^s$, in this case the considered deadline is $T_{max} = t_{max}^b = t_{max}^s$;
- $t_{max}^b < t_{max}^s$, in this case the considered deadline is $T_{max} = t_{max}^b$.

For agent b, the proposal to offer or accept is within the interval [min^b, max^b], where max^b is the reservation price of the buyer in the negotiation thread, and min^b is the lower bound of a valid offer.

Similarly, for agent s, the proposal to offer or accept is within the interval [min^s, max^s], where min^s is the reservation price of the seller and max^s is the upper bound of a valid offer.

Initially a negotiator offers the most favorable value for himself: agent b (buyer) starts with min^b and agent s (seller) starts with max^s. If the proposal is not accepted, a negotiator concedes with time proceeding and moves towards the other end of the interval.

2.3 Negotiation Behaviors

The paces of concession depend on the negotiation behaviors of agent b and agent s which are characterized by negotiation decision functions. For negotiation strategies, time t is one of predominant factors used to decide which value to offer next. For this purpose, time-dependent functions are used as negotiation decision functions varying the acceptance value (price) for the offer depending on the remaining negotiation time (an important requirement in negotiation), i.e., depending on t and t^b_{max} for agent b and depending on t and t^s_{max} for agent s.

Thus, the proposal $x^b[t]$ to be offered by agent b and the one $x^s[t]$ to be offered by agent s at time t, with $0 <= t <= t^i_{max}$ belonging to [0; T - 1], are as follows:

$$x^b[t] = min^b + \alpha^b(t) (max^b - min^b),$$

$$x^s[t] = min^s + (1 - \alpha^s(t)) (max^s - min^s),$$

where $\alpha^b(t)$ and $\alpha^s(t)$ are time-dependent functions ensuring that: \hfill (1)

$$0 <= \alpha^i(t) <= 1,$$

$$\alpha^i(0) = K^i \text{ (positive constant) and } \alpha^i(t^i_{max}) = 1.$$

Such $\alpha^i(t)$ functions can be defined in a wide range according to the way in which $\alpha_i(t)$ is computed (the way they model the concession), e.g., polynomial:

$$\alpha^i(t) = K^i + (1 - K^i)(\frac{min(t, t^i_{max})}{t^i_{max}})^{\frac{1}{\beta}}. \hfill (2)$$

Thus, this family of functions represent an infinite number of possible behaviors, one for each value of β. Indeed, the constant $\beta > 0$ determines the concession pace along time, or convexity degree of the offer curve as a function of the time. By varying β a wide range of negotiation behaviors can be characterized. Two sets of β can be identified to characterize two classes of behaviors: Boulware (B) with $\beta < 1$ and Conceder (C) with $\beta > 1$ [5], and the particular case of Linear (L) with $\beta = 1$. With a Boulware strategy a negotiator tends to maintain offered value until time is almost exhausted, then he concedes to the reservation price quickly. With a Conceder strategy a negotiator goes to the reservation price rapidly and early.

2.4 Negotiation Strategies

A collection of strategies, one for each agent, determines what proposals agents will make. In other words, an agent's strategy determines which combination (or behavior when using behaviors individually) of behaviors should be used at any one instant.

This Chapter deals with the second type of combination inspired from empirical human negotiation research as explained in Section 1. The work developed in [13] suggested negotiation strategies based on this second type of combination. Indeed, Lopes *et al.* suggested a combination of opening negotiation behaviors (starting optimistic, starting realistic, and starting pessimistic) and concession behaviors (where the concession magnitude variation is achieved with a concession factor modeled as a constant and as a function of the total concession), resulting in three strategy families (starting high and conceding slowly, starting reasonable and conceding moderately, and starting low and conceding rapidly). With regard to the work [13], in this Chapter, experiment results of strategies using behaviors individually are given in order to compare them with those of suggested strategies using behaviors in combination. More experimental measures are used as utility product and utility difference in order to a fine evaluation of strategy results, and in order to perceive the improvement degree given using such combination (inspired from empirical human negotiation research) with regard to strategies using behaviors individually.

Thus, during a negotiation *thread* (the sequence of rounds with offers and counter-offers in a two-party negotiation), a negotiation strategy consists to define the way in which one or several behaviors are used. In this Chapter, two kinds of strategies are considered for both agent b and agent s:

- individual behaviors (one behavior for each strategy), where each strategy use individually the behaviors Boulware (B), Linear (L), or Conceder (C) during a negotiation *thread* ;
- behaviors in combination (inspired from empirical human negotiation research), where each strategy use two behaviors (among B, L, C) in combination during a negotiation *thread*, resulting in the combinations: BL, BC, LB, LC, CB, and CL.

2.5 Agreement Rule

An agreement rule that determines the reach agreements (deals) stopping the negotiation.

Agent b accepts an offer (or a deal) $x^s[t]$ from agent s at time t if it is not worse than the offer he would submit in the next step, i.e., only if:

$$\begin{cases} x^b(t+1) >= x^s(t) \\ t <= T_{max} \end{cases}. \tag{3}$$

Similarly, agent s accepts an offer (or a deal) $x_b[t]$ from agent b at time t only if:

$$\begin{cases} x^s(t+1) <= x^b(t) \\ t <= T_{max} \end{cases}. \tag{4}$$

3 Experiments: Environments, Measures, and Results

In this Section, the environments and experimental measures are presented and a set of experiments are carried out for short and long term deadlines for both strategies using different behaviors individually or in combination.

3.1 Experimental Environments

Environments are defined in bargaining bilateral negotiation between buyer(b) and seller(s), in single issue negotiation j = *price*.

The experimental environment is defined by the following variables [t^b_{max} , t^s_{max} , T_{max}, K^b, K^s, min^b, max^b, min^s, max^s].

The negotiation interval (the difference between the minimum and maximum values of agents) for price is defined using two variables: θ^i (the length of the reservation interval for an agent i) and Φ (the degree of intersection between the reservation intervals of the two agents, ranging between 0 for full overlap and 0.99 for virtually no overlap). In the experimental environment:

- and θ^i are randomly selected between the ranges [10, 30] for both agents,
- $\Phi = 0$.

The negotiation intervals are then computed, setting $min^b = 10$, by:

$$\begin{cases} min^b = 10 \\ max^b = min^b + \theta^b \\ min^s = \theta^b \Phi + min^b \\ max^s = min^s + \theta^s \end{cases} \qquad (5)$$

The analysis and evaluation of negotiation behaviors and strategies developed in [8], indicated that negotiation deadline significantly influences the performance of the negotiation. From this, the experimental environment is defined with short term deadlines or with long term deadlines such as:

- short time deadlines are defined from random selection of the round number within [1, 5] which corresponds to a random selection of t^i_{max} within [2, 10],
- long time deadlines are defined from random selection of the round number within [15, 30] which corresponds to a random selection of t^i_{max} within [30, 60].

The parameter β ranges are defined as shown in Table 1.

Table 1. Parameter β ranges

Time-Dependent Behaviors	β Ranges
Boulware (B)	$\beta \in [0.01, 0.2]$
Linear (L)	$\beta = 1$
Conceder (C)	$\beta \in [20, 40]$

The constants K^i are chosen as small positive $K^i = 0.1$, for both agents, in order to not constrain the true behavior of each time-dependent functions (negotiation decision functions).

The initiator of an offer is randomly chosen because the agent which opens the negotiation fairs better, irrespective of whether the agent is a buyer (b) or a seller (s). This is because the agent who begins the negotiation round reaches $\alpha^i(t) = 1$ before the other agent, hence deriving more intrinsic utility.

3.2 Experimental Measures

To produce statistically meaningful results, for each experiment, the precise set of environments is sampled from the parameters specified in Section 3.1 and the number of environments used is $N = 200$, in each experiment. This ensures that the probability of the sampled mean deviating by more than 0.01 from the true mean is less than 0.05.

To evaluate the effectiveness of the negotiation behaviors the three following measures are considered:

- *Average Intrinsic Utility*, the intrinsic benefit is modeled as the agent's utility for the negotiation's final outcome, in a given environment, independently of the time taken and the resources consumed [7]. This utility, U^i, can be calculated for each agent for a price x using a linear scoring function, i.e., when there is a deal (an agreement) for a price x:

$$U^b = \frac{max^b - x}{max^b - min^b} \text{, and } U^s = \frac{x - max^s}{max^s - min^s} . \tag{6}$$

If there is no deal in an environment (a particular negotiation), then $U^b = U^b = 0$. Then, the average intrinsic utility for each agent AU^i is:

$$AU^i = \frac{\sum_{n=1}^{N} U^i[n]}{N_D} , \tag{7}$$

where N is the total number of environments in each experiment, U^i the utility of each agent, for each environment with deal, and N_D is the number of environments with deals.

- *Average Round Number*, rounds to reach an agreement (a deal), a lengthy negotiation incurs penalties for resource consumption, thus shrinking the utilities obtained by the negotiators indirectly [28]. The average round number AR is then:

$$AR = \frac{\sum_{n=1}^{N} R_D[n]}{N_D} , \tag{8}$$

where N is the total number of environments in each experiment, R_D is the number of rounds, for each environment with deal, and N_D is the number of environments with deals.

- *Percentage of Deals*, the percentage of deals $D(\%)$ is obtained from the *Average Deal* AD:

$$AD = \frac{N_D}{N} , \tag{9}$$

where N_D is the number of environments with deals, and N is the total number of environments in each experiment. Then, the Percentage of Deals $D(\%)$ is:

$$D(\%) = AD.100\% , \tag{10}$$

where N_D is the number of environments with deals, and N is the total number of environments in each experiment.

Also, in order to analyze the performance of strategies, three measures are obtained during the experimentation:

- *average performance*, the Average Performance (AP^i) of each agent is an average evaluation measure implying the average negotiation deadline At_{max}^i and three (03) experimental measures, i.e., the average intrinsic utility U^i, the average round number AR, and the average deal AD:

$$AP^i = \frac{AU^i + (1 - \frac{AR}{At_{max}^i}) + AD}{3}, \text{ with } At_{max}^i = \frac{\sum_{n=1}^{N} t_{max}^i[n]}{N_D}, \tag{11}$$

where N_D is the number of environments with deals, and N is the total number of environments in each experiment ;

- *utility product*, UP, once an agreement is achieved, the product, $AU^b.AU^s$, of the utilities obtained by both participants is computed, this measure indicates the joint outcome [9]:

$$UP = AU^b.AU^s ; \tag{12}$$

- *utility difference*, UD, once an agreement is achieved, the difference, $|AU^b - AU^s|$, of the utilities obtained by both participants is computed, this measure indicates the distance between both utilities [9]:

$$UD = |AU^b - AU^s|. \tag{13}$$

There is an important relation between these product and difference utility measures and a compromise should be taken into account. Even though a high joint outcome is expected (for a relatively *win-win* results), it is also important that the difference between both utilities is low. For this reason, the analysis and evaluation of the obtained results is based not only on the utility product but also on the utility difference (a compromise should then be taken into account).

3.3   Results: Negotiation Strategies Using Behaviors Individually or in Combination

In this Section, experimental results of negotiation strategies using behaviors individually or in combination are presented and analyzed for short and long term deadlines.

Note that these results are regarded from the buyer(b), and the similar results are obtained which are regarded from the seller(s).

3.3.1   Results and Analysis for Short Term Deadlines

Results given in Table 2 are obtained from 9 experiments (B-B, B-L, B-C, L-B, L-L, L-C, C-B, C-L, C-C) where both agents b and s have same or different strategies (using behaviors individually) and where each experiment is an average of N=200 environments. These results show (only adequate strategies for buyer when using B, L, or C) that the adequate strategy when both agents have strategies using behaviors individually is the L(buyer)-L(seller) case with the best performance in term of utility

Table 2. Results, short term deadlines, agent b (behaviors individually) and agent s (behaviors individually)

Buyer	AU^b	AP^b	Seller	AU^s	AP^s	AR, D(%), UP, UD
B	0.68704	0.62354	L	0.33958	0.50772	1.39062, 64%, 0.23330, 0.34746
L	0.52776	0.66841	L	0.48263	0.64853	1.18072, 83%, **0.25471, 0.04513**
C	0.57640	0.80867	C	0.43733	0.76494	0.20652, 92%, 0.25207, 0.13907

Table 3. Results, short term deadlines, agent b (behaviors in combination) and agent s (behaviors in combination)

Buyer	AU^b	AP^b	Seller	AU^s	AP^s	AR, D(%), UP, UD
BL	0.50382	0.61043	BL	0.50488	0.61949	2.71264, 87%, **0.25436, 0.00106**
BC	0.50566	0.63930	LB	0.49786	0.63883	2.30681, 88%, **0.25174, 0.00780**
LB	0.53763	0.65098	BL	0.52999	0.63183	2.38823, 85%, **0.28493, 0.00764**
LC	0.50506	0.70727	LC	0.50556	0.71034	1.88888, 99%, **0.25533, 0.00050**
CB	0.54967	0.83203	CB	0.49033	0.81232	0.27000, 100%, **0.26951, 0.05934**
CL	0.57246	0.84075	CL	0.43971	0.79730	0.24000, 100%, **0.25171, 0.13275**

product UP (high) about 0.25471 and utility difference UD (low) about 0.04513, and percentage of deals D(%) = 83% and average round to reach an agreement AR = 1.18072.

Results given in Table 3 are obtained from 36 experiments (BL-BL, BL-BC, BL-LB, BL-LC, BL-CB, BL-CL, and BC-BL, BC-BC, BC-LB, BC-LC, BC-CB, BC-CL, and LB-BL, LB-BC, LB-LB, LB-LC, LB-CB, LB-CL, and so on...) where both agents b and s have same or different strategies (using behaviors in combination) and where each experiment is an average of N=200 environments. These results show (only adequate strategies for buyer when using BL, BC, LB, LC, CB, or CL) that the adequate strategy when both agents have strategies using behaviors in combination is the LC(buyer)-LC(seller) case with the best performance in term of UP (high) = 0.25533 and UD (low) = 0.00050, and D(%) = 99%, and AR = 1.88888. From these results, since with incomplete information, an adequate strategy could be also here a strategy which imitate the other agent since, when the buyer is using BL, the adequate strategy for seller is BL (with UP=0.25436 and UD=0.00106), and when the buyer is using BC, the adequate strategy for seller is LB (with UP=0.25174 and UD=0.00780), and when the buyer is using LB, the adequate strategy for seller is BL (with UP=0.28493 and UD=0.00764), and when the buyer is using LC, the adequate strategy for seller is LC (with UP=0.25533 and UD=0.00050), and when the buyer is using CB, the adequate strategy for seller is CB (with UP=0.26951 and UD=0.05934), and when the buyer is using CL, the adequate strategy for seller is CL (with UP=0.25171 and UD=0.13275).

3.3.2 Results and Analysis for Long Term Deadlines

Results given in Table 4 are similar to those in short term deadlines, so same conclusions can be drawn.

Table 4. Results, long term deadlines, agent b (behaviors individually) and agent s (behaviors individually)

Buyer	AU^b	AP^b	Seller	AU^s	AP^s	AR, D(%), UP, UD
B	0.58207	0.46872	B	0.45639	0.42755	19.04615, 65%, 0.26565, 0.12568
L	0.50727	0.69729	L	0.50815	0.69197	9.50000, 100%, **0.25776, 0.00088**
C	0.57521	0.85068	C	0.40875	0.79525	0.52000, 100%, 0.23511, 0.16646

Table 5. Results, long term deadlines, agent b (behaviors in combination) and agent s (behaviors in combination)

Buyer	AU^b	AP^b	Seller	AU^s	AP^s	AR, D(%), UP, UD
BL	0.52948	0.66484	BL	0.47824	0.64405	12.62000, 100%, **0.25321,0.05124**
BC	0.48556	0.66266	LB	0.48916	0.66324	11.71717, 99%, **0.23751, 0.00360**
LB	0.51092	0.65374	LB	0.51428	0.66003	11.20652, 92%, **0.26275, 0.00336**
LC	0.49018	0.69880	LC	0.49401	0.69604	9.49000, 100%, **0.24215, 0.00383**
CB	0.53408	0.83706	CB	0.45521	0.81114	0.53000, 100%, **0.24311, 0.07887**
CL	0.52774	0.83700	CL	0.49164	0.82505	0.39000, 100%, **0.25945, 0.03610**

Results given in Table 5 show more performance (AP, AU, AR, UP, and UD) and they still similar to those in short term deadlines. With an important precision on the adequate strategy when both agents have strategies using behaviors in combination is the LC(buyer)-LC(seller) case with the best performance in term of UP (high) = 0.24215 and UD (low) = 0.00383, and D(%) = 100%, and AR = 9.49000, and with AP^b = 0.69880 and AP^s = 0.69604 (which are better than LB-LB and better than BC-LB). In fact, there is an important relation between these product and difference utility measures and a compromise should be taken into account. Even though a high joint outcome is expected (for a relatively *win-win* results), it is also important that the difference between both utilities is low. For this reason, the analysis and evaluation of the obtained results is based not only on the utility product but also on the utility difference (a compromise should then be taken into account). More, in order to help in such compromise, the Average Performance we suggested is useful, helping for a better compromise, as demonstrated by these results.

4   SOcial and COgnitive SYStem (SOCOSYS) for Learning Negotiation Strategies

The technology of multi-agent systems which facilitates the negotiation at operative level of the decision-making [29] is used. It allows agents to embody a notion of autonomy, in particular, to decide for themselves whether or not to perform an action on request from another agent. More, in order to satisfy their design objectives, agents are designed to be intelligent, i.e., capable of flexible behavior [2], [18], [19]: able to perceive their environment, and respond in a timely fashion to changes that occur in it (reactivity), able to exhibit goal-directed behavior by taking the initiative (proactiveness), and capable of interacting with other agents and possibly humans (social ability) ; such abilities are guaranteed in this work by psychological aspects of the personality, multi-agents, as well as Q-Learning and Sarsa-Learning.

4.1 SOcial and COgnitive SYStem (SOCOSYS)

In order to elaborate such negotiation strategy, SOcial and COgnitive SYStem (SOCOSYS) which is presented in Fig. 1, is developed. It is built of three main parts: a multi-agent system representing the environment model, an intelligent agent, and the simulation environment (SISINE[1] software). Thus, SOCOSYS is designed and developed as a social and cognitive system for learning negotiation strategies from interaction (human-agent and agent-agent).

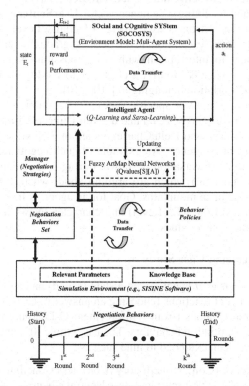

Fig. 1. SOcial and COgnitive SYStem (SOCOSYS) for learning negotiation strategies from interaction (human-agent or agent-agent).

[1] SISINE project, funded by European Union Leonardo Da Vinci Program, aimed to develop an innovative teaching methodology of negotiation skills exploiting an integrated system platform of simulation (http://www.sisine.net).

4.2 Integrated System of Simulation for Negotiation

In SISINE (Integrated System of Simulation for Negotiation) project, simulation environments enable a participant to interact with a virtual entity called "bot" (a software agent), through a communicative exchange: texts (one among three), voice (tone and volume), facial expression, and gesture. The objective of such simulation is to allow participants to directly experience the basic elements of negotiation, through:

- a "standard bot", an agent with reactions from simple rules,
- a "smart bot" an agent with reactions from a first intelligence level.

The suggested negotiation approach relies on a theoretical model of the negotiation process which is mainly based on the negotiator personality, i.e., characters Con, Neu, and Agg which define a 'psychological' aspect of the negotiator agent personality. Also, in this theoretical model, the character of a negotiator agent is defined by a character vector [Con, Neu, Agg] where each component belongs to the interval [0, 100] in percentage (%) such as: Con + Neu + Agg = 100 %. During a negotiation round, each negotiator agent is defined by its internal current state Current-State(CurrentEscLevel, **CurrentChar**, **UserSentence**), having its CurrentEscLevel and **CurrentChar**, and receiving a sentence vector **UserSentence** = [DeltaEscLevel, CharToModify, DeltaChar] from a user where: - DeltaEscLevel, an escalation level variation from [-60, +60], escalation level defines gradually different negotiation stages from agreement to interruption with seven possible stages modeled by EscLevel from [0, 60] ; - CharToModify, a character to modify (Con, Neu, or Agg) ; DeltaChar, a character variation belonging to [-10, +10].

In fact, during a round (a given specific state) of a negotiation as illustrated in Fig. 2, an agent has a given CurrentEscLevel and a given **CurrentChar**, and receives DeltaEscLevel, CharToModify, and DeltaChar extracted in SISINE software from a user (human or another agent) sentence. An example of such negotiation session is illustrated in Fig. 2, where a conversation is shown, during a negotiation round, between two agents (woman and man) which are arguing about whether they should spend their holidays at seaside or in mountains. Goal for an agent, is from a user sentence to update its escalation level (NewEscLevel) and character (**NewChar**), and to choose an answer based on these new values.

4.3 Negotiation Strategy: Case Study

In this work, for an *intelligent* agent, given a state (round) and a given strategy (basic buying strategy), the agent reactions (choosing an answer in each state) must be based on the updating of the Qvalues (from reinforcement Q-Learning or Sarsa-Learning) corresponding to the negotiation behaviors. The basic three (03) negotiation behaviors (Con, Neu, Agg) have been extended to nine such as:

- more Conciliatory (+Con), Conciliatory (Con), less Conciliatory (-Con),
- more Neutral (+Neu), Neutral (Neu), less Neutral (-Neu),
- less Aggressive (-Agg), Aggressive (Agg), more Aggresive (+Agg).

These behaviors correspond to the actions such as: +Con (Action0), Con (Action1), -Con (Action2), +Neu (Action3), Neu (Action4), -Neu (Action5), -Agg (Action6), Agg (Action7), and +Agg (Action8).

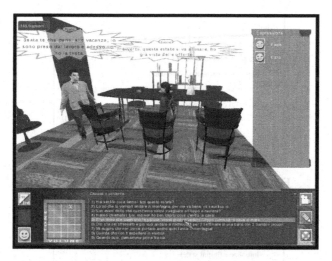

Fig. 2. Negotiation session (two agents) from SISINE software

Note that, in this work, + Neu is considered belonging to positive socio-emotional area than - Neu.

The answers are linked to actions such as for instance, the answer corresponding to +Con(Action0), is considered as a more conciliatory answer, or the answer corresponding to Con(Action1), is considered as a conciliatory answer, and so on.

Then, after learning, the answer corresponding to the character with the max Qvalue is chosen. Rewards given in Table 1 for the suggested basic buying strategy, are used for training. Note that, for most part of everyday negotiations, a completely conciliatory agent strategy, or completely neutral agent strategy, or completely aggressive agent strategy are insufficient. Then, an interesting strategy is necessarily a combination of these three basic conciliatory, neutral, and aggressive strategies. Such interesting combination is given through a basic buying strategy which could be neutral (+Neu) for some first states, then aggressive (-Agg) for some second states, and conciliatory (-Con) for the last states, suggesting for buying something with less price, it is interesting to negotiate at first neutrally, then aggressively, and finally conciliatory to conclude (buy).

4.4 Reinforcement Learning Approaches

One of the most important breakthroughs in reinforcement learning was the development of an off-policy temporal-difference control algorithm known as Q-learning and Sarsa-learning [21], [22], [23].

Reinforcement learning allows an agent (the learner and decision-maker) to use its experience, from the interaction with an environment, to improve its performance over time [21], [22], [23]. In other words, from the interaction with an environment an agent, can learn, using reinforcement Q-learning, to maximize the reward r leading to an optimal behavior policy. Indeed, in this on-line reinforcement learning, the agent incrementally learns an action/value function $Q(s, a)$ that it uses to evaluate the utility

```
Initialize Q(s, a) arbitrarily
Repeat (for each episode):
    Initialize s (initial state)
    Repeat (for each step of episode):
        Choose a from s using policy derived from Q
        Take action a, observe r, s'
        Q(s, a) ← Q(s, a) + α[r + γmaxₐ· Q(s', a') – Q(s, a)]
        s ← s'
    until s is terminal (final state)
```

Fig. 3. Q-Learning: an off-policy learning paradigm

```
Initialize Q(s, a) arbitrarily
Repeat (for each episode):
    Initialize s (initial state)
    Choose a from s using policy derived from Q
    Repeat (for each step of episode):
        Take action a, observe r, s'
        Choose a' from s' using policy derived from Q
        Q(s, a) ← Q(s, a) + α[r + γQ(s', a') – Q(s, a)]
        s ← s' ; a ← a'
    until s is terminal (final state)
```

Fig. 4. Sarsa-Learning: an on-policy learning paradigm

of performing action a while in state s. Q-learning leads to optimal behavior, i.e., behavior that maximizes the overall utility for the agent in a particular task environment [18]. The used Q-learning paradigm [21], [23] is shown in Fig. 3, while the used Sarsa-learning paradigm [23] is shown in Fig. 4.

Table 6. Strategy rewards (basic buying strategy).

	States From S0 to S8				States From S9 to S16				States From S17 to S24			
+ Con	- 60	- 60	...	- 60	- 60	- 60	...	- 60	- 2	- 2	...	- 2
Con	- 50	- 50	...	- 50	- 50	- 50	...	- 50	- 1	- 1	...	- 1
- Con	- 40	- 40	...	- 40	- 40	- 40	...	- 40	+ 1	+ 1	...	+ 1
+ Neu	+ 1	+ 1	...	+ 1	- 30	- 30	...	- 30	- 10	- 10	...	- 10
Neu	- 1	- 1	...	- 1	- 20	- 20	...	- 20	- 20	- 20	...	- 20
- Neu	- 2	- 2	...	- 2	- 10	- 10	...	- 10	- 30	- 30	...	- 30
- Agg	- 40	- 40	...	- 40	+ 1	+ 1	...	+ 1	- 40	- 40	...	- 40
Agg	- 50	- 50	...	- 50	- 1	- 1	...	- 1	- 50	- 50	...	- 50
+ Agg	- 60	- 60	...	- 60	- 2	- 2	...	- 2	- 60	- 60	...	- 60

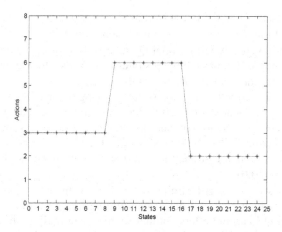

Fig. 5. Strategy results (Q-Learning: γ = 0.01).

Table 7. Strategy results (Sarsa-Learning: γ = 0.86).

S0	S1	S2	S3	S4	S5	S6	S7	S8
-136	-115	-134	-146	-141	-136	-138	-131	-117
-96	-88	-91	-100	-102	-101	-98	-94	-86
-166	-151	-155	-144	-125	-127	-113	-107	-102
-3.5	-2.2	-9.4	-3.5	-7.0	-6.8	-15	-22	-36
-0.7	**-1.5**	**-2.2**	**-2.6**	**-3.1**	**-5.1**	**-9.3**	-17	-32
-9.9	-11	-13	-14	-12	-11	-15	**-14**	**-20**
-30	-34	-51	-59	-72	-81	-77	-65	-38
-141	-141	-136	-134	-132	-136	-139	-144	-24
-60	-87	-98	-105	-103	-109	-100	-91	-59

S9	S10	S11	S12	S13	S14	S15	S16
-115	-119	-111	-115	-110	-110	-93	-59
-87	-88	-85	-75	-76	-74	-72	-51
-99	-96	-92	-97	-89	-83	-67	-40
-58	-49	-51	-47	-51	-52	-48	-35
-51	-43	-42	-38	-49	-37	-37	-35
-17	-15	-20	-18	-20	-21	-25	-33
-0.8	-1.2	**-1.1**	-4.7	-9.0	**-8.9**	**-18**	**-26**
-1.3	**-0.6**	-1.4	**-2.3**	**-4.7**	-10	-24	-47
-7.9	-5.8	-8.7	-4.5	-6.3	-11	-25	-48

S17	S18	S19	S20	S21	S22	S23	S24
-5.4	-2.3	-1.5	-1.8	-2.1	-3.1	-0.5	-0.9
-3.1	-2.9	**-0.7**	**-1.0**	-0.9	**-1.0**	0.0	-0.4
-2.9	**-0.7**	-2.3	-6.3	**-0.7**	-1.9	**1.8**	**1.0**
-14	-9.5	-6.1	-2.8	-6.5	-6.6	-4.7	-2.7
-29	-21	-17	-19	-21	-15	-9.0	-8.1
-47	-34	-16	-5.4	-16	-7.7	-17	-14
-58	-39	-29	-37	-28	-18	-26	-20
-79	-65	-52	-40	-29	-27	-23	-20
-83	-61	-51	-57	-65	-46	-42	-34

The parameter settings of the initial Q values, constant step-size parameter ($0 <= \alpha < 1$), and discount rate ($0 <= \gamma < 1$) have been done following the choice approaches given in [21] and [23] resulting in: initial Q values = 0.5, $\alpha = 0.1$, and $\gamma = 0.01$. The results of Q-Learning of the basic buying strategy, from the strategy rewards given in Table 6, with $\gamma = 0.01$ are globally following the rewards.

Q-Learning succeeded in learning the strategy giving the maxQvalues (1.0) for Action3 in the states (or rounds) from S0 to S8, then giving the maxQvalues (1.0) for Action6 in the states from S9 to S16, and finally giving the maxQvalues (1.0) for Action2 in the states from S17 to S24, as shown in Fig. 5. Note that similar results of Sarsa-Learning (basic buying strategy), from the strategy rewards given in Table 6, are obtained with $\gamma = 0.01$.

For other values of γ, Sarsa-Learning offers the advantage of safe solutions (not optimal solutions which could be dangerous offered by Q-Learning) [30], [31]. Indeed, in this case, Q-Learning offers optimal solution (similarly to the rewards shown in Table 6) corresponding to Action3. However, in on-line learning some results could be dangerous in the sense that, when learning Action3 (+Neu), the result could be related sometimes to Action2 (-Con) or Action4 (Neu). In the case of Action2 (-Con), this is dangerous because of the complete changing of the character under learning from neutral character to conciliatory character. Thus, Sarsa-Learning is offering more desirable properties to the suggested character learning than Q-Learning.

In this work, such safe solution (Sarsa-Learning) is obtained with $\gamma = 0.86$, as shown in Table 7 and Fig. 6. In fact, for state 0 to 8, action under learning is Action3 (+Neu) an offered solutions are Action4 (Neu) and Action5 (-Neu) ; for state 9 to 16, action under learning is Action6 (-Agg) an the offered solutions are Action6 (-Agg) and Action7 (Agg) ; and finally for state 17 to 24, the action under learning is Action2 (-Con) an the offered solutions are Action1 (Con) and Action2 (-Con).

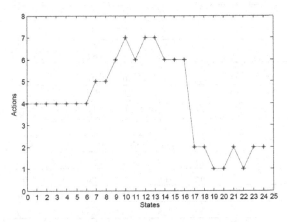

Fig. 6. Strategy results (Sarsa-Learning: $\gamma = 0.86$).

4.5 Fuzzy ArtMap Neural Networks (FAMNN)

Neural network (fuzzy artmap neural network in this work) implementation of reinforcement Q-learning offers the advantages of learning and generalization (essential traits of intelligent behaviors), robustness, massively parallel computations and distributed memory [25], [26], [27] and limited memory requirement for storing the knowledge [32], [33].

One of the main features of human memory is its ability to learn many new things without necessarily forgetting things learned in the past. The ability of humans to remember many details of an exciting movie is a typical example of fast learning. Such learning can be achieved exploiting the Adaptive Resonance Theory (ART) in combination with neural networks and fuzzy logic. Indeed, ART can be used to design a hierarchical artmap neural networks that can rapidly self-organize stable categorical mappings between i-dimensional input vectors and n-dimensional output vectors [34]. To better reflect human reasoning, fuzzy inference is incorporated to the basic architecture leading to Fuzzy ArtMap Neural Networks (FAMNN) [26], [34]. FAMNN are capable of *fast* and *stable* learning of recognition categories in response to arbitrary sequences of input patterns. Therefore, they achieve a synthesis of fuzzy logic and ART neural networks by exploiting a close formal similarity between the computations of fuzzy subset hood and ART category choice, resonance, and learning.

FAMNN classifier is then trained to acquire the suggested strategy (a basic buying strategy), from the learning algorithm detailed in [35], from one hundred (100) examples of the training set (4x9 examples for Action3, 4x8 examples for Action6, and 4x8 examples for Action2) with a normalization of the inputs (between 0 and 1) either by incorporating an example into an existing output category node or creating a new output category node for it. This classifier sprouted the output category node number $OCNN = N$ with $N = 9$ and yields convergence in well under the cycle number $CN = 1$ with the learning rate $\eta = 1$, small positive constant $\lambda = 0.000001$, baseline of the vigilance $\sigma = 0.4$. FAMNN results demonstrate a fast and stable learning (all adaptive weights only decrease in time).

After learning, the resulting FAMNN classifier is built of raw input layer (complement coder), input layer, output category layer, and category layer shown in Fig. 7.

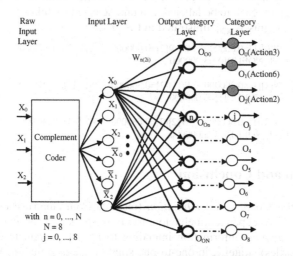

Fig. 7. Resulting Fuzzy ArtMap Neural Network (FAMNN) architecture.

Raw Input Layer: with three inputs corresponding to components of the input vector **X**: strategy (X_0), character (X_1), and state or round (X_2).

Input Layer: with six input nodes, constituting the input vector **I** in Eq. (14), obtained from the three inputs implemented in a complementary coding form by a complement coder from an input i-dimensional vector to 2i dimensional vector.

$$\mathbf{I} = [X_1, X_2, X_3, \overline{X}_1, \overline{X}_2, \overline{X}_3]. \tag{14}$$

Output Category Layer: starts with no output category nodes (no weights) and grows reaching finally nine output nodes.

For each input **I**, the output activation function (choice function), T_n, is defined by Eq. (15):

$$T_n(\mathbf{I}) = \frac{\left|\mathbf{I}_{(2i)} \wedge \mathbf{W}_{n(2i)}\right|}{\lambda + \left|\mathbf{W}_{n(2i)}\right|}, \tag{15}$$

where λ is a small positive constant $0 < \lambda \ll 1$.

Fuzzy AND operator $\wedge$ is defined by Eq. (16):

$$(\mathbf{p} \wedge \mathbf{q})_m \equiv \mathrm{Min}(p_m, q_m), \tag{16}$$

with m = 1, ..., M ; norm | . | is defined by Eq. (17):

$$|\mathbf{p}| \equiv \sum_{m=1}^{M} |p_m|, \tag{17}$$

for any m-dimensional vectors **p** and **q**.

Then, category choice in Eq. (18) is indexed by N:

$$T_N = \mathrm{Max} \{ T_n: n = 1, ..., N \}. \tag{18}$$

Category Layer: with nine category nodes O_j, the maximum number of categories the network can learn, each to be labeled as a unique category (class).

The resulting weights $W_{n(2i)}$, illustrated in Fig. 7, are:

W[0][0]=0.320000   W[0][1]=0.000000   W[0][2]=0.000000   W[0][3]=0.680000   W[0][4]=1.000000
W[0][5]=0.680000   CAT_NOEUD[0]=3.000000 ;

W[1][0]=0.320000   W[1][1]=0.200000   W[1][2]=0.360000   W[1][3]=0.680000   W[1][4]=0.800000
W[1][5]=0.360000   CAT_NOEUD[1]=6.000000 ;

W[2][0]=0.320000   W[2][1]=0.400000   W[2][2]=0.680000   W[2][3]=0.680000   W[2][4]=0.600000
W[2][5]=0.040000   CAT_NOEUD[2]=2.000000.

5   Discussion and Conclusion

In first part of this Chapter, a *combination* (inspired from empirical human negotiation research) of time-dependent behaviors has been suggested for autonomous agent systems with *incomplete* information intending to find the adequate (*win-win* solutions for both parties) strategy, in one-to-one single issue negotiation. With regard to strategies using behaviors individually, the obtained results demonstrate that the suggested combination allows agents to improve the negotiation process in terms of agent utilities, round number to reach an agreement, and percentage of agreements. Another interesting conclusion is that, since with incomplete information, an adequate strategy could be a strategy which imitate the other agent (adequate strategies have been **BL-BL**, BC-LB, LB-BL, **LC-LC**, **CB-CB**, and **CL-CL**).

With regard to the work developed in [13]:

- the strategies of the class starting reasonable and conceding moderately lead, on average, to superior outcomes (in our case this is corresponding to LB-BL and LC-LC),
- the strategies of the class starting high and conceding slowly lead, on average, to fewer and slower agreements (in our case this is corresponding to BL-BL and BC-LB),
- the strategies of the class starting low and conceding rapidly lead, on average, to more and faster agreements but with inferior outcomes (in our case this is corresponding to CB-CB and CL-CL).

An interesting alternative for future research works is with regard to type of combinations (linear) suggested in [7], which is expected to improve our suggested combination if it is used in our combination particularly for time-dependent with behavior-dependent behaviors.

In second part of this Chapter, a negotiation strategy essentially based on the psychological aspects of the human personality (negotiation behaviors: +Con, Con, -Con, +Neu, Neu, -Neu, -Agg, Agg, +Agg) is suggested. Such negotiation behaviors are acquired by reinforcement Q-learning and Sarsa-Learning approaches in order to guarantee the agent reactivity, agent proactiveness, and agent social ability. Analysis and comparison show that Sarsa-Learning offer the advantage of safe solutions with regard to Q-Learning. The strategy (a basic buying strategy) is then acquired by FAMNN exploiting particularly their *fast* and *stable* learning capabilities (essential trait of intelligent behavior), massively parallel computations, and limited memory requirement for storing the knowledge. Thus, suggested strategy displays the ability to provide agents, through a basic buying strategy, with a first intelligence level in SOCOSYS for learning negotiation strategies from interaction (human-agent and agent-agent).

In addition, this first intelligence level provided to an intelligent agent through negotiation behaviors acquired by Q-learning and FAMNN allows to interact, during a negotiation session, with another agent or with human through SISINE software. More, with such interactions, different training scenarios are possible with on-line learning (Q-learning) and on-line learning (FAMNN) exploiting one or the other, or exploiting them together. This first intelligence level still not enough to allow an agent to learn a negotiation strategy, for instance, from a human with a high negotiator quality. For this, a second intelligence level is under work handling problems of strategy, cooperation, competition.

Another interesting alternative for future works is to investigate the research works developed by Jim Camp "Start with No" (http://startwithno.com/) with regard to win-win solutions.

Finally, we believe that an intelligent negotiation system, win-win with incomplete information, should combine in an efficient way such behaviors developed in the first and second parts of this Chapter with fuzzy and prediction behaviors.

References

1. Rosenschein, J., Zlotkin, G.: Rules of Encounter. MIT Press, Cambridge (1994)
2. Wooldridge, M.: An Introduction to MultiAgent Systems. John Wiley & Sons, England (2002)
3. Jennings, N.R., Faratin, P., Lomuscio, A.R., Parsons, S., Sierra, C., Wooldridge, M.: Automated negotiation: prospects, methods, and challenges. Int. J. of Group Decision and Negotiation 10(2), 199–215 (2001)
4. Gerding, E.H., van Bragt, D., Poutré, J.L.: Scientific Approaches and Techniques for Negotiation: A Game Theoretic and Artificial Intelligence Perspective. CWI, Technical Report, SEN-R0005 (2000)
5. Li, C., Giampapa, J., Sycara, K.: Bilateral negotiation decisions with Uncertain Dynamic outside options. IEEE Trans. on Systems, Man, and Cybernetics, Part C: Special Issue on Game-Theoretic Analysis and Stochastic Simulation of Negotiation Agents 36(1), 1–13 (2006)
6. Lomuscio, A.R., Wooldridge, M., Jennings, N.R.: A classification scheme for negotiation in electronic commerce. Int. J. of Group Decision and Negotiation 12(1), 31–56 (2003)
7. Faratin, P., Sierra, C., Jennings, N.R.: Negotiation decision functions for autonomous agents. International Journal of Robotics and Autonomous Systems 24(3-4), 159–182 (1998)
8. Wang, K.J., Chou, C.H.: Evaluating NDF-based negotiation mechanism within an agent-based environment. In: Robotics and Autonomous Systems, vol. 43, pp. 1–27. Elsevier, Amsterdam (2003)
9. Ros, R., Sierra, C.: A negotiation meta strategy combining trade-off and concession moves. In: Auton. Agent Multi-Agent Sys., vol. 12, pp. 163–181. Springer, Heidelberg (2006)
10. Pruitt, D.: Negotiation Behavior. Academic Press, London (1981)
11. Hamner, W.: Effects of bargaining strategy and pressure to reach agreement in a stalemated negotiation. Journal of Personality and Social Psychology 30(4), 458–467 (1974)
12. Carnevale, P., Pruitt, D.: Negotiation and Mediation. In: Rosenzweig, M., Porter, L. (eds.) Annual Review of Psychology, vol. 43, pp. 531–581. Annual Reviews Inc (1992)
13. Lopes, F., Mamede, N., Novais, A.Q., Coelho, H.: A negotiation model for autonomous computational agents: Formal description and empirical evaluation. Journal of Intelligent and Fuzzy Systems 12, 195–212 (2002)
14. Bales, R.F.: Interaction Process Analysis: A Method for the Study of Small Groups. Addisson-Wesley, Cambridge (1950)
15. Rubin, J.Z., Brown, B.R.: The Social Psychology of Bargaining and Negotiation. Academic Press, New York (1975)
16. Silver, D., Sutton, R., Müller, M.: Reinforcement Learning of Local Shape in the Game of Go. In: Int. Joint Conference on Artificial Intelligence, pp. 1053–1058 (2007)
17. Miglino, O., Di Ferdinando, A., Rega, A., Benincasa, B.: SISINE: Teaching negotiation through a multiplayer online role playing game. In: The 6th European Conference on E-Learning, Copenhague, Danemark, October 04-05 (2007)
18. Pfeifer, R., Scheier, C.: Understanding Intelligence. MIT Press, Cambridge (1999)

19. Chohra, A.: Embodied cognitive science, intelligent behavior control, machine learning, soft computing, and FPGA integration: towards fast, cooperative and adversarial robot team (RoboCup). Technical GMD Report, No. 136, ISSN 1435-2702, Germany (June 2001)

20. Zeng, D., Sycara, K.: Benefits of learning in negotiation. In: Proc. of the 14th National Conference on Artificial Intelligence (AAAI 1997), Providence, RI, July 1997, pp. 610–618 (1997)

21. Watkins, C.J.C.H.: Learning from Delayed Rewards. PhD Thesis, King's College (1989)

22. Whitehead, S.D.: Reinforcement Learning for the Adaptive Control of Perception and Action. Technical Report 406, University of Rochester (February 1992)

23. Sutton, R.S., Barto, A.G.: Reinforcement Learning. MIT Press, Cambridge (1998)

24. Dietterich, T.G.: Hierarchical reinforcement learning with the MAXQ value function decomposition. Journal of Artificial Intelligence Research 13, 227–303 (2000)

25. Anderson, J.A.: An Introduction to Neural Networks. The MIT Press, England (1995)

26. Patterson, D.W.: Artificial Neural Networks: Theory and Applications. Prentice-Hall, Simon & Schuster (Asia) Pte Ltd, Singapore (1996)

27. Haykin, S.: Neural Networks: A Comprehensive Foundation, 2nd edn. Prentice-Hall, Englewood Cliffs (1999)

28. Lee, C.-F., Chang, P.-L.: Evaluations of tactics for automated negotiations. In: Group Decision and Negotiation. Springer, Heidelberg (2008)

29. Sandholm, T.W.: Distributed Rational Decision Making. In: Multiagent Systems: A Modern Introduction to Distributed Artificial Intelligence, pp. 201–258. MIT Press, Cambridge (1999)

30. Coggan, M.: Exploration and Exploitation in Reinforcement Learning., Research supervised by Prof. Doina Precup, CRA-W DMP Project at McGill University (2004)

31. Takadama, K., Fujita, H.: Lessons learned from comparison between Q-learning and Sarsa agents in bargaining game. In: North American Asso. for Computational Social and Organizational Science Conference (2004)

32. Lin, L.-J.: Self-improving reactive agents based on reinforcement learning, planning and teaching. In: Machine Learning, vol. 8, pp. 293–321. Kluwer Academic Publishers, Dordrecht (1992)

33. Touzet, C.F.: Neural reinforcement learning for behaviour synthesis. Robotics and Autonomous Systems 22, 251–281 (1997)

34. Carpenter, G.A., Grossberg, S., Rosen, D.B.: Fuzzy ART: Fast stable learning and categorization of analog patterns by an adaptive resonance system. Neural Networks 4, 759–771 (1991)

35. Azouaoui, O., Chohra, A.: Soft computing based pattern classifiers for the obstacle avoidance behavior of Intelligent Autonomous Vehicles (IAV). Int. Journal of Applied Intelligence 16(3), 249–271 (2002)

Intelligent Knowledge-Based Model for IT Support Organization Evolution

Jakub Chabik, Cezary Orłowski, and Tomasz Sitek

Gdańsk University of Technology, Department of Management and Economics,
ul. Narutowicza 11/12, Gdansk, Poland
jakub.chabik@ebit.pl,
{cezary.orlowski, tomasz.sitek}@zie.pg.gda.pl

Abstract. The goal of the paper is building the knowledge-based model for predicting the state of the IT support organization. These organizations are facing the problem of their transformation. The complexity of the processes, the difficulty adjusting the operations and limited ability to control the evolution implies the need of the solutions supporting the decision-makers to make the change happen. The solutions are based on systems gathering and processing the knowledge, built in the research units for the support of business organizations. The paper is the example of the applying this research for the banking sector. The result of the research is the intelligent, knowledge-based system used for the assessment of the organization and support the evolution of the organizations based on fuzzy modeling and mechanisms of reasoning using uncertain and incomplete knowledge and classification according to the ITIL (IT Infrastructure Library) model. The sources of the knowledge are the authors' experiences in managing the IT support organization in the banking sector. The paper presents the results in two areas. One of them is the financial sector, for which we propose the prognostic model for the IT support organization. The second is the selection of reasoning in social systems with uncertain and incomplete knowledge. In the paper we present the mechanisms of reasoning based on the knowledge and data in the research in the evolution of the IT support organization in the financial institution. The publication contains the description of the assumptions of the experiment, the experiment itself, the analysis of its results and conclusions.

Keywords: intelligent systems, IT support organization, ITIL, fuzzy modeling.

1 State and Processes of the IT Support Organization

1.1 The Problem of the Organization Evolution

Traditionally, IT organization is the supplier of technology and systems. IT supplies to its customers (internal or external) technology 'boxes', e.g. desktop computers, off-the-shelf systems or other pieces of IT infrastructure. Its purpose is defined as the maintenance of existing technology pieces (as network links, servers, desktops, applications) and purchasing (together with deployment) of new technology pieces for

E. Szczerbicki & N.T. Nguyen (Eds.): Smart Infor. & Knowledge Management, SCI 260, pp. 177–196.
springerlink.com
© Springer-Verlag Berlin Heidelberg 2010

the core business units. From that point of view, the main objectives of such organization are:

- Relevant procedures for fulfilling the customers' requirements through projects,
- Development or purchase of the information systems for these projects.

However, in the last years the market of IT services is quickly changing through saturation of the modern enterprise with information systems and technologies. The result of these changes is the need of different approach to IT organizations. The mature business organization expects that its IT will supply services rather than 'technology boxes'. First, it will understand the business goals and needs. Then it will select appropriate tools and technology. Last, it will supply it continuously according the agreed service level, supporting the business processes.

This approach requires fundamental rebuilding (or transformation) of the IT organization: from its structures, through processes and services, redefinition of the systems and technologies, rebuilding the competencies to change in every working workplace (fig. 1).

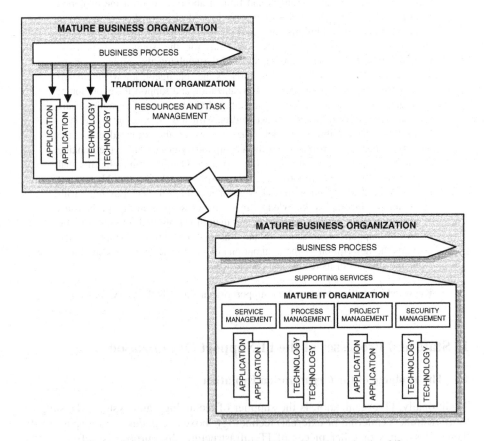

Fig. 1. Characteristic of the state of the IT Support Organization [4]

With this approach, the new objectives for the IT organization should be defined as

- Building and alignment of the strategy,
- Analysis of the business processes and their maturity,
- Identifications of customer requirements and needs and satisfying them though projects deploying systems and services,
- Building, agreeing and delivering IT services suitable for the business processes of the customers.

The crucial role in this transformation is played by the standards, in particular two of them: Capability Maturity Model (CMM), Capability Maturity Model Integration (CMMI) [2, 5] and IT Infrastructure Library (ITIL) [1]. They set up the direction based on the market's best practices and they build up a framework for the organization processes, at the same time providing the tools for IT organization assessment. The application of this assessment schema for the organization and its processes is the important metrics for the IT organization transformation and its current maturity.

1.2 Application of the ITIL Standard for the Assessment of the Maturity of the IT Support Organization

The ITIL standard can be used or assessing the maturity of the IT support organization [7] (fig. 2). It describes the process's framework, goals and its typical artifacts (input, output, intermediate products). It defines the roles, e.g. the role of the Problem Manager, and the typical actions and tasks of these roles, for instance, the Root Cause Analysis (RCA). ITIL, like CMM or CMMI, is the reference model and provides the maturity assessment for the processes.

Therefore, it can be used to assess (compare) existing processes to the framework or template model. The organization itself decides which of the processes are more important, which activities it selects and to what extent it is going to deploy the model. ITIL doesn't support or promote any kind system as 'supported' or 'preferred', leaving the initiative in the hands of the software companies.

The selection of the ITIL [8] model was based on the market's best practices, opinions and experiences and it wasn't by the structured framework selection process.

The ITIL version 2 was built by the Office of Government Commerce (OGC) [12] and divided into:

- Service Delivery,
- Service Support,
- Planning to Implement Service Management,
- Application Management,
- ICT Infrastructure Management,
- Security Management,
- Business Perspective.

The experiences for using the ITIL standard in the existing organization show that the Service Delivery and Service Support are dominating parts [9]. From that point of

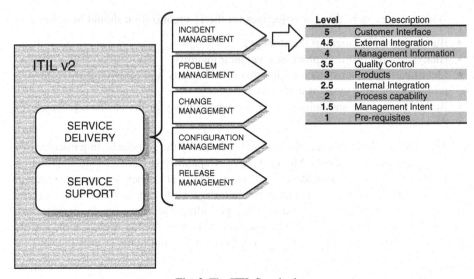

Fig. 2. The ITIL Standard

view, the assessment from these two perspectives is crucial for identifying the state of the organization.

The Service Support covers 5 processes and one function:

• Configuration Management,
• Incident Management,
• Problem Management,
• Change Management,
• Release Management,
• Service Desk function.

These processes cover the daily maintenance and operational management of the services. The second area is the Service Delivery and its covers the following processes [11]:

• Service Level Management,
• Capacity Management,
• IT Service Continuity Management,
• Availability Management,
• IT Financial Management.

The Service Delivery area describes the IT services from outside of the organization and in a more tactical perspective. On the other hand, the Service Support provides a more internal view and operational perspective. These two perspectives were chosen as a representation for the variables in the prognostic model. Using the questionnaire (used for self-assessment or external assessment) the IT organization, can decide to what extent it fulfills the requirements of a certain level of maturity. The questionnaire also helps to choose the reference maturity model (typically between 60% and 80%) as 'pass' level.

As the Service Support and Service Delivery define 6 management processes, together with 9 maturity levels for each process, it all brings a complete environment to assess the maturity of IT support organization.

1.3 The Evolution of the IT Support Organization with ITIL

The ITIL standard describes the state of the IT support organization (in particular, the maturity of the processes in this organization), but it doesn't describe the transition processes from the traditional, technology-oriented organization to the modern, service-oriented organization that focuses on customers' needs and processes.

In the earlier publications, we used the word 'transformation' to describe this strategic shift; however, now, the word 'evolution' seems to be more descriptive. 'Transformation' suggests somehow finished boundaries in time and relative stability of the organization in the start and end state. In reality, the change in the organization is a continuous process and the organization is never really stable in any of its states. The strategic shift is more iterative. The organization can capture the current state through the assessment, but it doesn't mean that it freezes the state – in the same way as taking a photo of the moving train doesn't stop the train itself, even if it seems not to be moving in the picture [4].

The literature on ITIL [15] presents a wide spectrum of the solutions related to the state, but it doesn't offer clear recipes on transition towards higher maturity. The objective of this research is to build up an addition for the model and to increase its usability in this way. Also, the goal is to use the environment of assessing the organization.

In order to cover a possibly complete spectrum of evolution, the research was carried out in six fields: processes, services, projects, knowledge management, technologies and organization culture. The dynamics of the evolution of the organization in these fields can be different, but the balance between them is essential. If the balance is not kept, the 'organizational tension' can be observed, which can result in a lack of stability. For instance, the organization that builds up the internal processes, but doesn't offer services needed by the customer, can face the opinion „we don't care how much you control your own organization as long as you don't support our business" form the customer.

The changes in the organization can be divided into three main groups:

- **Management processes** – are spread along a period of time, change the structure of the organization, the way it works, its tools and qualifications, e.g. the deployment of the software tool.
- **Management decisions** – are one-time actions that institutionalize and 'legalize' earlier processes. The example of such decisions is announcement of the official internal regulation that formally obliges the employees to follow the process Change Management.
- **Management artifacts** – are products – items, documents, etc. related to the management processes or decisions. The example of the management artifact is the detailed description of the RCA form in the Problem Management process.

It seems that the good example of all three kinds of changes is the service in the Service Level Management process. The **management process** is the negotiation of the shape and the parameters of the service with the customer. The **management decision** is signing off the agreement with the customer. The management artifact is the detailed description of the service, defining its availability, capacity, agility, hours of service, etc.

Having described the organization states and the transition processes it is important to show the relationship between them. The experience shows 4 kinds of such relationships (correlations) (fig. 3):

1. Very strong – observed in different organizations as well as different processes.
2. Strong – observed either in different organizations or in different processes, but not both at the same time.
3. Weak – observed partially in organizations as well as processes, but not consistently and repeatedly.
4. Inconsistent – observed in organizations as well as processes, but in different ways or directions.

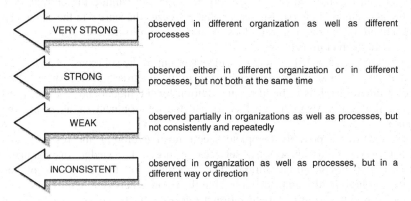

Fig. 3. Types of relations between the organization states and the transition processes

All correlations can be either positive (certain changes support higher maturity) or negative (certain changes disable higher maturity or decreases maturity). All in all, we have 7 combinations of correlations:

- Positive – very strong
- Negative – very strong
- Positive - strong
- Negative - strong
- Positive - weak
- Negative - weak
- Inconsistent

The relationship between the action and the maturity can be represented in the knowledge base as discrete or fuzzy rules. This knowledge base can be used for predicting the organization state or decision support in managing the evolution of this organization.

1.4 Methods of Reasoning Based on Uncertain or Incomplete Knowledge

The description presented above based on the ITIL standard can be used for assessing both maturity of IT support organization and the strength of relationships between state and transition. It implies the use of reasoning algorithms suited for uncertain or incomplete knowledge. Therefore, before building the model, the preliminary assessment was done in order to choose the right way of reasoning methods. The methods chosen are very important as it comes to the correctness of reasoning – they are based on the assumption that the information in the knowledge base is true and therefore the reasoning is correct and the results are true.

This assumption can be too hard to take, because the knowledge gathered from experts – even if is certain – is possibly incomplete and therefore not perfect. Using the methods normally suited for complete and 'perfect' knowledge for incomplete and imperfect knowledge, there is a risk that the results will be incorrect. Therefore, it is very important that the reasoning system be supplied with additional mechanisms for preprocessing this knowledge to increase the overall certainty. In this way it will be possible to characterize the uncertainty level of the knowledge about the organization state and transition processes. This approach can be used for the new knowledge inferred from the existing one by the reasoning system.

There are three basic kinds of uncertainty in knowledge about the IT support organization state and transition:

- The uncertain knowledge about the transition processes for which the knowledge about relationship between the change and state is certain,
- The incomplete knowledge about the organization state some of the certain states are unknown, but we cannot assume that 'unknown' means 'incorrect',
- The imprecise knowledge about the transition and state, in which some of the relationship is not known exactly.

In case of uncertain knowledge the results cannot be clearly labeled as true or false. The methods are needed for characterizing the level of the knowledge certainty – both this from the preliminary knowledge base and this from experts' assessment. Incompleteness of the knowledge means that the status of the results inferred is unknown – they may be true as well as false, which may require taking the assumptions about the certainty or leaving it open to repeat (and – potentially – change) the reasoning in case of getting the new knowledge that would invalidate the previous results. Uncertainty is the consequence of an inability to distinguish the knowledge in the area, for which we are researching, or the objects in the certain relationship to this knowledge, from the objects in different area, unrelated.

For the purpose of building the model, the analysis was carried out in the reasoning methods for different kinds of uncertainty [10]. It was carried out in such a way that the reasoning led to the results whose quality could be assessed. These characteristics would allow answering the question how uncertain or incomplete results can be inferred from uncertain or incomplete knowledge. The following methods were considered [6]:

- Fuzzy logic – methods for processing the incomplete knowledge, for which partial relationship between the item and set is considered.
- Dempster-Schaffer algorithm – for processing the uncertain knowledge, for which the probability of correctness is attributed to base statements.

- Certainty levels – the method for processing the uncertain knowledge, for which each statement has the certainty level attributed, expressing the expert's conviction about the correctness of this statement.
- Non-monotonic logic – methods for processing incomplete knowledge, in which it is allowed that the new statements will appear invalidating earlier inferred rules.
- Probabilistic reasoning – the method for processing unsure knowledge based on direct application of the probabilistic calculus; in this approach the probability level of correctness is attributed to results.

2   Stages of Building the Model

The previous part of the paper presents the concept of building the model for predicting the evolution of the IT support organization.

This chapter concentrates on the detailed model in the proposed and used mechanisms and algorithms. The analysis of the knowledge gathered in the project was performed. We identified the variables (parameters) relevant for the changes in the IT support organization and, from that point, we present the model that is responsible for processing this knowledge and generating conclusions.

The analyzed case shows exactly that it is not possible to gather complete, sure knowledge. The authors decided to use the means for processing the unsure and incomplete knowledge.

2.1   Input and Output Variables

The project of predicting changes in the IT support organization was initiated by the thorough analysis of the current state. Both interviews and the analysis of the documentation were intended to identify the key factors that induced the current state. It seems to be obvious that such factors can be found in different areas of the organization's activities. It was assumed that the acquisition of the knowledge relevant for this project will be done in all such areas (called threads) by the dedicated member of the research team. This kind of division of responsibilities determined also the division of the overall goal into the partial goals for the team members. The division was done according to the components of organization evolution (fig. 4). It should be noticed that the overall success of the evolution is the result of the orchestration of the following components:

- Design and deployment of processes,
- Harmonization of structures,
- Knowledge gathering and distribution,
- Tool support,
- Cultural change.

Obviously, these components cannot be treated separately. If one wants to achieve an overall result, the organization should be perceived in a holistic way, e.g. as in corporate architecture framework TOGAF (The Open Group Architecture Framework).

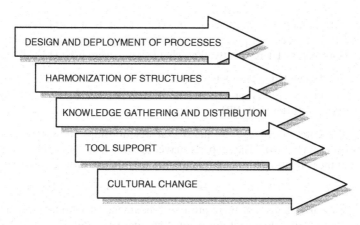

Fig. 4. Components of the successful evolution [3]

From that point, the following threads were chosen in order to structure the knowledge about the state of the organization and transition processes (fig. 5):

- Processes,
- Projects,
- Services,
- Knowledge,
- Technologies,
- Organization culture.

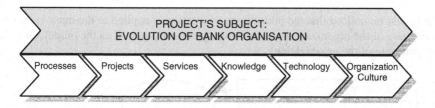

Fig. 5. Evolution areas of the organization [3]

According to the assumptions, the result of this research was supposed to be the knowledge base suitable for further processing. It should be ready to infer the relevant change actions given the planned states. For building up this database, the researchers chose the rule-based representation, because this kind of representation creates the possibility of easy modeling / processing the knowledge. It seems also the most relevant for reasoning about the relationships between the end state of the process as a consequence of the begin state and given changes. Additionally, the experts' knowledge was represented by facts that are going to be used in the inference process [14].

Assuming the initial assessment of the current state in all of the six areas, it was agreed that the products of the first stage of the project, will be informal description. To describe this state in a formal way, it was necessary to identify the input and output parameters of the model. Unfortunately, this knowledge turned out to be incomplete. Not all the threads received precise identification of the key factors influencing the results of the research. Therefore, two knowledge bases were used [13]:

- Fundamental knowledge base KB_POD,
- Knowledge base for initial preprocessing of the input variables for the fundamental knowledge KB_PRE1, KB_PRE2.

The fundamental knowledge are the overall relationships showing the changes in the organization (fig. 6). The experts have agreed that the appropriate way of representing these relationships is the following way (in the Prolog notation)

```
End_state :- Begin_state, Transition_processes
```

This rule clearly states that there are two fundamental factors that impact the end (pr-edited) state of the organization: begin state and transition processes. Obviously, this kind of statement, describing the evolution of the organization, is a significant simplification. At this point it cannot be different (i.e. more detailed) as we are thinking of the common knowledge base for all areas which are different from one another. Therefore, for each variable we have found out the dedicated set of rules making up a preliminary processing database. For instance, for the area „Technology", the transition processes are the consequence of the number of factors. The rule describing these processes can be the following:

```
Transition_Processes:-
            Compliance_with_technology_stack,
            The_cost_of_technology,
            Vendor_support,
            Team_competencies
```

It should be noticed that the preliminary processing is applied to the input variables (begin state of the organization and transition processes) as well as the output variable (and end state of the organization).

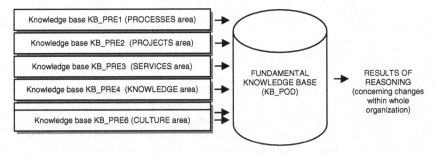

Fig. 6. The scheme for processing knowledge and reasoning in the model

2.2 The Proposal for Model

The previous subchapter presents the relationship between the post-transition state of the organization and the pre-transition state. This is one of the stages on the way to build the full prediction mode, which would allow simulating the evolution thus becoming the tool of the decision support for the managers running the evolution of the organization.

The goal is to build the mechanism that would be supplied within existing knowledge. For this, the following notation was chosen:

```
End_state :- Begin_state, Transition_processes
```

where:
the variable: End_state has values from the set [1..5]
the variable: Begin_state: has values from the set [1..5]
the variable: Transition_processes has values from the set [1..5]

It was assumed that all the variables will have the numeric values representing the achieved level. This assumption is obviously based on the previously presented concept of the organization assessment using ITIL and CMMI framework. At the same time, it is the way to formalize the knowledge when this knowledge is in some way imperfect (e.g. incomplete or unsure). However, it should be noticed that the built solution doesn't represent technical matters (where knowledge is typically precise and concrete), but is attributed to the social system (as the organization culture is the research area).

Is the fuzzy model the only correct solution to the difficulty of incomplete knowledge in the analyzed problem? Even if its application seems to be correct, it cannot be the only method of reasoning for the knowledge coming from different sources and represented in many ways. The fuzzy model will be implemented only in the stage of processing the preliminary knowledge (KB_POD).

The situation becomes different for the knowledge bases dedicated to different project threads. In each case the relevant input variables will be different, while the output variables will be the same (for state of the organization and state of the transition processes). Still we are facing the imperfect (incomplete, unsure) knowledge, in which case the appropriate method for reasoning is necessary.

During the interviews with the experts, it was observed several times that the knowledge is based only on the memory of the expert, without being backed up by any documentation. Some statements (facts about the state of the organization as well as rules describing the relationship between these states) were simply stated as 'probable'. The only way to cope with this situation could be the consultancy in the bigger group – but even then it wouldn't be completely certain. Therefore it was necessary to introduce a certainty factor (CF) in knowledge processing. For most statements the subjective score representing the level of certainty in the given fact or rule was given. The example of the fact from this knowledge can be represented in the following way:

```
Exists_in_the_technology_stack (MS SQL Server, 2004,
0.5)
```

where the last argument is the certainty factor and can have the following values: $CF \in [-1..1]$. We should set the minimum level of CF, for which any reasoning from so uncertain facts is acceptable.

All in all, for such complex problem applying the right reasoning method seems the most crucial task. There are several such methods, as shown in the previous subchapters, but each of them has some limitations. In this case, the proposed method is the hybrid.

2.3 Relationship between Variables (Strength of Impact)

During the analysis of the gathered knowledge the hypothesis, which wasn't considered before, appeared, suggesting that not in all cases the variable End state is related to both input variables in the same way. Generally, it seems that this relationship is applicable to all areas in this research, but the strength of influence should be considered. The issue is to identify which input variable has more impact (influence) on the end state of the organization after the evolution. It may happen that in some cases the end state will be strongly influenced by the begin state, while in the other cases it will be strongly be the result of transition processes.

At the time of building the model, we adopted the simplest solution – the strength of influencing was represented by multipliers (weights) as the modifiers of reasoning processes, which are arbitrarily set up by the expert.

The authors are planning to use the reasoning method based on Dempster-Schaffer method in this context, which allows reasoning from the statements (in our case: rules) based on their relationship with other rules. Are such relationships present in the KN_POD knowledge base? During the knowledge acquisition process it was assumed that the bases will be separate and independent. During the analysis of the documentation we noticed an explicit relationship. It seems that the evolution in the area of e.g. technology influences the organization as a whole and it influences the evolution in other areas. Therefore, it would be useful to be able to reason about the results of the evolution in all areas. At this point the authors are facing the problem of finding the right means for representing this kind of knowledge and relationship.

2.4 The Reasoning Processes - The Experiment Plan and Its Execution

The above considerations should be treated as the guidelines and assumptions to practical activities that are carried out to achieve the goal of the research project. This goal, building the prediction model in the area of evolution of the IT support organization, requires a plan. This subchapter presents the plan of the experiment.

It should be noticed that some of the stages of the project are already fulfilled and earlier in this paper we described their results and problems that we have faced. The list of stages shown below should be treated as some kind of structuring this information. The most important part of the plan is the part about next steps, as they show the directions of future research.

– **Stage 1:** Knowledge acquisition
The system will be the knowledge-based solution. In the first step it is crucial to gather this knowledge. The analysis of a problem so complex as the evolution of the IT support organization in many areas doesn't allow to choose only one method of gathering the knowledge. It is assumed that it will be knowledge of a different quality, gathered in different ways. So the product of the first stage is the semantic description of the current state, divided into six areas (threads) of the research. For this project,

both the people from the inside of the organization were involved (as experts and thread leaders) and the students from Gdansk University of Technology. There were realizing their graduation theses in this project. The principal methods for gathering the knowledge were the interviews and analysis of the documentation.

– **Stage 2:** Knowledge formalization

The way that the knowledge is gathered in the first stage is not perfect. However, it is not possible to perform the knowledge formalization earlier, i.e. during acquisition, because at this stage its characteristic (i.e. which variables, discrete or linguistic, etc.) is unknown. The analysis of artifacts gathered in the first stage would allow setting the rules of the knowledge formalization.

As it was described in the earlier subchapters, it is assumed that the knowledge will be divided into several separate knowledge bases, including the fundamental knowledge base and the knowledge bases for preprocessing (for each thread). It is worth to mention at this point that there isn't any reasoning at this stage, only 'translation' of the gathered specifications to the rule-based knowledge. This stage is realized by the employees of the Information Technology Management Team who are knowledge engineers at this stage.

– **Stage 3:** Knowledge verification and reasoning

The product of the previous stage will be the structure of the knowledge base using the proper combination of variables and relationships. In case of the preliminary knowledge, what should be expected are the conclusions inferred from the input data. During the interviews this knowledge should be structured, but it is not possible to build up the preliminary reasoning for the whole KB_POD knowledge base. The question arises how to carry out the reasoning in the fundamental knowledge base. Two possibilities are considered:

- Identification of rules considered on few examples that would be the generalization for the knowledge; this approach would allow automatic reasoning,
- The session with the expert, presentation of the preliminary structure of the knowledge base and performing full reasoning (for each rule) solely with his or her knowledge.

The first solution would be more effective (mostly, because automatic reasoning is more scalable). This is the objective that we are setting. There are several threats, i.e. the incompleteness of the knowledge and incomplete understanding of the problem domain may cause the ambiguity during the definition of such algorithms.

Because of that we propose the experiment that would mutually verify both approaches. We want to attempt building the algorithms, but at the same time to run a session with an expert. The expert wouldn't know the algorithm; therefore he will be using only his subjective knowledge. In both cases the goals is the same – to get the complete set of rules with conclusions. Both results will be then compared and then, again with the expert's help, we will point out the conclusions that are significantly different (in particular, conflicting). It cannot be assumed that any of the sources will generate the results more credible. The expert has more experience, but this might be an apparent strength. In case of larger number of rules, the human might be unable to formulate conclusions for all possible combination of input variables or even can make multiple mistakes. The experiment's objective is to identify and eliminate the weaknesses of the algorithms and its improvement in such a way that it is applicable to any size of the knowledge base.

During this stage it is necessary to consider the activities related to the completing the knowledge base. It could happen that the gathered knowledge is so unsure or incomplete that it will be necessary to perform the sessions with experts in all threads. Even if it will make the whole experiment longer, such a step should be included in the schedule.

– **Stage 4:** Building the model, implementation of the reasoning algorithm

This step is purely technical. At this stage we don't need to analyze the quality of the knowledge or verify the correctness of reasoning. The product of the 4th stage should be the implemented model allowing making predictions related the IT organization evolution. This step should be divided in the smaller steps allowing the incremental checking of the functions of the system.

– **Stage 5:** Fine-tuning the model, final verification

This is the last stage of the projects that ends up with the acceptance of its products by the all beneficiaries. It is assumed that it is possible to fine-tune the model in the full application area or even – in case some major issues are found – the return to the previous stage. At this stage the knowledge will not be analyzed, only the mechanisms for its processing and management.

2.5 Verification of the Applicability of Prepared Model in the Evolution of the IT Support Organization

The next stage is the verification of the applicability of the IT support organization evolution model. At the beginning we have presented the example of using the model for selected organization. Then the model was assessed both for the given example as well as its applicability to other organizations. At the end we pointed out the future development possibilities taking into account better availability of input variables and reasoning methods.

The prognostic model was verified against incomplete and unsure knowledge. Therefore two levels of the solution verification were prepared.

The first level covered the verification for the knowledge for the output variables. We assumed five values classifying these variables. This approach was used because of existing assessment framework (aggregation of ITIL process maturity framework). For other threads (projects, services, knowledge management, technologies and organization culture) such assessment framework is not present or cannot be applied and therefore the solution presented in this paper should be verified. A similar solution was applied for assessing transition processes.

The second level of the verification covered the procedures of reasoning for the model. The verification process was executed in two steps. In the first step we assessed the initial values for input variables used in the reasoning algorithms. The selection of the reasoning algorithms was incremental. They were selected according to the possessed knowledge. The more knowledge the expert had, the higher the level of applied solution was. At the beginning, we used the approaches based on fuzzy modeling, then the approach using certainty levels, later the Dempster-Schaffer algorithm, at the end the proposed hybrid approach. The proposed scenarios were then verified by the experts.

The experimentation process:

– Stage 1: Selection of the areas for IT support organization assessment
We assumed that the research will cover 5 previously chosen areas (threads). During the research process we have found out that the most difficult will be the cultural organization area. The difficulty was the consequence of fact that the we were missing the knowledge as well as specialists (psychologists and sociologists) that would be able to assess current status of the organization culture and identify relevant management processes. That's why this area was excluded from the research but in the further development of the model this area will be included again.

– Stage 2: Acquisition of the knowledge in other areas
The knowledge describing the state of other areas was acquired with the research in the company involving the employees of the departments of the IT support organization. These specialists were supported by the students realizing their MSc thesis. The knowledge acquisition was performed incrementally. In the first iteration the knowledge was acquired and preliminarily assessed. We considered the completeness of this knowledge in terms of organization state and transition processes. Initial assessment has shown incompleteness of the knowledge. Therefore, we analyzed its applicability and we prepared the plan for the next iteration.

– Stage 3: Initial formalization of the knowledge
For the knowledge acquired and preliminarily assessed we assumed that the reasoning will be carried out with the fuzzy approach. The choice of these methods was impacted mostly by the experiences of the authors using this approach [6]. Therefore at this stage we picked the input variables and the levels of their granularity. The knowledge was assigned to the levels of granularity.

– Stage 4: Full formalization of the knowledge
After the choice of variables and their values' range we started to build the model selecting the areas for the needs of the model. As we said before, the organization culture was excluded because of the lack of the data. For other areas (threads), the fuzzy model was applied as presented above.

– Stage 5: Reasoning processes
The processes of reasoning were performed in two streams. The possible values of the variables were initially assessed. It appeared that we faced the problems during the assignment of the results. Therefore this approach was assessed as initial and the scenario-driven analysis was performed. There were three scenarios: the first, following the initial results; second, where the levels of certainty in the fuzzy model were chosen and the third using the Dempster-Schaffer algorithm.

– Stage 6: Verification of the picked scenarios
The scenarios were verified. The verification processes included the expert's assessment of the presented scenarios in the company. During this assessment it appeared that the initial assumptions were corrected in the practice. The experiment (assessment of the scenarios) has shown that the approach selected in the experiment plan cannot be carried out without an expert.

The experiment plan included that during the building of the scenarios the preliminary analysis will be carried out for the informal knowledge. Based on this, we prepared the full set of rules (for each scenario) including all combination of initial variables implemented according to the Prolog structures:

```
End_state :- Begin_state, Transition_processes.
```

The verification processes were carried out in one thread (processes) with an expert (the IT manager of the company). It covers two scenarios:

- Scenario one: the expert carries out the reasoning processes about the target state of the organization according to the initial conclusions prepared during the automatic rule generation in the knowledge base. Then, the expert verifies the rules, builds alternative conclusions (and comments). During that process the engineers do not interfere.
- Scenario 2: the expert, using the same set of rules, carries out the reasoning processes. This time he considers additionally the levels of correlation between the input parameters and their impact of the value of the output variables. The objective of this scenario is the attempt to pick the appropriate algorithm of reasoning, its formal statement and the building the relation to the algorithms covered in this paper. Using this statement gives a strong foundation for using the approach for similar knowledge bases and rules.

The reasoning processes, carried out according to the first scenario, made the authors understand the complexity of the process and the need for verification by the expert. It appeared also to be very difficult to use formalities in the processes of automatic reasoning. The initial assessment of the process by the knowledge engineer would rather cause the rejection of several rules because of low probability of their applicability in practice, for example the companies with very low maturity realizing the transition that would immediately transform them into high-maturity organization. However, even if these considerations had low applicability in practice, it appeared to be fruitful in the experiment. It led to discovery of other, previously unknown, mechanisms of reasoning, they were very valuable in the attempting the automation of the results generation. The same time, the verification process allowed to discover the weaknesses of the model and pointed out the directions of the future research in the reasoning processes.

The second scenario was applied for building up the algorithm with the objective of supporting the reasoning process (reasoning without an expert). Being independent of an expert at each stage of the reasoning process would significantly simplify the whole system, e.g. because if the errors that could cause an inconsistency in the reasoning. During the verification process, the knowledge engineer suggested the solution based on assigning the weighs for variables of the model (the mechanism applying this approach was prepared for the session with expert). Using this approach would increase the effectiveness of the reasoning by expert, even in case of using the weighs introduced by the engineer. This approach seemed to be more useful for the knowledge engineer. The Dempster-Schaffer algorithm was proposed as the next stage of the model verification. It will be used as a mechanism supporting the reasoning processes based on the initially processed knowledge.

Using this approach, the results appeared to be achievable and usable by the organization. These results were related to the assessment of the ability of the organization to increase its maturity. It was clear the achieving the higher maturity of the organization is possible only when it carries out the transition processes suitable for its maturity level or – which is not as clear – a slightly more challenging. If the same organization would have been carrying out processes suitable for the lower maturity organization, its state would remain the same (but it will not deteriorate). This reasoning allows predicting the state of the organization. What would happen to the organization that would carry out the transition processes suitable for significantly higher maturity that it actually is on? These cases were identified as the most destructive for the companies. As they are not ready for such advanced changes, they would spin off the chaos and – additionally – make a negative impact on the morale of the IT support organization personnel.   These results significantly limit the mechanisms of reasoning and bring up the need of finding the intermediate solutions.

The presented scenarios have shown that the incomplete understanding of the domain is the factor that effectively prevents understanding of the reasoning for the key rules. The experiments have confirmed the theses stated initially:

- The construction of the knowledge base at the assumed level of generality a the given level of preprocessing is generally correct,
- It is possible to reason automatically, but the expert must verify the results.

- **Stage 7:** Verification of the model base on the survey research in other organizations

The data was gathered in the on-line survey. It spanned about 100 of the support organizations and its goal is to answer the question: to what extent the company's IT support organization is typical for the market as a whole and therefore – to what extent the conclusions from the deep case study can be generalized to other organizations. It is also trying to answer the question to what extent the model of the evolution is generic for all IT support organization and to what extent it is specific for the support organization of the financial institution.

3   Analysis of the Results

It is clear that the evolution processes are one of the key aspects of the development of the organization. It is difficult to find out the reason. Possibly, this is the consequence of the crisis in the financial sector and the need to adjust this organization the needs of the customer. Surely, these processes are (and will be) going on in the majority of the financial institutions and large organizations in which supporting processes are key for their efficiency. The pressure arises from the representatives of the banking sector for the solutions that support the transition processes. This pressure has the key importance for building the business models, but it doesn't automatically mean the availability of the knowledge about the evolution. On one hand the organizations would like to know about the processes of evolution. On the other, they are reluctant to share the knowledge about these processes. It is even more difficult to process this knowledge. That's why the support organization of the bank uses the convenient procedure of supporting their evolution with the ITIL model.

The standard allows describing the organization with five key processes (for both Service Support and Service Delivery) and enables to assess the maturity of these processes according to the framework. This approach lets us describe the complete organization which is very important in the evolution. The representative of the organization as well as the modeling person have agreed that the ITIL model can make up a solid base for the preparation (about the knowledge of the evolution, from the organization's side) and the formal description, using semantic model (from the modeler side). The commitment to use the model was mutual and so the modeler as well as the representative of the organization agreed on using it. When they started modeling they were sure about the availability of the complete knowledge from all five areas. Actually, quickly it appeared that the knowledge availability is limited. It was difficult to find out in the organization the people responsible for the about the organization culture. This area is rather specific because of the need to aggregate the psycho-socio-technical knowledge and therefore the need for particular knowledge arises. Neither inside the IT support organization, nor on the modeler side such competency existed. That's why it was agreed that this area will be researched in the next phase of the research project. The lack of the data from this area raises the question about the correctness of the reasoning with so incomplete knowledge. The same question arises in other areas as well. In the paper it was defined which knowledge is crucial from the point of view of building the model and its verification. It appeared that it wasn't acquired entirely. What's more, the acquired knowledge isn't completely certain. That's why on the modeler side it was assumed that the methods for reasoning suited for processing incomplete and uncertain knowledge will be applied. Because in such cases several algorithms are used, it was proposed to verify which of them will be adequate to this level of uncertainty in reasoning. The selection was preceded by the experiment described in the paper.

Because of the specifics of the IT organization (in the researched case, IT support organization) there is a need to carry out specific experiments with the objectives of verifying the selected approach and the assessment of the model for predicting the state of the organization. Because of that it was proposed to:

- Pick up relevant scenarios of reasoning based on existing knowledge,
- Carry out the preliminary reasoning processes,
- Assess used mechanisms of reasoning,
- Assess the credibility of the predictions and its applicability to the organization.

4   Conclusions

The goal of the paper was the presentation of the model for the assessment of the IT support organization evolution and its application in the predicting of the transition of the organization. In the conclusions three mains aspects of this paper are covered: processes of the evolution of the organization, the ITIL model (formalizing the evolution processes) and the model solution. The solution is presented in the context of the provided knowledge about the evolution processes, used formalization methods and reasoning methods and the possibility of extending the model.

The proposed experiment was complex both in terms of designing and executing. It was clear that the knowledge obtained from the expert completely verified the applicability of the reasoning methods. It allows picking the fuzzy model, adequate both in terms of reasoning and prepared model. However, the knowledge available was assessed as insufficient. This opinion applies mostly to the sources of the knowledge about the initial state of the organization. It appeared that in the IT support organization in the financial companies the documentation about the state of the organization is not complete.

The knowledge about the state of the organization is typically informal and hard to discover by the modeler. It seems that this is not specific to the bank; in most cases the status of the processes of the organization is not a subject to the research. That's why the concepts presented in this paper can be intermediate solution in which the knowledge about the IT support organization doesn't require complete documentation, only the preprocessing of the knowledge for to suit the model.

References

1. Cannon, D.: Service Operation ITIL, Version 3 (ITIL). Stationery Office, 3rd edn. Norwich (2007)
2. Carnegie Mellon University, Software Engineering Institute, The Capability Maturity Model: Guidelines for Improving the Software Process. Addison-Wesley Professional, Boston (1995)
3. Chabik, J., Orłowski, C.: Ewolucja dojrzałości informatycznych organizacji wsparcia. In: Orłowski, C., Kowalczuk, Z., Szczerbicki, E. (eds.) Zastosowanie technologii informatycznych w zarządzaniu wiedzą, pp. 179–184. Pomorskie Wydawnictwo Naukowo-Techniczne, Gdańsk (2009)
4. Chabik, J., Orłowski, C.: Knowledge-Based Models of the Organization Transformation. In: Information Systems Architecture and Technology: Application of Information Technologies in Management Systems.Oficyna Wydawnicza Politechniki Wrocławskiej, Wrocław (2007)
5. Chrissis, M.B.: CMMI®: Guidelines for Process Integration and Product Improvement, 2nd edn. Addison Wesley Professional, Boston (2006)
6. Czarnecki, A., Orłowski, C., Sitek, T., Ziółkowski, A.: Information Technology Assessment Using a Functional Prototype of the Agent Based System. In: Foundations of Control and Management Sciences, vol. 9, pp. 7–28. Poznań University of Technology, Poznań (2008)
7. Office for Governmental Commerce: ITIL Lifecycle Publication Suite, Version 3. The Stationery Office, London (2007)
8. Office for Governmental Commerce: Introduction to ITIL. The Stationery Office, London (2005)
9. Office for Governmental Commerce: Service Support. The Stationery Office, London (2005)
10. Orłowski, C., Szczerbicki, E.: Application of Blackboard Architecture in Concurrent Environment for IT Project Management. In: Zarządzanie, K.Z., Knosala, R. (eds.), vol. 2, pp. 249–254. Wydawnictwo Naukowo-Techniczne, Warszawa (2004)
11. Pultorak, D.: Microsoft Operations Framework (MOF): A Pocket Guide. Van Haren Publishing, Norwich (2005)
12. Service Support Book 12th Impression. Office of Government Commerce (OGC), London (2001)

13. Sitek, T., Orłowski, C.: Model of Management of Knowledge Bases in the Information Technology Evaluation Environment. In: Wilimowska, Z. (ed.) Information Systems Architecture and Technology: Models of the Organization's risk Management, pp. 221–231. Oficyna Wydawnicza Politechniki Wrocławskiej, Wrocław (2008)
14. Sitek, T., Orłowski, C.: Evaluation of Information Technologies – Concept of Using Intelligent Systems. In: Wilimowska, Z. (ed.) Information Systems Architecture and Technology: Application of Information Technologies in Management Systems, pp. 217–224. Oficyna Wydawnicza Politechniki Wrocławskiej, Wrocław (2007)
15. Steinberg, R.: Implementing ITIL: Adapting Your IT Organization to the Coming Revolution in IT Service Management. Trafford Publishing, Trafford (2005)

Modeling Context for Digital Preservation

Holger Brocks[1], Alfred Kranstedt[2], Gerald Jäschke[3], and Matthias Hemmje[1]

[1] FernUniversität in Hagen, Chair of Multimedia and Internet Applications,
Universitätsstrasse 1, 58097 Hagen
`{Holger.Brocks,Matthias.Hemmje}@FernUni-Hagen.de`
[2] Deutsche Nationalbibliothek, Adickesallee 1, 60322 Frankfurt am Main
`A.Kranstedt@d-nb.de`
[3] Globale Informationstechnik GmbH, Julius-Reiber-Strasse 15a, 64293 Darmstadt
`gerald.jaeschke@globit.com`

Abstract. Digital preservation can be regarded as ensuring communication with the future, that means ensuring the persistence of digital resources, rendering them findable, accessible and understandable for supporting contemporary reuse as well as safeguarding the interests of future generations. The context of a digital object to be preserved over time comprises the representation of all known properties associated with it and of all operations that have been carried out on it. This implies the information needed to decode the data stream and to restore the original content, information about its creation environment, including the actors and resources involved, and information about the organizational and technical processes associated with the production, preservation, access and reuse of the digital object. In this article we propose a generic context model which provides a formal representation for capturing all these aspects, to enable retracing information paths for future reuse. Building on experiences with the preservation of digital documents in so-called memory institutions, we demonstrate the feasibility of our approach within the domain of scientific publishing.

1 Introduction

There exist diverse notions of context in various communities and environments, hence a clear understanding of context in a complete, concise and unambiguous way remains a challenge even within computer science (cf. [18], [4]). However, across the board context is defined by the *interrelated conditions in which something exists or occurs*[1]. This implies for digital resource management, that the context of a digital object is complex, possibly containing concepts which are shared with other objects. This might be the process environment in which they are created, the associated actors, resources and information objects. Or the preservation environment in which they are stored. Hence it is feasible to regard context representations as separate entities describing particular perspectives, which maintain references to concepts and instances within a particular universe of discourse.

The preservation of digital content, especially in the long term, covers periods of time, during which the nature of digital resources as well as their usage settings change

---

[1] Context. (2009). In Merriam-Webster Online Dictionary.
http://www.merriam-webster.com/dictionary/context. Cited 30 Apr 2009.

E. Szczerbicki & N.T. Nguyen (Eds.): Smart Infor. & Knowledge Management, SCI 260, pp. 197–226.
springerlink.com   © Springer-Verlag Berlin Heidelberg 2010

[30]. An important goal of digital preservation is to foster reuse of digital content. Exploitation with a novel purpose and verification are important variants of reuse. Production and consumption of digital content from the archive do not coincide. Preservation of digital content, especially in the long term, bridges gaps in time, space, culture, expertise, interests, and other dimensions. As consumers cannot refer back to the creators, reuse depends on proper descriptions of preserved digital objects provided through the archive. Understanding the nature of the digital content and its origin can support interested parties in identifying relevant elements in the archive collection and interpreting them correctly.

1.1 Digital Preservation

Digital preservation can be regarded as ensuring communication with the future, that means ensuring the persistence of digital resources, rendering them findable, accessible and understandable for supporting contemporary reuse as well as safeguarding the interests of future generations.

Online publications are becoming increasingly important: be them electronic journals, domain-specific databases, abstract collections, independent publications such as digital dissertations or electronic books. Internet resources appear in parallel to paper publications, even superseding them in certain disciplines.

Results from scientific research which are based on primary data are preferably published through digital platforms. Paper publications that are retroactively digitized with enormous efforts create vast amounts of digital information. Archives are confronted with collections of electronic records, which have to be transferred to digital archives to be preserved for the future. Museums which incorporate digital artifacts in their holdings are, according to their legal mandate, important actors in the creation of digital surrogates.

All digital resources with enduring relevance need to be accessible for long-term in the light of rapid changes in technology and despite outdated software environments. The extensive range of individual problem statements in digital preservation spans from the situation, discussing basic issues such as appraisal criteria towards the provision of appropriate hardware and software which guarantees that a digital document can be opened in the future, towards the search for scalable storage infrastructures allowing to persistently manage large amounts of data and metadata.

The preservation of digital resources as part of the cultural heritage enables the representation of social, historical, technical and artistic developments, but also safeguards access to existing knowledge. Since scientific research always requires the reflection of past achievements, the long-term preservation of digital resources is essential for the competitiveness of education, science, and businesses.

[17] provides a good starting point for the characterization of digital preservation and related concepts, some of which are also discussed in [21] from a slightly different perspective.

1.2 Problem Statement

The preservation of their environmental context has a huge impact on rendering digital holdings accessible, understandable and reusable in the future. Special care has to be taken to capture, represent and preserve the information about the context in which the digital objects have been created. Hence a context model for digital preservation needs to provide an infrastructure-independent representation of the attributes associated with and (implied) relations between digital objects. Context is not only defined by the digital objects themselves, but also by the processes, in which they were created, assembled, ingested, preserved, accessed, adapted, and reused. Context is thus defined as a model for resources and their spatial, temporal, structural and semantic relationships within a particular environment. It is suitable for adopting various perspectives, including document-centric and process-centric views on preservation.

The Open Archival Information System (OAIS) [8] reference framework specifies the relevant terms and concepts for archiving digital information. The OAIS standard is established as the predominant, widely accepted conceptual frame in digital preservation. It comprises a functional model, an environment model, and an information model. While the OAIS reference model focuses on representation information (cf. Sect. 3), it only offers a limited characterization of (preservation) processes and their effects.

The scope of OAIS is limited by its archival-centric view, largely ignoring anything outside the archival environment. OAIS delegates full responsibility for what happens before a digital object enters the archive and for everything that happens after a digital object leaves the archive to abstract stereotypes denoted as Producers and Consumers. OAIS Producers and OAIS Consumers are defined by their relationship with the archive system as the role played by those who provide digital information or those who access digital information, respectively. However, more often than not, Digital Objects are not created for the purpose of archiving! In the domain of scientific publishing, for example, typical business comprises processes like abstract writing, reviewing, editing, indexing and publishing and involves actors like researchers, scientific associations and publishing houses. If context like reviewing criteria or review reports is not captured within these phases in the life cycle of a digital object, most of its imprint is lost for archival. Institutions with the aim to bring OAIS into application cannot ignore these facts and have to implement mechanisms to capture, store, and preserve relevant context information.

Therefore, a comprehensive context model is needed that can be smoothly aligned with the OAIS framework and that provides means for modeling the dimensions of context explicitly throughout the whole life time of digital objects.

1.3 Approach

Following the definition of the context of a digital object as *the representation of known properties associated with it and the operations that have been carried out on it* [27], this article proposes a generic model for representing the required context information.

Sect. 2 introduces the domain of application: document production, archival, access and reuse within digital archives and libraries. We will show and exemplarily illustrate that the rapid changes in digital publishing and archiving, the rising number of stakeholders and their complex interactions demand a detailed analysis of the processes affecting objects before archiving with respect to preservation and future reuse.

As a prerequisite to the development of the context model, Sect. 3 discusses necessary extensions to the OAIS reference model. At first, the OAIS environment model is complemented with an extended life cycle model that identifies and differentiates the phases which are relevant for capturing context, i.e. before a digital document enters an archive (Pre-Ingest) and after it leaves the archive (Post-Access). To accommodate our notion of representing context as a separate object, the OAIS information model will be extended by a specialized context information package.

In Sect. 4 we suggest the OAI-ORE model as a feasible starting point for representing complex objects. Exemplarily, we demonstrate the interoperability by mapping the popular packaging and exchange format METS to OAI-ORE. The object model establishes the basis for embedding complex digital objects within a sophisticated representation of context.

After setting the baselines above, we develop the details of our context model in Sect. 5. The proposed formalism has contact points to ontologies as well as business process models and is based on the ABC Ontology and Model. The context model itself is lightweight, but provides a yet powerful enough framework to express the concepts (actors, activities, events, states) required for adopting process-centric as well as data/resource-centric perspectives.

Sect. 6 illustrates the developed formalisms with a comprehensive example taken from the domain of scientific publishing. Lastly, Sect. 7 briefly summarizes the contributions of this article and discusses some of its main outcomes.

2   Scenario - Document Production, Archival, Access and Reuse

This section describes the scenarios for demonstrating the feasibility of our proposed context model. Starting with a brief characterization of archiving and preservation in libraries and archives, we more closely inspect the domain of scientific publishing, which is used as a example in Sect. 6.

2.1   Digital Archiving and Preservation within Libraries and Archives

Digital archiving and preservation is first of all pushed forwarded by depot libraries and digital archives which have the mission to provide their digital holdings for future reuse. As may be expected from a domain that has its roots in paper based publications, the most widespread digital object formats currently archived proved to be document-style formats containing text and images, with dedicated image, sound and video formats proving to be important for particular collections.

In the majority of the depot libraries and archives processes for digital archiving are already under development and partly operational in day to day work practices. Furthermore, these institutions have begun to make decisions on the preservation functionality they are intending to implement and the requirements the preservation tasks will have on collecting and preparing digital documents for archiving. For an overview of the current state of discussion in the National libraries see, e.g., [33] or [30].

Archiving and preservation processes are seen as generally embedded within the business processes of these organizations, such as acquisition, cataloging, and search and retrieval according to their objectives, obligations, and business models.

Most libraries started collecting and archiving digital material with the focus on text documents very similar to printed publications adopting the processes for indexing and cataloging they had at hand for printed material. So-called book-like digital publications are publications which are in content as well as in layout and structure very similar to printed books. The same holds true for journal-like digital publications. Both can be published electronically in parallel to a print version or as digital born publications without a print version. A typical example of book-like publication holdings are the collections of electronic thesis that are currently build at several university libraries as well as at the national libraries.

Besides cataloging data, administrative and - if remaining capturing mechanisms are realized - preservation metadata are generated and stored with the digital documents. For organizing digital content and metadata the Metadata Encoding and Transmission Standard (METS)[2] becomes more and more popular. An example of an online thesis and the relating METS file can be found in the appendix, the METS standard will be described in more detail in Sect. 4.2.

In addition to this primary group of digital object types a wide variety of other digital object types can be identified that libraries and archives have a remit to preserve in future, e.g., database files, emails and newsletters, computer game binaries, or web sites.

Digital publishing evolved very quickly beyond the scope of traditional publishing. The production, distribution and use of documents are being subjected to fundamental change. Digitization and interconnectivity fuel liberalization, in which new structures emerge that challenge traditional institutionalized structures with their well-defined, in-part exclusive roles and tasks. As to be described in the following scenario, besides publishing houses, research organizations and universities start to operate publication repositories and also individual authors make use of the Internet in order to bring their work to the public.

2.2 Scientific Publishing

Congresses serve the mediation of information and the establishing of contacts. Within a short period of time, the important heads of a field of science meet in order to present their work and ideas, to discuss new products, to determine trends, to socialize, and to initiate co-operation and collaboration. The scientific associations make efforts to create the ideal environment, possibly supported by a professional congress organizer. Special attention is paid to the assembling and provision of scientific contributions. For the conference, scientific contributions are accepted, indexed and reviewed, speakers get invited, a scientific program is set-up, categorized and linked thematically. Congresses are an integral part of work and discourse in the field of science.

Traditionally, printed abstract books document the scientific contributions of the congress. For some time now, congress organizers have been giving electronic copies of these abstract books to congress participants on CD-ROM that feature full-text search. Moreover, the scientific contributions are made available online in the congress web site. Embedded in the scientific program of the congress, the interactive presentation allows for structured access along date and time, type of presentation, topic, and presenter. Taking into account ongoing changes in the field of scientific publishing,

---

[2] http://www.loc.gov/standards/mets/mets-home.html

another publishing packaging and other publishing formats are feasible and are likely to be developed in the future. Such formats could be, for example, single abstract net publications.

Single conference editions scale up to 8000 participants, 2000 abstracts, and 500 presentations and beyond. The data collection incurred and managed in the course of one conference and/ or consecutive editions of one conference features heterogeneous material with metadata and multitude of relationships. Entities in the collection comprise, amongst others, abstracts, papers, posters, presentations, authors, sessions, and topics. Data and document collections comprise two general types: self-contained documents that can be considered complete and well-established (for example, presentation slides, posters, the printed conference abstract book), and the multitude of data, texts, images, and document parts gathered or produced in the course of the conference. This material includes, amongst others, organizational data including conference participants, presenters, events, sessions, talks, and topics, as well as structured text information from conference contributions, especially abstracts and their tables and figures.

Scientific contributions are characterized by the following features exceeding the regular scope of traditional publishing:

- A scientific contribution can be represented by several publications of different type, for example abstract, slides, poster or full paper.
- Each of these manifestations can be disseminated using different media such as printed proceedings, a conference web site, an interactive CD-ROM or a net publication.
- A scientific contribution can be embedded in the world of scientific discourse via several mechanisms as citation nets, interest and competence profiles of persons and organizations, and discussion threads.
- A scientific contribution holds direct and indirect connections to other documents like review reports and conference reports.

Spanning the whole process from production to archiving, this scenario focuses on the interactions of the stakeholders before and in the course of ingest processes and the organization of context capturing within these complex interactions. A special challenge will be to preserve persistent links between the different manifestations of the scientific publications to be archived.

This scenario aims for providing an experimental application scenario that takes into account the changing publication practices leading to complex publication and archiving processes with several stakeholders and diverse but highly interlinked outcomes. We chose scientific publishing and archiving in the context of scientific conferences as concrete instance of this scenario.

3   Contextualizing OAIS

With the Open Archival Information Systems (OAIS) Reference Model (ISO 14721, [8]) a conceptual framework for the standardization of systems for archiving information was provided[3]. It defines common terminology and comprises several models

---

[3] http://public.ccsds.org/publications/archive/650x0b1.pdf

which prescribe a minimal set of responsibilities required for the preservation of digital information. The *Functional Model* (see Sect. 3.1) distinguishes between seven separate functional entities and their related interfaces. The *Information Model* (see Sect. 3.4) recognizes three types of information packages, namely *Archival Information Package (AIP)*, *Submission Information Package (SIP)*, and *Dissemination Information Package (DIP)* which address the encapsulation of information objects within an OAIS. Information packages include content information composed from the data object and *Representation Information* and *Preservation Description Information (PDI)* and additional packaging information. With the Environment Model (see Sect. 3.2), OAIS differentiates between three roles: *Producers*, *Consumers* and *Management*. OAIS is intended to identify the necessary features of an archival information system rather than recommend any particular type of implementation.

3.1   OAIS Functional Model

The OAIS functional model aims for structuring the whole functionality provided within compliant preservation systems into functional entities and to specify the interfaces and the major information flow between them. Doing this, the functional model provides the internal structure of OAIS systems from a functional point of view, independent of concrete system architectures and implementations.

The entities defined within this model and their naming has become the most known and widely accepted part of OAIS definitions used in all discussions on digital preservation.

Beside various Common Services dedicated to basic functions concerning operation systems, network and security the OAIS Functional Model identifies the following six entities (for reference see [8]):

Ingest provides all functions to accept submission information packages from Producers[4], to ensure quality assurance of the received data, to enrich the representative information if needed, and to prepare the material for storage and management within the archive including the generation of archive information packages.

Archival Storage provides all functions for the storage, maintenance, and retrieval of archive information packages.

Data Management provides all functions for populating, maintaining, and accessing the archive database holding both descriptive metadata which identifies and documents archive holdings and administrative data used to manage the archive.

Administration provides all functions for the overall operation of the archive system including auditing and monitoring, maintaining hardware and software, and maintaining standards and policies.

Preservation Planning provides all functions to ensure that the information stored remains accessible over the long term, even if the original computing environment becomes obsolete.

Access provides all functions that support Consumers in determining the existence, description, location and availability of information stored and address consumers requests for information products.

---

[4] For the definition of the OAIS roles Producer and Consumer, see Sect. 3.2.

3.2 OAIS Environment Model

The OAIS Environment Model embeds OAIS systems into their overall environment by defining, on a very high level of abstraction, generic roles interacting with OAIS systems. OAIS identifies Producers, Consumers, and Management as these external roles [8]:

Producer is the role played by those persons, organizations, or systems, which provide the digital information to be preserved.

Consumer is the role played by those persons, organizations, or systems that address the OAIS systems to find and acquire preserved information.

Management is the role played by those persons or organizations who set the overall OAIS policies in the sense of Management's responsibilities, not in day-to-day archive operations.

A Designated Community is an identified group of potential consumers, which might also be composed of multiple user communities, who should be able to understand a particular set of information.

While there exists a correlation between the environmental management role and the administrative management of an archival, (producers and) consumers are outside the OAIS, only vaguely characterized by the Designated Community umbrella. This separation of concerns is adequate from a purely conceptual, archival perspective which has only the concern to draw - together with the functional model - the borderline of OAIS.

Nevertheless, the environment around OAIS affecting the digital holdings and their information context before ingest and after access is much more complex. The following section depicts that a further differentiation of this environment within a Life Cycle Model of digital archive holdings offers the chance to identify the relevant steps in the life time of objects feasible for capturing representative context information needed for long-term archiving and future reuse.

3.3 Information Life Cycle Phases

From the perspective of long-term preservation, Digital Objects undergo during their life-span the phases Creation, Assembling, Archiving, Adoption, and Reuse.

Creation is the initial phase during which new information comes into existence. More often than not, Digital Objects are not created for the purpose of archiving. Their shape aims at use within regular business models of the creator. Creation processes can be rather complex and involve a multitude of stakeholders until chunks of information result which are worth considering for archival. Use means the exploitation of information according to the original purpose the object was created for. Traditionally, objects are archived right after Creation. From the perspective of the Archive, Use is a concurrent thread in the life of a Digital Object that also starts with completion of Creation.

Assembly denotes the appraisal of objects relevant for archival and all processing and enrichment for compiling the complete information set to be sent into the future, meeting the presumed needs of the Designated Community. Assembly requires in-depth

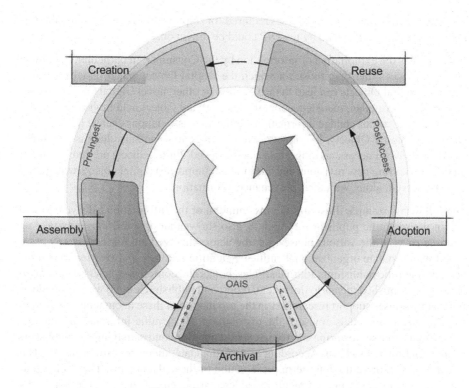

Fig. 1. Information Life Cycle Phases

knowledge about the Designated Community in order to determine objects relevant for long-term preservation together with information about the object required for identification of relevant archive objects and their Reuse some time later in the future. Assembly generates SIPs for ingest.

Archival addresses the life-time of the object inside the archive. During Archival, Digital Objects are managed in the form of AIPs. In most archives, policies prohibit irreversible deletion of content. Hence, preserving is a perpetual activity. The Archiving phase is open-ended - unless Digital Objects are to be irrevocably removed from the archive. As a matter of principal, information disseminated by the archive must enable the Designated Community to use that information.

Adoption encompasses all processes by which accessed Digital Objects, received as DIPs from the archive, are examined, adapted, and integrated to enable understanding and Reuse. It is not the mandate of the archive to accommodate arbitrary consumer scenarios and the corresponding requirements of their respective Designated Communities. Adoption makes efforts to re-contextualize Digital Objects and accompanying information for the respective Consumer Environment for effective exploitation within regular operational infrastructure and working processes. Altogether, the Adoption phase might be regarded as a mediation phase, comprising transformations, aggregations, contextualization, and other processes required for re-purposing data. Additional information beyond that provided by the DIP could be drawn on.

Reuse means the exploitation of information by the Consumer. In particular, reuse may be for purposes other than those for which the Digital Object was originally created. Reuse of Digital Objects can lead to the Creation of other, novel Digital Objects. Reuse also may add or update metadata about the base Digital Object held in the archive. For example, annotations change information content and associations change relationships with other Digital Objects. In collaborative working environments, there is a continuous flow between access and ingest, in that retrieved digital objects are reused and/or modified, yielding new revisions and additional (composite) digital objects which have to be preserved, along with their provenance information.

The following example illustrates the instantiation of the Information Lifecycle Model within the scientific publishing domain. It also shows that organizations may take up multiple roles in the various phases, i.e. the actors involved are not necessarily associated with separate organizational entities. Scientific findings (*Information*) are at the heart of scientific publishing. Journal articles, books and conference papers (*Digital Objects*) are established means to document, report and publish scientific works and ideas. Publishing houses support researchers in the production of these documents (*Creation*). Regular business models envisage the distribution of scientific information, enabling scientific discourse, fostering the researchers reputation, and satisfying the publishers commercial interests (*Use*). According to their mandate, memory institutions like National Libraries drive the long-term preservation of these documents. They either purchase that material or obtain it by universal legal arrangements like legal deposit or by agreements with single publishers. Being one single organization, memory institutions often act both as a library and as an archive, providing library services as well as preservation services. But this may change in the future. However, publishers never directly interact with the archive system. It is the memory institution who appraises and indexes the documents, collect and organize the necessary context information for preservation, and prepare the material for archiving (*Assembly*) before it is ingested into the archive and preserved from one generation to another (*Archival*), primarily for research, education, cultural and legal purposes (*Designated Community*). Also, memory institutions take care that Consumers can understand and reuse Digital Objects accessed from the archive. Provision systems offer to end-users interfaces for search and retrieval. They unpack, examine, adapt, transform, integrate and display archive packages (*Adoption*). If required, this processing also includes emulation or migration by access, respectively, as well as linkage to related information. Building on previous findings as one ultimate principle of science, leads to the generation of novel insights documented in other Digital Objects (*Reuse*).

3.4 OAIS Information Model

The *OAIS Information Model* describes the types of information that are exchanged and managed within the OAIS [8]. Information objects that are exchanged and managed within the OAIS are encapsulated as *Information Packages* [3]. Information packages include content information (the data object and *Representation Information*),

Preservation Description Information PDI, and additional packaging information, the following two definitions are derived from [8].

Representation Information accompanies a digital object, or sequence of bits, and is used to provide additional meaning. It typically maps the bits into commonly recognized data types such as character, integer, and real and into groups of these data types. It associates these with higher-level meanings that can have complex inter-relationships that are also described.

Preservation Description Information includes information that is necessary to adequately preserve the particular Content Information with which it is associated. It is specifically focused on describing history of the Content Information, ensuring it is uniquely identifiable, and ensuring it has not been altered.

OAIS specifies three representation information types, which are referred to as the *Structure Information, Semantic Information*, and a *Representation Network*. It further distinguishes between four different PDI types, namely *Reference Information* (assigned identifiers for the content information), *Context Information* (relationships with the environment), *Provenance Information* (history of the content information), and *Fixity Information* (data integrity checks).

The OAIS Information Model recognizes three types of information packages which get deployed at different stages of archiving. With *Submission Information Packages (SIP)*, Producers hand-over content for preservation to the archive. Internally, the archive manages content as well as additional data required for executing preservation in *Archival Information Packages (AIP)*. The archive responds with *Dissemination Information Packages (DIP)* to Consumer requests. These variant types differ in their mandatory content, the complexity of their internal structure, and their function. Especially, these information package variants differ even if applied to one and the same digital object, e.g., submissions might not contain all information necessary for preservation, whereas not all content information needs to be disseminated to consumers. The OAIS specifications for representation information and preservation description information are vague, providing only basic guidelines for the information types to be contained within an information package. The OAIS information model is document-centric, providing a logical model for archival information. Hence its purpose is to preserve information for a designated community over time by providing additional metadata which is attached to a content object. While cross-references between representation information are admissible, the resulting representation network is constrained by the containing information package.

3.5   Extended Information Model

To accommodate our notion of context, the OAIS information model has been extended, see Fig. 2. In order to avoid overloading, OAIS concepts remain unaltered. Information packages are *either* aggregated *traditionally* (red dotted fringe), *or* constitute (complex) context representations (blue dashed fringe). Links between layers or across information packages require persistent identification of the referenced objects, possibly delimited by logical namespaces. Adhering to the principle of separation of concerns, the Extended Information Model expresses context within a distinct, separate information

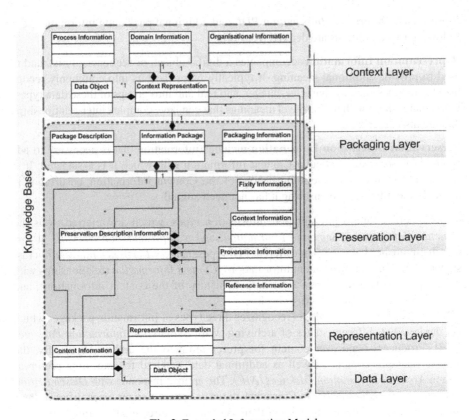

Fig. 2. Extended Information Model

package. Consequently, OAIS representation information and OAIS preservation description information constitute mandatory segments of an individual digital object's context, possibly complemented by references to one or more external shared context information packages.

The elements of the extended information model can clustered in five layers, where each layer corresponds to particular roles within the model. The *traditional* OAIS information package (red dotted fringe, bottom to top) is composed of the following layers:

The *Data Layer* is responsible for preserving the bitstream through time, whereas the *Representation Layer* describes the logical structure of the digital object being transported, but also the information necessary to decode its content. The *Preservation Layer* manages the properties related to the archival of a digital object within its preservation environment, i.e. it documents past and present states of the digital object being preserved, provides persistent identification of its content, and information about the integrity of its data. The *Packaging Layer* wraps the various data and metadata objects, together with corresponding manifests and packaging information. The external *Knowledge Base* expresses the intellectual assumptions which characterize a particular designated community.

The context information package (blue dashed fringe, top to bottom) has a slightly different structure: The *Context Layer* contains the representation of the context (and its serialization), which is further decomposed into three main constituents (see Sect. 5.2). It represents different a information type, which is wrapped similar to above as an information package in the *Packaging Layer*.

Conceptually, a context information package does not constitute a new type of information package, but follows the regular flow of SIP, AIP, and DIP. Context representation and digital objects are explicitly linked through logical namespaces, associating persistent state information at instance level with conceptual references describing the surrounding environment. However, context information packages do not need to maintain their own preservation description information. Their evolution is documented through the PDIs of the digital objects which are linked to them, i.e. provenance, reference and context information are updated with the unique identifier of the new (or revised) context information package.

The OAIS community knowledge base could in principle be used to store context information, but its manifestation, preservation and evolution are not formally specified (cf. [8]). OAIS does not prescribe the knowledge base as an information technology component, but a conceptual entity external to the preservation environment. Pragmatically, the knowledge base might be implemented, and accessible/usable for the preservation environment. If versioned properly (and provided with unique persistent identifiers), the externalization of context information as first class objects allows for modeling changes of concepts and terminology over time, characterizing production and (potential) reuse environments, and facilitates transfer to different communities by providing mappings of the underlying structured representations of concepts and relations.

The interaction policies between context representation and the community knowledge base need to be defined. In principle, the context layer operates at instance level, with the corresponding semantics and schema information being stored in the knowledge base (if implemented as information technology component) or in a separate context representation. From a preservation perspective, referential integrity and ontological commitment have to be maintained through ongoing evolutions. Alternatively, context information packages might also be self-contained, i.e. complete representation of schemas and instances at the cost of storage (and processing) overhead.

Within the extended information model, context constitutes a manageable segment of the world. It is represented partly within the information package wrapping the digital objects, as separate context information packages which might be shared across digital objects, and expressed through the background on production, archival and reuse provided by the community knowledge base.

3.6 Context Components

Production and Reuse Context corresponds to the producer and (anticipated) consumer environment, i.e. the respective designated communities creating and accessing digital objects. The creation environment includes the actors and resources involved, but also a formal representation of the organizational and technical processes carried out in the production of a digital object. To re-trace information paths the representation of the production context has to be maintained during the transition from the production into the preservation environment. Reuse depends on proper descriptions of the significant properties of preserved digital objects and the associated domain-specific knowledge, allowing for efficient access and usage even from outside the original creating community. The producer and consumer environment are not explicitly part of an OAIS, hence only fragments of those are preserved through representation information. In this sense, the context information package provides sophisticated models of the production and future reuse context, complementing the representation information of the objects being contextualized.

The scope of the *Preservation System Context* is to capture the way organizations address their issues related to information, processes and technology. From the preservation perspective, this implies the documentation of an implemented archival system with respect to the assumptions, principles, and rules defined by a corresponding reference architecture. The virtualization of processes for the explicit representation of policies, preservation processes, operations and their effects within logical name spaces is theoretically covered by preservation description information (PDI context information and PDI provenance information). Digital audit trails have to document service instantiations, including the particular revision of a service that has been invoked. Hence the evolution of an information system has to be tracked over time, unveiling limitations of the (current interpretation of) PDI. Correspondingly, the context information package (complementing/extending PDI) might be more adequate for explicitly representing the underlying enterprise architecture.

There is a fluent passage between *Production and Reuse* and *Preservation System*, since both notions of context can be expressed through concepts of an Enterprise Architecture Framework. Only the denoted purpose and perspective differs, whereas the context representation might exhibit and share similar structures and patterns on various levels of abstraction.

4   Object Model

An *Object Model* is an abstract specification of the structure of a digital object, its constituting components, and associated resources. In essence, object models provide platform-neutral and language-neutral interfaces for implementing digital objects across system boundaries. Therefore, the object model of choice provides the underlying framework to organize digital documents and the relating representation information throughout the processing of the objects in their life cycle.

A *Packaging Format* is a specific instantiation of an object model, e.g., through a *XML Serialization* and corresponding *URI*s (Uniform Resource Identifiers) needed for data exchange between the functional components of a preservation system. *Application Profiles* specify further application-dependent configurations and semantics.

The OAIS reference model does not specify any particular object model or packaging format for information (cf. [3]). Hence, various implementations have been proposed for serializing OAIS information packages. In the following we will describe briefly the relatively new OAI-ORE abstract object model and illustrate it with a simple example taken from the scenarios described in Sect. 2. Subsequently we discuss shortly the alignment between OAI-ORE and METS to demonstrate the compatibility of our approach with existing standards. The Metadata Encoding and Transmission Standard METS is widely used to describe digitized library material, but becomes more and more popular in the preservation community for born digital material.

4.1 OAI-ORE

The *OAI-ORE*[5] is a relative new development and build on the web architecture as defined by the W3C. It defines standards for the description and exchange of web resources, for the definition and identification of composed digital objects.

The OAI-ORE abstract data model specifies *Resource Maps*, which contain assertions about *Aggregations*, *Aggregated Resources* and optional additional properties, i.e. metadata about the resource map and aggregation. An aggregation is a set of other resources, whereas aggregated resources denote the constituents of an aggregation. Additionally, a *Proxy* might be used to contextualize aggregated resources, i.e. it asserts relationships which are specific to a particular aggregation.

Digital objects described as resource maps constitute connected RDF graphs, which are constrained by a number of structural restrictions[6]. The nodes of these RDF graphs are resources identified by URIs, and the edges are predicates expressing relationships between these resources.

Fig. 3 illustrates the formalism with a simple example taken from the book-like digital holdings described in Sect. 2.1. The representation depicted on the left represents a digital PhD thesis encoded in several files. The related METS file organizing the content files together with all remaining metadata can be found in the appendix. As it can be seen there, the structure of the content files is very simple, 32 files are organized in a flat hierarchy within one unit. Therefore, also in the OAI-ORE representation only one *Resource Map* with one *Aggregation* is needed. Metadata elements like persistent identifier, authors name, transfer checksum, and package creation date are assigned directly to the *Resource Map*.

Resource Maps can be expressed in different syntaxes to generate exchange data. OAI-ORE does not define a specific serialization for that but offers RDF/XML[7], RDFa[8] or ATOM XML[9].

4.2 Relating OAI-ORE to Metadata Schemata and Exchange Formats Like METS

The Metadata Encoding and Transmission Standard (METS)[10] is a container format for metadata and content files maintained by the Library of Congress (USA). METS aims for the management of objects within a repository as well as the exchange of objects

---

[5] Open Archives Initiative Object Reuse and Exchange [20], http://www.openarchives.org/ore/

[6] http://www.openarchives.org/ore/1.0/datamodel.html

[7] http://www.w3.org/TR/rdf-syntax-grammar/

[8] http://www.w3.org/TR/xhtml-rdfa-primer/

[9] http://www.atompub.org/

[10] http://www.loc.gov/standards/mets/mets-home.html

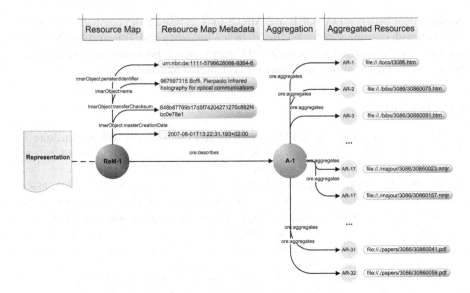

Fig. 3. Simplified ORE Example representing a digital phd thesis

between repositories. Therefore, METS is not a pure object model, but a standard to define complex hierarchically organized container structures containing content files as well as their remaining metadata usually serialized in XML.

A METS file for a digital object could contain the following sections:

Descriptive Metadata describes the content of the object.
Administrative Metadata holds data about the object itself including technical metadata, rights metadata, source metadata and digital provenance metadata.
File Section lists all the files of the object.
Structural Map defines the hierarchical structure of the resources listed in the File Section.
Structural Links record hyperlinks between items of the Structural Map.
Behavior Section associates executable behaviors with content in the METS object.

The internal structure of resources is described in the Structural Map that organizes all components in a strict hierarchical manner. In addition, the Structural Links provide the opportunity to connect elements independent of the hierarchical structure and, therefore, allows for representing general graph structures essential to capture the linking of, e.g., web content.

Currently, there are ongoing efforts to align OAI-ORE with the METS world pushed forward by members of both communities. [24] discusses issues that arise in mapping METS to OAI-ORE and makes some suggestions for best practices in METS implementation to ensure OAI-ORE compatibility. T. Habing and T. Cole present two approaches for describing OAI-ORE aggregations in METS and illustrate them by providing XSLT transformations for both approaches[11]. The first approach suggests a mapping in METS

---

[11] Currently available as preliminary draft only, http://ratri.grainger.uiuc.edu/oremets/

such that a specific METS structMap could be thought of as corresponding in essence to an OAI-ORE Resource Map. The second approach suggests a mapping where the METS document itself is an alternative serialization of the OAI-ORE Resource Map.

5 Context Representation

The context of a digital object to be preserved over time comprises the representation of all known properties associated with it and of all operations that have been carried out on it. This implies i) the information needed to decode the data stream and to restore the original content, ii) information about its creation environment, including the actors and resources involved, and iii) information about the organizational and technical processes associated with the production, preservation, access and reuse of the digital object. To retrace information paths for future reuse, a formal representation of all these aspects of context is a fundamental prerequisite.

In this section, we propose a generic model for representing the required context information, which specifies high-level concepts and activities for documenting the control flow (process-centric), and data/resource flow (document-centric) of production, working, and preservation processes. It provides a unifying framework for context-aware, knowledge-based process support using domain, enterprise and process ontologies. It integrates conceptual descriptions of process definitions with typed representations of the entities, roles and information objects involved. It supports the characterization of processes to support the explicit representation of policies applied, services and operations carried out and their effects within logical name spaces, with digital provenance trails documenting specific instantiations.

5.1 Contact Points

Our proposed context model has contact points to ontologies, enterprise modeling as well as business process models. Adopting an information technology perspective, this section established a brief thematic and terminological foundation for these topics, covering the relevant aspects which are necessary for the definition of the context model.

Ontologies - To improve ease of access, information has to be organized and made available for search and browsing. Hence, if it is connected in a meaningful way it becomes knowledge. *Ontologies* represent *concepts* and their *relations* to one another. Concepts can be associated with facts (*instances*) and in consequence, relations can be defined concept-concept, concept-instance and instance-instance. Ontologies are formal models of a specific domain (see, e.g., [16]), or, as defined by Gruber in [14]:

An ontology is a specification of a conceptualization.

Here conceptualization denotes an abstract, simplified view on the world which is represented to serve a particular purpose. Ontologies are used to establish a common understanding about knowledge existing within a domain. They describe a shared vocabulary and a set of logical propositions which define the semantics of the concepts and relations described (cf. [22]).

One important aspect of ontologies is that they formally express the semantics of each element contained, enabling Humans and machines alike to access and process the knowledge represented. Rules and inference (or reasoning) mechanisms can be employed to derive new insights, i.e. making so far implicitly existing knowledge explicit. Essentially, ontologies allow the knowledge sharing and communication, re-use, derivation and analysis of knowledge about a domain.

There are several standardized languages for ontologies, e.g., *RDF* (Resource Description Framework[12]) and *OWL* (Web Ontology Language[13]) represent more recent specifications from the *Semantic Web community*[14].

Enterprise Modeling - Enterprise Modeling [13] provides instruments for managing current and future operations and developments in organizations. Enterprise models address business issues as well as technological aspects, such as the improvement of communication between persons, increased interoperability of tools, a higher flexibility for reacting to changing circumstances and appropriate means for supporting the employees in performing their tasks. Often, enterprise models comprise a collection of integrated ontologies formalizing particular aspects of an organization (cf. [32], [12], [13]).

Business Process Models - A *business process* consists as a set of coordinated *activities* that are embedded into a net of *control-flow* constructs that define their order of execution. Activities represent the tasks and resources which are required to accomplish a certain business goal. They can either be atomic or aggregated from other activities as a semantic unit (see, e.g., [1], [23]). A business process is described by an (abstract) *process model*, a formal definition of its constituting activities, with distinguished start and end points, and constructs prescribing the flow of control (e.g., sequential or parallel execution, choice) between the activities. Each activity is characterized by its input and output parameters, called *preconditions* and *postconditions* respectively. Activities are performed by *actors*, which are Humans or other resources. A *business case* describes the concrete instance of a business process.

There exist a large variety of different representation formalisms for process models, some are graphically oriented, others are structured textual representations mostly based on XML[15]. Graphical notations are easier to understand for Humans (especially for non-experts), whereas XML-based representations can be processed directly (without serialization) by machines. Mappings between the various representation formats are possible, exploiting shared process concepts. But due to mismatches in the intended purposes such mappings are not always sound and complete, i.e. information is lost during the transformation. Generally, commercial interests undermine ongoing standardization efforts, resulting in often proprietary tool suites with limited interoperability support.

---

[12] http://www.w3.org/RDF/

[13] http://www.w3.org/TR/owl-ref/

[14] http://www.w3.org/2001/sw/

[15] eXtensible Markup Language, http://www.w3.org/XML/

5.2 Context Model

Our notion of context comprises an infrastructure-independent representation of the attributes associated with and (implied) relations between digital objects and collections. Nevertheless, context is not only defined by the digital objects themselves, but also by the processes in which they were created, preserved, accessed and reused. Domain-specific groundings provide interfaces to the relevant concepts and topics of the designated communities addressed, in addition to formalizations of the organizational structures involved, including associated role assignments. Our context model is based on previous work on the representation and management of knowledge-intensive processes (cf. *Extended Process Model* in [5], [6], [7]).

Fig. 4 illustrates a high-level overview of the context model, which is explained in more detail below.

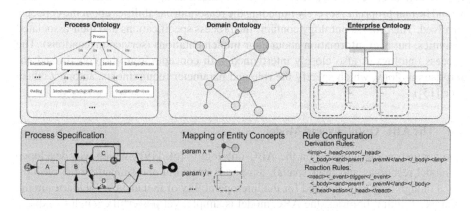

Fig. 4. High-Level Context Model

Process specifications are created by selecting appropriate activities from the process ontology and embedding them within a control-flow structure. Domain and enterprise ontologies are used to specify supporting entity concepts within logical namespaces, and appropriate rule configurations determine context management policies. The rule configurations are beyond the scope of this article and not discussed here.

Adopting a functional, knowledge-based perspective, processes are expressed as a sequence of recurring activities, which represent instantiation templates for concepts defined within the process ontology. A consistent terminology represents a fundamental requirement for the effective integration of the unified context representation, i.e. there are cross-references and inter-dependencies between (instantiated concepts of) the domain, enterprise and process ontologies.

Domain Ontology - The domain ontology defines the concepts and topics, but also their relations which are relevant for a particular application domain (designated community). For the domain of scientific publishing, for instance, possible concepts include *Abstract*, *Abstract Book*, *Presentation* or *Supplement* (additional data completing a document publication), but also the *Conference Topics* or *Review Criteria*. An accepted abstract might be included in the published abstract book (Abstract *is_Part_Of* Abstract Book), other relations between concepts within the domain ontology are admissible as well. Even though the superclass/subclass relationships might be dominant, there are other relation types spanning vertically.

Enterprise Ontology - The enterprise ontology models the structural layout of organizational environments, such as *Affiliations*, *Persons*, or *Roles* for describing a set of relevant concepts for scientific publishing. The enterprise ontology is neither strictly hierarchical nor completely self-contained, e.g., an *instance_Of* person *affiliated_With* a particular *instance_Of* institution might have the role (*has_Role*) of reviewer for a set of given abstracts, the latter are instantiated concepts from the domain ontology.

Process Ontology - The semantic classification of processes and activities as their building blocks requires their formal, hierarchical representation and description within an ontological structure (see [29]). A purely syntactical approach based on data types would be inappropriate, severely impeding the reasoning capabilities about the concepts involved. The process ontology contains the process specifications and their associated activities, but also information about their implementations (service invocations). The process ontology is also closely intertwined with concepts from the domain and enterprise ontology, which specify the role and parameter requirements, amongst others (see [15]).

5.3   The ABC Ontology and Model

The context model is based on the ABC ontology[16] (cf. [19]), which was developed to model resources and their spatial, temporal, structural and semantic relationships. Even though originally defined for the interoperability of metadata vocabularies in the context of digital libraries, the ABC model is simple, yet powerful enough to express the concepts (actors, activities, events, states) required for adopting a process-centric as well as a data/resource-centric perspective. Strongly influenced by the RDF(S)[17] data model (Resource Description Framework, http://www.w3c.org/RDF/), the concepts of the ABC Ontology which are most relevant for the context model are events, situations, actions, agents and manifestations. According to the specification in [19], the (abbreviated) definitions of the core classes are as follows:

- **Situation** - a situation is a context for making time-dependent or existential assertions about actualities
- **Event** - an event marks a transition between situations, one that is associated with the event through a precedes and another through a follows property
- **Actuality** - a primitive category for physically existing entities
- **Agent** - an actuality that is present during an event or is the party of some action.
- **Action** - an activity or verb performed by some agent or agents in the context of an event
- **Manifestation** - a form of an artifact that stands as the sensible realization of some work

---

[16] http://www.metadata.net/harmony/

[17] http://www.w3.org/RDF/, http://www.w3.org/TR/rdf-schema/

ABC classes are interlinked by property relationships, which may have restrictions with respect to their admissible source and destination classes (denoted *range* and *domain* in RDF(S)). The most relevant property relationships for the context model are:

- **precedes** - binds a situation within its context as existing before an event
- **follows** - binds a situation within its context as existing after an event
- **hasAction** - an event can have one or more actions
- **hasParticipant** - associates an agent as an active participant in an event or action
- **hasPatient** - the actuality involved is transformed by the action or event
- **creates** - some actuality is coming into existence as the result of an event or action
- **involves** - expresses the involvement of an actuality in the performance of an action or an event

The next section describes the specific interpretation of the ABC classes and property relationships for modeling context.

5.4   Classes and Relations

The classes and relations of the ABC model are used to model the context of production, preservation, and working processes. They are used to embed the activity concepts from the process ontology within a formal control-flow structure and parameterize the resulting process specification. The ABC property relationships are used to provide the references to concepts from the domain and enterprise ontology. Hence, the semantics of processes and activities is defined on a conceptual level, i.e. it is generic and independent of a specific environment.

Within the context model, the operational semantics of the ABC classes specifying control-flow is formed by the concepts from business process management (see, e.g. [2], [23], [28]). The derived interpretation for these classes is described as follows:

- **Situation** - describes a particular state within a process which is reached after a particular sequence of activities.
- **Event** - defines the transition from one state to another, i.e. it corresponds to (the effect of) an activity
- **Action** - referring to the components of an activity
- **Agent** - corresponds to the actor role, i.e. the resource (e.g., human, web service) performing an activity
- **Manifestation** - models input and output parameters of processes and activities, which correspond to different resource concepts

Furthermore, the semantics of ABC property relationships within the context model are defined as:

- **precedes, follows** - order activities and states
- **hasAction** - elaborates on the structure of activities, if necessary
- **hasParticipant** - assigns an actor to an activity or its subcomponents
- **hasPatient** - associates resources with activities

- **creates** - specifies the resources which are created during the execution of an activity
- **involves** - expresses the involvement of a resource in an activity

To sum up, the context model is composed of *events* and *situations*, which are structured by *precedes* and *follows* property relationships. The associated actors and other resources are modeled as *agents* and *manifestations* respectively, connected by the *hasParticipant*, *hasPatient* and *creates* relations. Activities might be further decomposed by *action/hasAction*. In the next paragraphs, we will apply the context model to the scenario of scientific publishing. Please note that the usage of ABC classes and property relationships in the following examples is not exhaustive, but sufficient for the representation for illustrating the overall approach.

5.5   Related Work

The context model in this article is based on the ABC Ontology and Model [19]. Other representation models might also be used to express the provenance of digital objects, such as extensions to CIDOC CRM in [31], the Open Provenance Model, or the PREMIS Data Dictionary for Preservation Metadata.

The *CIDOC Conceptual Reference Model (CRM)* is an extensible semantic framework describing concepts and relationships in the domain of cultural heritage [10]. Its primary goal is to enable information exchange and integration (semantic interoperability) between heterogeneous sources of cultural heritage information. CRM is a formal ontology which declares 90 classes and 148 properties, which are derived from the underlying semantics of database schemas and document structures found in typical cultural heritage and museum documentation. Since 2006 CIDOC CRM is official ISO standard 21127:2006.

The *Open Provenance Model (OPM)* has been designed as a format and associated semantics for the interchange of provenance information [25]. It defines three primary entities (Artifact, Process, Agent) and five dependencies (used, wasGeneratedBy, wasTriggeredBy, wasControlledBy, wasDerivedFrom) between these entities. The resulting provenance graph captures causal dependencies between entities, some of which (used, wasGeneratedBy, wasControlledBy) can also be annotated with differentiating roles. The formal model underlying OPM can be exploited to draw inferences about processes.

The *PREservation Metadata: Implementation Strategies (PREMIS)* Data Dictionary defines a (core) set of semantic units, which are mapped (as properties) to the entities of the Data Model (Rights, Objects, Agents, Events, Intellectual Entities) [9]. Some semantic units are defined as containers, grouping related semantic units. Relationships express associations between instances of entities. PREMIS semantic units have a direct mapping to the metadata elements defined in the PREMIS XML Schema[18].

As noted in [26], the ambiguity and complexity of CIDOC CRM might yield inconsistent models, rendering practical usage (e.g., the formulation of queries) without prior harmonization difficult. There has been a mediation process to harmonize the ABC and CIDOC CRM ontologies [11], some results of which have been integrated in later

---

[18] http://www.loc.gov/standards/premis/premis.xsd

versions of the CIDOC CRM. PREMIS focuses on preservation functions within digital repositories, but does not provide support for anything outside the archival phase. Uptake of OPM currently appears to be limited to so-called provenance challenges[19], which are used to evaluate and further refine the specification. While there might be slight incongruencies at the conceptual level, general mappings/transformations (of admissible subsets) from one representation model into another can be performed, possibly with some loss in expressiveness.

6   Example from Scientific Publishing

This section takes up the models presented in Sect. 4 and Sect. 5 and illustrates their main concepts based on a comprehensive example from the domain of scientific publishing (see Sect. 2.2). We first present the control-flow of a scientific publishing process and then go into more details regarding the actors, resources and information objects involved. Finally, we illustrate the OAI-ORE object model based on the simplified example of an abstract book.

6.1   Control-Flow

The following example depicts the simplified control-flow of a publishing process, inspired by the organization of large scientific congresses. Fig. 5 shows the control-flow in full. The circles correspond to *Situations*, with *Start* and *End* respectively indicating the start state and end state of the process. The rounded rectangles denote activities (*Events*), which are connected to situations by *precedes* (solid line) and *follows* (dashed line) relations.

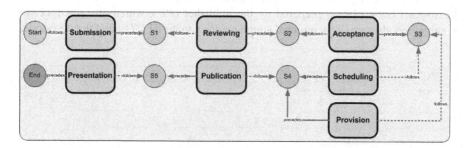

Fig. 5. Simplified Scientific Publishing Process - Control Flow

The process starts with the *Submission* of an abstract to the congress organizer, which then has to undergo a peer-reviewing process (*Reviewing*). In case the submitted abstract is accepted (*Acceptance*), its presentation during the congress is scheduled (*Scheduling*). In parallel, the *Provision* of the relevant documents for the accepted contribution is coordinated, possibly including technical quality control. Then the abstract is published as part of an abstract book (*Publishing*), and finally the scientific contribution is presented during the congress (*Presentation*).

---

[19] http://twiki.ipaw.info/bin/view/Challenge/WebHome

6.2 Resource/Data-Flow

From the perspective of resource and data flow, the publishing process is a bit more complex, as illustrated in Fig. 6. Here, activities (*Events*) are again depicted by rounded rectangles, but the control-flow structure is omitted. References to the domain ontology (*Topics, Reviewing Criteria*) are illustrated by the symbols in the upper left corner, whereas a reference to the enterprise ontology (Speakers, Moderators) is indicated by the symbol on the middle right-hand side. *Agents* are marked by a yellow circle, and *Manifestations* are denoted by the green shape. Relations to existing *Actualities* (here: actors and manifestations) are marked by a solid line (*hasParticipant, involves, hasPatient*) and new *Actualities* (here: manifestations) are marked by a dotted *Create* line. The abbreviations for *Actors* and *Manfestations* are explained in the two legend boxes in the lower-left corner.

Each abstract (*AB*) is submitted by at least one author (*AU*) and is categorized according to a set of predefined *Topics*. Additional supplementary material (*SU*) might also accompany the submission. During submission, the author has to provide descriptive metadata (*DM*) about her contribution. Each submission is then evaluated by one or more reviewers (*RV*), according to the *Reviewing Criteria* provided by the congress organizer. The organizer (*OR*) consults the review report (*RE*) in order to take a decision (*RU*) whether the submission should be accepted or not. The organizer creates the schedule for the scientific program (*SP*) based on the list of accepted submission. The information published in the scientific program (*SP*) is based on the descriptive metadata (*DM*) provided by the author (*AU*). Additionally, the organizer (*OR*) specifies a list of (invited) speakers and moderators (*Speaker, Moderators*) for the presentation sessions.

The author (*AU*) provides the demonstration material (*PP*) for her contribution. The abstract book (*AB*) is prepared by the publisher (*PU*), which also has received the submitted abstract (*AB*), optional supplementary material (*SU*), and the descriptive

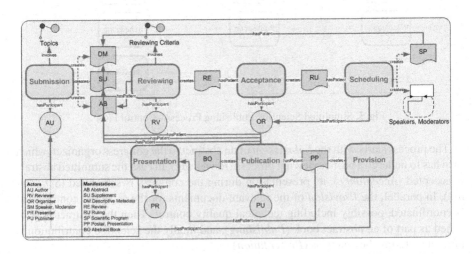

Fig. 6. Simplified Scientific Publishing Process - Resource and Data Flow

metadata (*DM*). During the scientific congress, the presenter (*PR*) uses the demonstration material (*PP*) to give her talk. Note that presenter and author roles coincide in many cases.

6.3   Object Model

Similar to the example in Sect. 4.1, complex objects in scientific publishing can also be expressed as an OAI-ORE aggregation. Fig. 7 illustrates a possible structure of the abstract book (*BO*) manifestation from above.

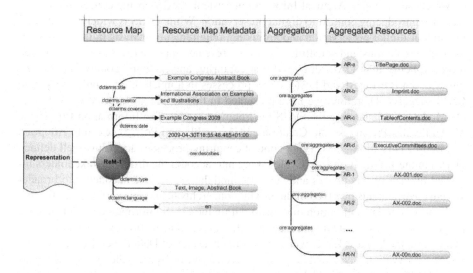

Fig. 7. Scientific Publishing Process - Example Object Model

Here, the abstract book for *Example Congress 2009* comprises an imprint, which might list the editor-in-chief and the editorial assistants, assistant managing directors, associate editors, reviews editors, contributors, editorial board, ethics committee and international advisory board. The imprint is followed by the table of contents, and another listing of the executive committees of the organizer association and associated organizations. The accepted abstracts constitute the remaining aggregated resources (*AG-1...AG-N*). The abstracts are modeled as simple files which are identified by their local filenames (which should be translated into URIs). In real-life, each abstract might as well be complex, comprising various text sections (such as objectives, methods, results, and conclusions) and be complemented with additional supplements (e.g., tables, diagrams, photos). Like the abstract book aggregation, the abstracts themselves might also be provided with descriptive metadata (omitted here).

7   Conclusions

In this article we have discussed the notion of context in digital preservation. We have identified relevant components of context and the interrelationship between the different aspects of context and we have proposed a generic model to represent the required context information.

This article has presented an approach that defines the context of a digital object as *the representation of known properties associated with it and the operations that have been carried out on it* [27]. As stated in the introduction, implied aspects of context are i) the information needed to decode the data stream and to restore the original content, ii) information about its creation environment, including the actors and resources involved, and iii) information about the organizational and technical processes associated with the production, preservation, access and reuse of the digital object.

We chose the Open Archival Information System (OAIS) as the conceptual framework for introducing context to digital preservation. While OAIS is widely accepted in the area of digital preservation and archiving, requirements analysis in the domain of memory institutions and scientific publishing revealed that context has to be modeled more explicitly than OAIS does and that capturing and handling context has to take into account Pre-Ingest and Post-Access processes in a much more detailed way than OAIS suggests. Therefore, we have presented in this article an Information Life Cycle Model complementing the OAIS environment model. The Information Life Cycle Model divides Pre-Ingest into Creation and Assembly and Post-Access into Adoption and Reuse. This allows classifying context capturing processes according well defined phases in the Information Life Cycle taking into account the different stakeholders involved in each of this phases and supports modular solutions embedded into broader system environments involved in the production, archival, and reuse of digital objects.

To accommodate our notion of context within the OAIS information model we have proposed to express context within a distinct, separate information package. Novel elements form the Context Layer which builds upon a stack of Data Layer, Representation Layer, Preservation Layer, and Packaging Layer, as we interpreted the existing elements of the OAIS information model.

The Information Life Cycle as well as the Extended Information Model extend OAIS concerning the needs of digital preservation within memory institutions and scientific publishing, but for compliance, all basic OAIS concepts remained unaltered. The extensions were aligned carefully with the OAIS functional model, the OAIS environment model, and the OAIS information model.

After having discussed adequate object and packaging formats, we proposed OAI-ORE as a feasible starting point for a flexible and generic abstract object model which is capable to represent the internal structure of any kind of digital material, content data as well as context information. We have shown that OAI-ORE representations can be aligned with popular packaging and exchange formats like METS (Metadata Encoding and Transmission Standard) to ensure compatibility with existing metadata schema.

Based on these prerequisites we have detailed the formalism of our context model. The proposed formalism has contact points to ontologies as well as business process models and is based on the ABC Ontology and Model. Itself slight, ABC provides a yet powerful framework to express the concepts (actors, activities, events, states) required for adopting a process-centric as well as a data/resource-centric perspective for context modeling.

With exemplary illustrations of our formalisms by means of two scenarios in the domain of document production, archival, access and reuse within digital archives and libraries we could show that the context comprising stakeholders, resources, and their complex interactions before archiving adds eminent information to emerging objects with respect to preservation. This context information modeled in an adequate manner and preserved together with the content data is an essential prerequisite for future access, understanding and reuse of data.

Acknowledgements. The research described is related to and partly funded by the European Commission's 7th Framework Programme in the integrated project *SHAMAN* (*Sustaining Heritage Access through Multivalent ArchiviNg*)[20]. SHAMAN targets a framework integrating advances in the data grid, digital library, and persistent archives communities in order to attain a long-term preservation environment which may be used to manage the storage, access, presentation, and manipulation of potentially any digital object over time. Based on this framework, the project provides application-oriented solutions across a range of sectors, including those of digital libraries and archives, engineering, and scientific repositories.

References

1. van der Aalst, W., van Hee, K.: Workflow management: models, methods, and systems. MIT Press, Cambridge (2002)
2. van der Aalst, W.M.P., ter Hofstede, A.H.M., Weske, M.: Business process management: A survey. In: van der Aalst, W.M.P., ter Hofstede, A.H.M., Weske, M. (eds.) BPM 2003. LNCS, vol. 2678, pp. 1–12. Springer, Heidelberg (2003)
3. Ball, A.: Briefing paper: the oais reference model. Briefing paper, UKOLN, University of Bath (2006), http://www.ukoln.ac.uk/projects/grand-challenge/papers/oaisBriefing.pdf
4. Bolchini, C., Curino, C.A., Quintarelli, E., Schreiber, F.A., Tanca, L.: A data-oriented survey of context models. SIGMOD Rec. 36(4), 19–26 (2007), http://www.sigmod.org/sigmod/record/issues/0712/p19.cesar-curino.pdf
5. Brocks, H., Begger, C., Kamps, T.: Process ontologies for IT-Services. In: Zimmer, F., Mohammadian, M., Khadraoui, D. (eds.) Proceedings of the international conference on Advances in Intelligent Systems - Theory and Applications, AISTA 2004 (2004)
6. Brocks, H., Meyer, H., Kamps, T., Begger, C.: The extended process model - unifying knowledge and process. In: Abramowicz, W. (ed.) Proceedings Business Information Systems, 8th International Conference on Business Information Systems, pp. 242–257. Wydawnictwo Akademii Ekonomicznej w Poznaniu, Poznan (2005)
7. Brocks, H., Meyer, H., Kamps, T., Begger, C.: The extended process model - transforming process specifications into ontological representations. Cybernetics and Systems 37(2-3), 261–281 (2006)
8. CCSDS: Reference model for an open archival information system (oais). Blue Book 1, Consultative Committee for Space Data Systems (2002), http://public.ccsds.org/publications/archive/650x0b1.pdf, Recommendation for Space Data Systems Standards, adopted as ISO 14721:2003

---

[20] http://www.shaman-ip.eu

9. Committee, P.E.: Premis data dictionary for preservation metadata version 2.0. Tech. rep., Library of Congress (2008), http://www.loc.gov/standards/premis/v2/premis-2-0.pdf

10. Crofts, N., Doerr, M., Gill, T., Stead, S., Stiff, M.: Definition of the cidoc conceptual reference model. Tech. rep., ICOM/CIDOC CRM Special Interests Group (2009), http://cidoc.ics.forth.gr/docs/cidoc_crm_version_5.0.1_Mar09.pdf, ISO 21127:2006

11. Doerr, M., Hunter, J., Lagozw, C.: Towards a core ontology for information integration. Journal of Digital Information 4(1) (2003), http://www.itee.uq.edu.au/~eresearch/papers/2003/JODI/_Oct2002.pdf

12. Fox, M., Barbuceanu, M., Gruninger, M., Lin, J.: An organization ontology for enterprise modelling. In: Prietula, M., Carley, K. Gasser, L. (eds.) Simulating Organizations: Computational Models of Institutions and Groups, pp. 131–152 (1997)

13. Fox, M., Gruninger, M.: Enterprise modelling. AI Magazine, 109–121 (1998)

14. Gruber, T.R.: A translation approach to portable ontology specifications. Knowl. Acquis. 5(2), 199–220 (1993)

15. Grüninger, M., Pinto, J.: A theory of complex actions for enterprise modelling. In: Symposium Extending Theories of Action: Formal Theory and Practical Applications, pp. 94–99 (1995)

16. Guarino, N.: Formal ontology and information systems. In: Guarino, N. (ed.) Proceedings of the 1st International Conference on Formal Ontologies in Information Systems, FOIS 1998, pp. 3–15 (1998)

17. JISC: Digital preservation - continued access to authentic digital assets. Briefing paper, Joint Information Systems Committee (2006), http://www.jisc.ac.uk/media/documents/publications/digitalpreservationbp.pdf

18. Kaenampornpan, M., O'Neill, E., Ay, B.B.: An integrated context model: Bringing activity to context. In: Workshop on Advanced Context Modelling, Reasoning and Management - UbiComp (2004), http://www.itee.uq.edu.au/~pace/cw2004/Paper17.pdf

19. Lagoze, C., Hunter, J.: The ABC Ontology and Model. Journal of Digital Information 2(2) (Article No. 77, 2001-11-06) (2001)

20. Lagoze, C., de Sompel, H.V., Nelson, M.L., Warner, S., Sanderson, R., Johnston, P.: Object re-use & exchange: A resource-centric approach. CoRR abs/0804.2273 (2008)

21. Lavoie, B., Dempsey, L.: Thirteen ways of looking at..digital preservation. D-Lib Magazine 10(7/8) (2004), http://dlib.org/dlib/july04/lavoie/07lavoie.html

22. Mädche, A., Staab, S., Studer, R.: Stichwort ontologien. Wirtschaftsinformatik, WI-Schlagwort 4 (2001)

23. Marinescu, D.C.: Internet-Based Workflow Management: Toward a Semantic Web. Wiley Interscience, Hoboken (2002)

24. McDonough, J.: Aligning METS with the OAI-ORE Data Model. In: Proceedings 9the ACM/IEEE-CS Joint Conference on Digital Libraries, ACM, New York (2009) (to be published), https://www.ideals.uiuc.edu/handle/2142/10744

25. Moreau, L., Plale, B., Miles, S., Goble, C., Missier, P., Barga, R., Simmhan, Y., Futrelle, J., McGrath, R.E., Myers, J., Paulson, P., Bowers, S., Ludaescher, B., Kwasnikowska, N., den Bussche, J.V., Ellkvist, T., Freire, L., Groth, P.: The open provenance model (v1.01). Tech. rep., University of Southampton (2008), http://eprints.ecs.soton.ac.uk/16148/1/opm-v1.01.pdf

26. Nussbaumer, P., Haslhofer, B.: Putting the cidoc crm into practice. experiences and challenges. Tech. rep., University of Vienna (2007), http://www.cs.univie.ac.at/upload/550/papers/putting_the_cidoc_crm_into_practice.pdf

27. Partners, D.: DPE digital preservation research roadmap. Public Deliverable D7.2, Dig-italPreservationEurope (2007), http://www.digitalpreservationeurope.eu/ publications/reports/dpe_research_roadmap_D72.pdf
28. Reijers, H.A.: Design and Control of Workflow Processes: Business Process Management for the Service Industry. In: Reijers, H.A. (ed.) Design and Control of Workflow Processes. LNCS, vol. 2617. Springer, Heidelberg (2003)
29. Schwarz, S.: Task-Konzepte: Struktur und Semantik für Workflows. In: Reimer, U., Abecker, A., Staab, S., Stumme, G. (eds.) WM 2003: Professionelles Wissesmanagement - Erfahrungen. GI (2003)
30. Schwens, U., Liegmann, H.: Langzeitarchivierung digitaler Ressourcen. In: Kuhlen, R., Seeger, T., Strauch, D. (eds.) Handbuch zur Einführung in die Informationswissenschaft und -praxis. Grundlagen der praktischen Information und Dokumentation, ch. D9, vol. 1, 5, völlig neu gefasste Ausgabe edn, pp. 567–570. Saur, München (2004), http:// nbn-resolving.de/urn:nbn:de:0008-2005110800
31. Theodoridou, M., Tzitzikas, Y., Doerr, M., Marketakis, Y., Melessanakis, V.: Modeling and querying provenance using cidoc crm. Tech. rep., Institute of Computer Science, FORTH-ICS (2008), http://www.casparpreserves.eu/Members/metaware/ Papers/modeling-and-querying-provenance-using-cidoc-crm/at_ download/file Draft 0.94
32. Uschold, M., King, M., Moralee, S., Zorgios, Y.: The Enterprise Ontology. The Knowledge Engineering Review 13 (1998); Special Issue on Putting Ontologies to Use
33. Verheul, I.: Networking for Digital Preservation: Current Practice in 15 National Libraries. K.G. Saur, Munich (2006)

Appendix

The METS file below describes a typical example of a digital thesis taken from the holdings of the German National Library. This file captures the internal structure of the content files, in this case a flat hierarchy of 32 files listed in the *Structural Map*. The *amdSec* (Administrative Metadata) contains administrative and preservation metadata serialized using the LMER schema[21]. This includes the persistent identifier of the object.

The METS file presented was generated with the KoLibRi[22] tool, which is part of the digital preservation infrastructure of the German National Library developed within the Kopal[23] project. It constitutes together with the content files the submission information package to be ingested into the long-term archive.

```xml
<?xml version="1.0" encoding="UTF-8"?>
<mets OBJID="" TYPE="kopal SIP" PROFILE="DNB" xsi:schemaLocation="http://www.loc.gov/METS/>
  <metsHdr CREATEDATE="2007-08-01T13:22:31.192+02:00" RECORDSTATUS="TEST">
    <agent ROLE="ARCHIVIST" TYPE="ORGANIZATION">
      <name>Deutsche Nationalbibliothek</name>
    </agent>
  </metsHdr>
  <amdSec ID="AmdSec-0001">
    <techMD ID="TechMD-LMER-Object">
      <mdWrap ID="TechMD-LMER-Object-MdWrap" MDTYPE="OTHER" OTHERMDTYPE="lmerObject" LABEL="LMERobject">
        <xmlData>
          <lmerObject:name>Boffi, Pierpaolo Infrared holography for optical communications</lmerObject:name>
          <lmerObject:persistentIdentifier>urn:nbn:de:1111-5796628066-9364-6</lmerObject:persistentIdentifier>
          <lmerObject:transferChecksum>648b87769b17d5f74204271270c882f4bc0e78e1</lmerObject:transferChecksum>
          <lmerObject:masterCreationDate>2007-08-01T13:22:31.193+02:00</lmerObject:masterCreationDate>
          <lmerObject:metadataCreationDate>2007-08-01T13:22:31.193+02:00</lmerObject:metadataCreationDate>
          <lmerObject:metadataRecordCreator>KOPAL DIAS</lmerObject:metadataRecordCreator>
          <lmerObject:startFile>./tocs/t3086.htm</lmerObject:startFile>
          <lmerObject:numberOfFiles>33</lmerObject:numberOfFiles>
        </xmlData>
      </mdWrap>
    </techMD>
  </amdSec>
  ...
  <fileSec>
    <fileGrp ID="ASSET" ADMID="TechMD-LMER-Object">
      <file ID="FILE-0"><FLocat LOCTYPE="URL" xlink:href="file://./tocs/t3086.htm"/></file>
      <file ID="FILE-1"><FLocat LOCTYPE="URL" xlink:href="file://./bibs/3086/30860075.htm"/></file>
      <file ID="FILE-2"><FLocat LOCTYPE="URL" xlink:href="file://./bibs/3086/30860091.htm"/></file>
      ...
      <file ID="FILE-17"><FLocat LOCTYPE="URL" xlink:href="file://./majour/3086/30860023.nmjr"/></file>
      <file ID="FILE-18"><FLocat LOCTYPE="URL" xlink:href="file://./majour/3086/30860157.nmjr"/></file>
      ...
      <file ID="FILE-31"><FLocat LOCTYPE="URL" xlink:href="file://./papers/3086/30860041.pdf"/></file>
      <file ID="FILE-32"><FLocat LOCTYPE="URL" xlink:href="file://./papers/3086/30860059.pdf"/></file>
    </fileGrp>
  </fileSec>
  <structMap TYPE="ASSET">
    <div ORDER="1" LABEL="File list" TYPE="ASSET">
      <fptr FILEID="FILE-0"/>
      <fptr FILEID="FILE-1"/>
      <fptr FILEID="FILE-2"/>
      ...
      <fptr FILEID="FILE-17"/>
      <fptr FILEID="FILE-18"/>
      ...
      <fptr FILEID="FILE-31"/>
      <fptr FILEID="FILE-32"/>
    </div>
  </structMap>
</mets>
```

---

[21] http://www.d-nb.de/eng/standards/lmer/lmer.htm

[22] http://kopal.langzeitarchivierung.de/index_koLibRI.php.en

[23] http://kopal.langzeitarchivierung.de/index.php.en

UML2SQL—A Tool for Model-Driven Development of Data Access Layer

Leszek Siwik, Krzysztof Lewandowski, Adam Woś,
Rafał Dreżewski, and Marek Kisiel-Dorohinicki

Department of Computer Science
AGH University of Science and Technology, Kraków, Poland
{siwik,doroh,drezew}@agh.edu.pl

Abstract. The article is a condensed journey over UML2SQL: a tool for model-driven development of data access layer. UML2SQL includes an object query language and allows for behavior modeling based on UML activity diagrams, effectively linking structural and behavioral aspects of the system development. From the general idea of UML2SQL and its origins, we go through the details of its architecture and beyond the processes and schemes which make UML2SQL a distinct tool in the data access domain. Finally, an example of developing an application using UML2SQL is given as an illustration of its practical usage.

1 Introduction

The realm of Model-Driven Development (MDD) or, in general, Model-Driven Architecture (MDA) introduces the idea of a complete model-based application generator. Such a concept seems really attractive: it would allow concentrating only on an application's design, driving it to perfection. The process of coding and testing the application would be as simple as clicking a button and getting through some wizard dialogs. It is obviously a simplification, but indeed Model-Driven Development can be seen in such a light.

Unfortunately the research on existing projects leads to the conclusion that actually there is no fully functional MDD tool to facilitate application development in the domain of data access layer. Nevertheless, there are some CASE solutions which provide their users with various benefits of MDD. Mainly, these tools let one design and develop applications of the real-time genre (examples include Telelogic Rhapsody [21] and IBM Rational Rose [10]). There are also tools like LLBLGen Pro [12] or OrindaBuild [15], which let users generate code to support data access. But none of those tools can be regarded as really MDD ones, since the connection between application model and its development in general is rather scarce, if any.

The aim of UML2SQL is to fulfil the need to have a tool for data layer model-driven development. The ambition of being an MDD tool is only one aspect of UML2SQL—the other, equally important, is to provide a possibility to use only one integrated application model during all system engineering phases.

The paper is organized as follows. First, in section 2 a short discussion of existing MDD tools is given. Then, section 3 presents basic concepts of UML2SQL, and section 4—selected implementation details of the tool. In section 5 the process of

E. Szczerbicki & N.T. Nguyen (Eds.): Smart Infor. & Knowledge Management, SCI 260, pp. 227–246.
springerlink.com

modeling of a simple application is presented, which serves as a proof-of-concept of UML2SQL.

2 A Glance at Existing MDD Tools

There are many examples of tools that try to make use of Model-Driven Development ideas and approaches, ranging from simple code generation extensions in UML-based graphical tools, to advanced modeling, development, simulation, and testing platforms.

The first complete tools for MDD were introduced in the field of real-time systems, with the most widely known example of Telelogic Rhapsody. Rhapsody is targeted at embedded and real time systems, with its goal set to provide a fully generated application code. It is able to construct all necessary artifacts necessary to build the system from a platform-independent model by targeting C/C++, Java or Ada languages. Rhapsody uses, known since the 70s, the Shlaer-Mellor method [32], which utilizes state machines to model behavior. By creating UML state diagrams with transitions and states annotated with conditions and activities, a complex behavior of a system can be described. The ability of adding a custom code to all elements of the diagram, which is later included in generated code skeletons, allows for specifying a complete system in UML only.

The generation of a complete application code from an UML model was not unique for the real-time systems' domain for a long time. Nowadays many tools allow for creating code from a UML model, and the extent and range of this code depends on the tool in question. Basic functionality of generating high-level programming language skeletons of classes (packages, modules etc.) is currently present in almost all UML modeling tools. However, they can not be perceived as Model-Driven Development tools merely because of the fact that they are, after all, general purpose modeling tools with simple option allowing for transformation of UML model into its another representation—a programming language. Only the presence of integrated and embedded behavior modeling mechanisms means that a specific tool is actually a Model-Driven Development one.

There are numerous examples of tools that facilitate behavior modeling using UML models, and the most widely known are IBM Rational Rose, Blu Age or AndroMDA. All of them, apart from allowing for modeling complete logical models, support modeling business logic of an application. This is achieved by making the use of many Shlaer-Mellor method variations or by using activity and sequence diagrams. For instance, the Blu Age documentation states that the modeling is done with the use of three groups of components [5]:

- class diagrams integrate the business objects notions and the simple and complex business rules, such as the entities,
- services' modeling, the same as the controllers', is made by the means of activity and sequence diagrams,
- the roles and authorizations are expressed by means of use cases.

All these tools also offer the ability to insert user-defined code into the generated code skeletons, thus ensuring that a complete, working application can be generated solely from the model. The option of reverse engineering is also available, making them even more productive.

Table 1. Selected features of the most important MDD tools and platforms

Tool / Platform	Orientation	UML Oriented	Structural modelling	Structural modelling scope	Behavior modelling	Behavior modelling approach	Add custom code during modelling
Telelogic Rhapsody [16]	real-time systems	yes	yes	complete logical model	yes	Shlaer-Mellor state machines	yes
AndroMDA [1]	MDA framework	yes	yes (via cartridges)	depending on the cartridge	yes (via cartridges)	N/A	N/A
AndroMDA BPM4Struts [2]	business process	yes	yes	complete logical model	yes	activity diagrams	yes
OpenArchitecture Ware [14]	MDA framework	yes (integrated via Eclipse EMF)	yes	depending on plugins	yes	depending on plugins	N/A
Blu Age [5]	JavaEE/.NET	yes	yes	complete logical model	yes	activity/sequence diagrams	yes
Middlegen [13]	DB persistence layer	yes	yes	database persistence layer using EJB, JDO, Hibernate etc.	no	N/A	N/A
IBM Rational Rose Data Modeler [9]	database applications	yes	yes	object models, data models and data storage models; physical-logical mapping	no	N/A	N/A
IBM InfoSphere Data Architect [8]	data layer	yes	yes	logical, physical domain models for various RDMSes	no	N/A	N/A
LLBLGen Pro [12]	data layer	no	yes	database model	limited	can design queries using a complex query builder	can include linq expressions
UML2SQL	data layer	yes	yes	complete logical model	yes	stored procedures/triggers by activity diagrams	yes (using a query builder)

The extensive and "rich" abilities of MDD tools for general application development are unfortunately insufficient in the field of data layer modeling. Even though, there are tools that allow for generating a database schema or database persistence layer code from a UML logical model, however the lack of support for modeling business logic in databases is obvious at first glance.

Such tools as IBM Rational Rose Data Modeler is able to convert a UML logical model expressed as class diagrams into a physical storage model in Relational Database Management Systems (RDMS), expressed in terms of tables and relations comprising a complete database schema. There are also tools, such as Middlegen, that are able to generate a database persistence layer code which facilitates using the generated schema from inside a programming language such as Java or C#. There is still no way, however, for modeling stored procedures, triggers or any other business logic that can be included in a database, which sets this scenario of MDD usage light-years behind the mainstream of "Java application" scenario. The ability, available in probably the most advanced data layer MDD tool—LLBLGen Pro, to design queries using a complex graphic query builder and include linq queries [11] in the generated code is still nothing compared to the abilities of Telelogic Rhapsody or AndroUML.

A short comparison of tools mentioned above is presented in table 1. It is needless to say, that there is a need for a tool that in the context of data layer, apart from structure modeling, facilitates also behavior modeling. UML2SQL, which aims to be such a tool, is also included in the comparison, and will be presented in the following sections of this chapter.

3   UML2SQL Concept

What would be the requirements like for a tool which could aid model-driven engineering? How do existing systems approach this problem? It is an obvious practice to base an MDD (CASE) tool on the Unified Modeling Language [10,21]. Using this standard allows architect(s) to get accustomed with new functionality delivered by the tool quickly and painlessly. The power of UML [30,27] itself is well known and we do not want to focus on it in this article. It has to be stated that UML2SQL is based on the UML standard too—as it is explained later (in section 4.1), the tool is designed as a plug-in to an existing UML modeling application.

Consequently, the process of designing system structure is similar for all types of CASE tools. We can model the application using different types of diagrams. Automating application code generation from its structure description is not a problem either. Although mapping of the object-oriented application model to the relational data model is not straightforward and requires some consideration regarding the choice of mapping strategy, it is well described in literature ([18,22] and section 4.5) and many (if not all) UML modeling tools provide such functionality.

What is more, the structural model of an application can be used to produce method skeletons and CRUD-like (Create, Read, Update and Delete) operations which can be deduced from the model. Such model characteristics are also widely applied in various engineering systems. One may ask so, how does UML2SQL differ from the above-mentioned CASE tools? What is its added value?

3.1   Behavior Modeling

The big issue starts when one wants to model the behavior of an application. It applies not only to UML itself, but foremost to the later possibilities of application code generation and general MDD tool operability. UML provides the designer with several types of diagrams (e.g. state diagrams, collaboration diagrams, sequence diagrams, etc.) which can be used to present system's behavior. However, even the most accurate diagram cannot be simply translated into source code. What is more, it is not the diagram's goal to be a complete template for code generation! To enable automatic application creation, some kind of an extension to UML has to be used, that would allow one to describe precisely the actions taken by the system.

The most popular approach nowadays is to model system behavior using the concept of state machines. The discussion on this idea dates back to the 1970s, when "Shlaer-Mellor Method" was presented [32]. It is widely used in real-time and embedded systems domain. Some other methods described in literature include:

– "Interaction Based" [20] modeling based on collaboration and sequence diagrams with an example of generating Java code from collaboration diagrams [24];

- "Contract Based" [20] method which abstracts from behavioral models and uses strongly the structural model, additionally equipped with some pre- and post-conditions imposed on operations and expressed in Object Constraint Language [29,26];
- using Abstract State Machines with AsmL language to model application behavior and also to test and evaluate a running system [19];
- applying activity diagrams to represent application behavior, which is expressed by means of states and transitions [2]. Such an approach is used in AndroMDA powered BPM4Struts cartridge (Struts J2EE application generator) and it can be understood as some variation on state machines.

As a practical example of behavior modeling, Rhapsody by Telelogic [21] can be presented. The diagrams created to illustrate system functionality can be completed with Java or C++ code snippets which are later included in the generated code. In the case of modeling an "executable" application it is an acceptable approach. The high-level programming language code inserted into diagrams still holds the idea of developing the Platform Independent Model (PIM), which can be easily compiled to the platform specific application (PSM) [28]. But when it comes to model data access, things get complicated.

First of all, when accessing data there is no straightforward object collaboration or message passing that could be modeled by executing adequate methods. It is not a situation that can be described by changing application's or objects' states either. Moreover, it is hard to imagine having to insert handcrafted high-level programming language code, which would access the database into any type of diagram. Even inserting code based on object-relational mapping technology would be painful, as any model refactoring could make it obsolete and, additionally, translating such code into SQL could be difficult. The natural way of accessing data is SQL, but even if the MDD tool provided us with such a possibility, we would encounter another problem: the impedance mismatch [18] between object-oriented paradigm and relational databases. Finally, by using a dedicated SQL dialect we face the danger that changing the database vendor would require rewriting code which operates on and communicates with the database.

Proposed UML2SQL tool was designed in order to avoid the mentioned above problems. It lets one model procedures that access data with activity diagrams (see section 4.4). Additionally, a dedicated language (called *eOQL* for enhanced Object Query Language) was designed. eOQL is a simple extension of Hibernate Query Language (HQL) [6, Chap. 14] and can be understood as a middleware between the application's structural model and its behavioral model (more details may be found in section 4.3). With the assistance of a graphical editor one can create statements that are assigned to activity diagram nodes—each node has one statement—and later used to generate stored procedures in a required SQL dialect (described in details in section 4.5). The connection between structural and behavioral models as well as the whole system architecture will be elaborated in section 4. One thing worth of mentioning here is that when modeling procedures with the eOQL editor, one can focus on the structural (i.e. object) model only, and need not to worry about the low-level data model. In MDA terminology, one copes with PIM instead of PSM [28].

At this point, it has to be underlined also that proposed approach is one of many possibilities when it comes to generate code that operates on data. Stored procedures are taken under consideration since using them lets preserve the consistency of the whole generated system and integrates easily with other components which emerges as outcome of running UML2SQL generators. It can be imagined that instead of stored procedures for example high level programming language code is generated which connects to the database, queries data and runs according to the model data flow.

Questions may arise whether UML2SQL is something more than the already mentioned tools like LLBLGen Pro [12] or OrindaBuild [15]. These tools provide functionality of object-relational mapping and generation of database access code. Below it is argued that such functionalities are merely an outcome of UML2SQL, and not its essence.

3.2   Is This Endeavor Worth It?

"MDA is another small step on the long road to turning our craft into an engineering discipline." [28] Since designing a new system is often a challenge of adopting a new environment and a new technology, fast prototyping techniques are priceless. Still, the refactoring activities are very important from the quality point of view. Linking the application design directly with its code means the process of keeping both of them up-to-date is easy and can be done by clicking the proverbial button. It contributes not only to the quality of the code, but facilitates later system maintenance.

Apart from the purely engineering motives, the second equally important issue connected with UML2SQL is the aforementioned impedance mismatch between object-oriented approach and modeling data in a relational fashion. Relational databases are independent from programming languages and can be easily shared among multiple systems and users. It is also important that behind the relational model there is a highly reliable and mathematically proven theory, tested for a very long time. Moreover, the biggest database management systems vendors invested huge assets in their products and it is obvious that they will not easily give the stage to competitors providing other solution for managing data (e.g. object-oriented databases [25]). Even though coping with relational database engines introduces some problems in application development, it will not be abandoned in the near future. Consequently, the concept of keeping one common model describing relational data and object-oriented structure of the application seems crucial. The simplest solution is UML [30], but UML itself is not enough. Some kind of a connector between relational and object-oriented paradigms is necessary to enable using one model in both aspects.

4   UML2SQL Deep Dive

From the concept presented in the previous section, a standard usage scenario was created. It perfectly underlines the goal of UML2SQL. The list of steps to go through is as follows:

1. The user designs the application structural model in an UML modeling application.
2. The user also creates abstract procedures accessing data using activity diagrams (also in the modeling application).
3. Activity diagram nodes are populated by the user with eOQL statements, which are directly based on the model designed in the first step (see sections 4.3 and 4.4).
4. The tool generates Hibernate mappings, application source code, database schema and stored procedures (section 4.5).
5. Generated code is compiled against the database if necessary.

4.1 Background

As it was already mentioned, the ambition of UML2SQL is to become a complete MDD tool for data access layer development. Thus, the first thing to consider was the choice between delivering a proprietary UML editor or using an existing one, extending its functionality with plug-ins. We have chosen the latter. This choice was dictated by the potential to reach a wider audience (i.e. users of an existing tool), and of course by our limited resources. After thorough research we decided to use Visual Paradigm Standard Edition [17].

The second, even more important decision regarded the approach to the object-relational mapping (ORM). To achieve relational and object models consistency and ensure that UML2SQL user could focus on PIM [28] design, the following issues were taken under consideration:

- usage of a custom ORM implementation vs. usage of an existing ORM engine,
- providing a simple way to access designed and generated stored procedures,
- usage of a query language which would allow operating on object model,
- building a graphical query editor vs. parsing user input.

Our experience with first prototypes of UML2SQL let us decide to use an existing ORM engine. Hibernate [6] was determined to be a perfect solution for us. It not only supplies a recipe for mapping object model to relational model and the interface to use it, but it also provides a way to use stored procedures compiled against the database as any other query. This way the developer, equipped with specific Hibernate mappings generated directly from the application design by UML2SQL, can easily access data stored in the database and call procedures generated from the model.

As far as the query language is concerned, its origin is associated with the chosen ORM engine. The simple to learn and use Hibernate Query Language (HQL) [6, Chap. 14] was adapted for UML2SQL. In order to enable an architect to model the behavior precisely, HQL was extended to give birth to *enhanced Object Query Language* (eOQL), which is discussed in section 4.3.

4.2 UML2SQL Architecture

In Fig. 1 a component view of UML2SQL is presented. Two main components can be distinguished: "UML2SQL Core" and "UML2SQL Plugin". The main reason why

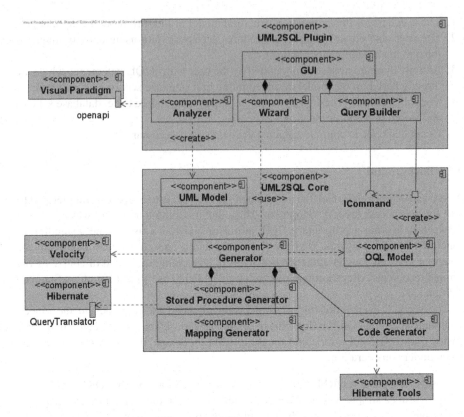

Fig. 1. UML2SQL component view

presented system is split into such parts is maximal independence from the UML editor and its interface. As it was already described, the first two steps of the scenario presented at this section's beginning are done using Visual Paradigm, but it can be imagined (and it was taken into consideration) that UML2SQL is plugged into another CASE tool when the "Plugin" component is replaced.

When the structural model of an application is ready, we can start playing with UML2SQL. When a request to build an eOQL statement for an activity diagram node is detected "Analyzer" is launched. It gathers the data from the UML editor and creates an internal UML model. Providing no errors or incompleteness of the UML model are detected, the "Query Builder" starts. This is a tool, an editor, which give an opportunity to build eOQL statements. From a software engineering point of view, it is worth to mention that the command design pattern was used here ([23]).

When activity diagrams are ready (i.e. populated with eOQL statements), application code can be generated. Using the "Wizard", the generation process can be customized in order to comply with application requirements. With the support of Velocity templates [4] and Hibernate [6] and using various generator types the application code is manufactured. The process is described in details in section 4.5.

4.3   eOQL—Enhanced Object Query Language

The question which for sure arises is whether it was necessary to introduce a new language to satisfy the undertaken requirements. We have already mentioned that a language to perform queries in an object-like fashion does exist! What is more, HQL [6, Chap. 14] is not the only one language to do that. There are also other languages like JPQL and OQL. When using object query languages it is possible to benefit from the power of their declarative approach known from SQL language and to stay in the realm of object-oriented programming simultaneously. Therefore, the problem of impedance mismatch is avoided as one does not have to be aware of relational model when writing queries. So, why have we decided to go for the eOQL then?

Again, we should look at UML2SQL from MDA perspective. "Fully-specified platform-independent models (including behavior) can enable intellectual property to move away from technology-specific code, helping to insulate business applications from technology evolution and further enable interoperability." [31] The procedure that accesses data is not only a collection of queries that select, update or delete it. To specify precisely the behavior of such a procedure some additional functions have to be used. For example, it may be necessary to create new objects (which can be understood as inserting data into a database), specify conditions deciding about the execution path, call other procedures or use helper variables or transactions.

Some of those things are not purely object-oriented, but one has to remember that the aim of such modeling is to generate SQL stored procedure's code. Its nature however is not object oriented, but conforms to a procedural programming paradigm. The essence of eOQL is therefore to provide well-known techniques for querying data in an object-like manner, at the same time giving an opportunity to use other types of statements to model application behavior (business logic) as accurately as it is possible. eOQL is not an extension of OQL language itself. It should be treated as an extension of object query language concept in general.

One more important issue regarding eOQL is that UML2SQL's user does not have to be aware of its concept or presence. From the user's perspective, eOQL is a middleware between structural and behavioral models of the application. eOQL is hidden behind the graphical editor which provides the interface to build eOQL statements. Consequently, the knowledge of any query language and some rudiments of computer programming are perfectly enough to build eOQL statements and use UML2SQL.

Types of eOQL statements and their relation to SQL expressions are described in the following sections (4.4 and 4.5). The eOQL "grammar" also includes instructions which are not available to the end user, but are used during code generation. These are mainly flow control statements and declarations.

4.4   Behavior Modeling

Since UML2SQL strongly depends on the provided UML model, the diagrams describing the application structure have to be adequately prepared. UML diagrams that are used for code generation must have some restrictions imposed on them. The following sections describe them in details.

Class diagrams preparation. In the case of class diagrams, our goal was to ensure maximum support for existing diagrams. Therefore, UML2SQL tool supports all class diagram elements by using the graceful degradation approach. That means that all unrecognized elements (that is, unnecessary as far as UML2SQL is concerned) of UML class diagrams are ignored. This is in contrast with the treatment of activity diagrams, which is described in the next section.

Bearing in mind that UML models can be constructed on different levels of abstraction, we decided to require an explicit indication of classes that should be included in the data model generated by UML2SQL. This allows the user to use an existing UML class model (with all the implementation-specific or utility classes) and only indicate to UML2SQL which classes should be included in the data model. In current implementation (under Visual Paradigm), this is achieved by adding the "ORM Persistable" stereotype to the class. Moreover, if system requirements demand having relations that keep the order of entities, one can use the "ordered" stereotype to mark the relations and have them generated appropriately later.

Apart from these requirements, there remain only a few rules. In general, a few details which are optional in UML class diagrams, should be defined. Among those are field names and association multiplicities. Without these details the engine would not be able to properly interpret class diagrams.

Building activity diagrams. Activity diagrams are treated in an entirely different way from class diagrams. The reason becomes apparent when one takes a look at a sample activity diagram presented in Fig. 2.

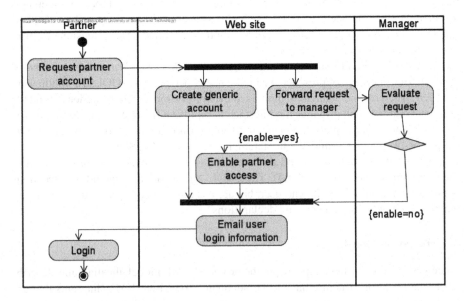

Fig. 2. An example of an activity diagram

From UML2SQL's point of view, there are several problems with such a diagram. First of all, several modeling constructs could not be translated to stored procedures because of lack of support in a declarative language that is SQL. These include, among others, elements related to object flow, constraints, synchronization elements (stored procedures are by design sequential and database engines do not provide any way of parallel execution inside a stored procedure) and exception handlers (some of the dialects provide exception handlers for stored procedures, but this is still far from the complexity that can be modeled in UML).

Moreover, even after restricting the structure of a diagram problems still remain mainly with the contents of the diagram. Human-friendly, textual information has little meaning for the procedure translation engine. Therefore UML2SQL requires that contents of activity diagrams are edited using an eOQL editor (see section 4.4).

To sum up, an activity diagram that can be translated by UML2SQL into SQL should meet the following criteria:

– it should contain only initial and final nodes, action nodes, decision nodes and loop nodes,
– each action, decision and loop node should have a statement assigned to it,
– each statement must be build using the eOQL editor (4.4).

Using eOQL editor. As it was already mentioned, the eOQL editor is a graphical tool to define eOQL statements which are assigned to activity diagram nodes. The classification of eOQL statements and their correspondence to the activity diagram nodes is presented in Table 2. The eOQL editor also provides a possibility to associate initial node with a list of variables which will be treated as procedure's arguments and to assign a variable to final node which would indicate that the user wants the procedure to return a value.

Since eOQL operates on object model only, eOQL editor provides user with the concepts defined on the class diagrams, i.e. classes and their attributes. One has no

Table 2. eOQL statements types

General statement genre	Description
Query	Well known query types (e.g. "SELECT", "DELETE") have their direct counterparts in eOQL, whose semantics are just the same. "Pure" queries can be assigned to action nodes and additionally "SELECT" query can be used to define subqueries in other statements (including logical expression attached to decision node) and to define an iterable collection for loop node.
References	There are two statements which allow modeling an action of adding and removing an object to/from a collection (which corresponds to a relation modelled on class diagram). An example can be seen in section 5.
Condition	Decision nodes have to be populated with a statement modeling the logical expression determining the execution path.
Other	There is also a possibility to assign values to variables, execute other procedures and handle transactions.

awareness of how the object model is mapped to the relational model in the process of SQL generation. Described editor delivers some handy, wizard-like panels which assist user in query building in a fashion known from MS Access interface.

It has to be stressed once again that the main concept of introducing eOQL together with its editor is to set an additional layer of abstraction on the top of the object-relational mapping process. In that way one can model the behavior of the application operating on object model only preserving one of the fundamental assumptions that the whole process of modeling is based on one model, i.e. object model.

4.5   Code Generation

Finally we can look at the output generated by UML2SQL. This is also a section which gives a basis for regarding UML2SQL as a complete MDD tool in the field of data access. After thorough design preparations, including building application structural and behavioral models and specifying concrete actions to be undertaken by created activity diagrams using eOQL editor, the code generation process can be launched and the delivered components can be used within an application.

Generation plan. The whole generation process is designed in a pipeline manner [23]. The initial input consists of the application class model, activity diagrams with assigned eOQL statements and database information (gathered by the "Wizard" mentioned in section 4.2). The process can be divided into three main stages, i.e.:

1. Hibernate mappings and configuration generation;
2. database schema and application source code generation and compilation (at the moment Java code generation is supported);
3. assembling stored procedures from activity diagrams and SQL code generation.

Hibernate mapping files. The first stage is done independently by UML2SQL itself. Using Velocity templates [4] the following Hibernate XML files are generated:

  – configuration file (e.g. storing database connection data) [6, Chap. 3],
  – mapping of persistent classes [6, Chap. 5-9],
  – mapping of non-persistent classes,
  – mapping of stored procedures [6, Chap. 16].

Two class mapping files are generated in order to distinguish classes from which SQL schema has to be generated. Both mappings are additionally equipped with some meta data describing the methods modeled on class diagrams and any other characteristics according to the Hibernate specification [6, Chap. 5-9]. As far as the object-relational mapping strategy is concerned, UML2SQL generates code which complies with "mapping each class to its own table" method [18]. Selected strategy provides intuitive operability at the object level and does not introduce any redundancy. However, performing operations at database level is sometimes a little bit complicated, especially when it comes to updating or deleting data.

Hibernate tools in action. During the second stage the outcomes of the first, presented above, are used intensively. Taking the advantage of Ant [3] and Hibernate tools [7], Java code for persistent and non-persistent classes is generated and compiled. The database schema and a Java interface with constants describing created procedures and their arguments are also created.

Stored procedures generation. The last stage is the most complicated one. The activity diagram itself is a graph describing possible execution paths. However, we do not cope directly with constructs known from programming languages. The naive approach to building procedures would include labelling every single statement and adding goto instructions when a "jump" is detected. Obviously, it is not an acceptable solution to have code full of the "hated" goto instruction. In UML2SQL, each diagram is subject to the loop detection algorithm as described in [33, Chap. 20], after which eOQL flow control instructions (like *DO-WHILE, WHILE-DO*) are added. *FOREACH* loop is generated for "Loop" nodes, "Decision" nodes are translated into *IF-THEN* statements. The variables that are used inside procedure's statements and are not its arguments are gathered to be listed on declaration list. After the procedures are assembled from the abstract eQOL statements, the SQL generator is launched. The generation process uses the following components:

- dialect specific *template provider*, whose responsibility is to feed the generator with an adequate Velocity template for a given eQOL statement;
- *Velocity templates* [4], which describe how given eOQL statement is translated into SQL code and, for statements other than queries, provide the recipe for direct generation;
- Hibernate "QueryTranslator" [6] which is used to transform "SELECT" queries;
- Hibernate "Executor" [6] which stores the execution plan for queries that modify data, i.e. "DELETEs" and "UPDATEs";
- compiled Java binaries (generated during the previous stages) which are used by components taking their origin in Hibernate.

All of the components mentioned above are dialect dependent and the generated code may differ when various SQL dialects ([6, Chap. 3]) are chosen. Finally, when the SQL code is ready it can be complied against the database if requested. There is also a possibility to create the database itself. The example of generated code can be found in section 5.

At the moment UML2SQL is equipped with generators which let one generate whole database system (i.e. database schema together with stored procedures) in MySQL and T-SQL. However, only code generated for MySQL is well tested. As far as database schema generation is concerned, all dialects supported by Hibernate library can be handled.

5   UML2SQL Test Case

This section serves as an illustration of the ideas introduced in the paper—an example of making a use of UML2SQL is presented. The goal is to show how UML2SQL can

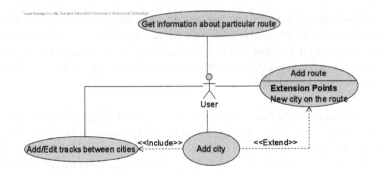

Fig. 3. Use case diagram describing requirements for the designed system

be applied in the subsequent stages of the application development process. Generally speaking, such a process can be divided into some well-defined phases, and UML2SQL covers some parts of the application structural and behavioral design, development, testing and post-deployment maintenance.

The proposed example is based on an application that can be described as a computer aided register of tourist routes or as an *ElectronicRoadAtlas*. After pointing out how UML2SQL supports development of *ElectronicRoadAtlas*, a short discussion presents what has to be conducted to create a similar system without UML2SQL's help.

5.1   Definition of Requirements

The first step of application development engaging fully software engineers is the process of defining and describing requirements for a created system. In figure 3 an use-case diagram is presented which conveys the requirements set for the application. Analyzing the diagram the following requirements can be distinguished:

1. Adding and editing routes between cities.
2. Defining a new city together with creating new links between existing cities and the one defined.
3. Adding new routes, which can require adding non-existent cities, which are present on the added route.
4. Retrieving the information about existing routes.

If the scope of the UML2SQL is taken under consideration, the requirements should be seen more from the data access layer domain perspective. Consequently, the problem of user interface will not be described here since the main stress is put on the way how selected functionalities are exposed in the programming interface.

5.2   Structural Modeling

Having the requirements' definition, one can focus on architectural-oriented activities. The first step is the preparation of a structural model. The form of the model itself is strongly determined by the requirements for the final application. All requirements have

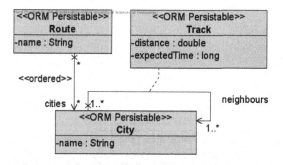

Fig. 4. Computer aided route database system data model

to be satisfied, therefore the structural model cannot restrict the development of some areas, which can make final product unacceptable. An exemplary model, which can be understood as a database of information regarding tourist routes, is presented on the diagram 4.

A particular tourist route (described by the Route class) is represented by a list of subsequent cities which appear on it. The association class Track describes data characterizing the link between two cities. The relation of self-aggregation for the City class can be understood as an extended representation of distance table which can be commonly found in the road atlas books. It is user responsibility to keep the data correct, i.e. the distance between cities A and B should be the same as the distance between B and A.

5.3   Business Logic Modeling

Taking into account the typical UML2SQL usage scenario the next step is the modeling of business logic, which can cover some of the application requirements. Some of the requirements will be satisfied thanks to the applying Hibernate library with delivered (generated) by the UML2SQL tool configuration and object-relational mapping files only. The requirement which would demand delivering a stored procedure is "Getting information about particular route". A model of an algorithm, which computes a distance and an expected travel time (two figures are returned as a result) is shown on diagram 5. The UML2SQL procedure model is constructed as an activity diagram. The algorithm computing mentioned data is based on iteration process, which step by step sums up the distances and expected travel times assigned to the subsequent cities' links appearing on the considered route.

5.4   Application Code Generation

Referring to the typical UML2SQL usage scenario again, the next step to take after having all modeling activities finished is to launch application code generation process. As a result UML2SQL gives the following artifacts which can be used in various ways at the subsequent development stages:

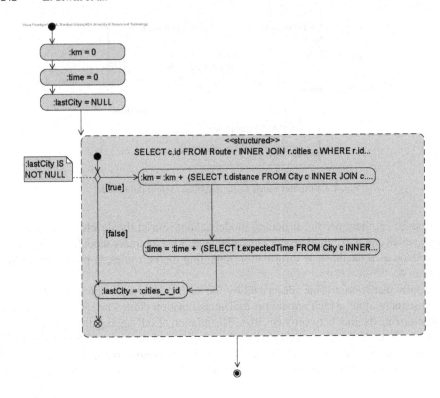

Fig. 5. Model of a procedure computing expected time and travel time of chosen route

- Hibernate configuration files,
- Hibernate object-relational mapping files,
- database schema (in a chosen SQL dialect),
- SQL stored procedures code (in a chosen SQL dialect),
- Java code of structural model and some additional interfaces facilitating further development.

The code of a stored procedure based on the activity diagram shown in fig. 5 is presented below:

```
CREATE PROCEDURE SP_getExpectedTimeAndDistance ( route bigint )
BEGIN
 DECLARE km double precision   ;
 DECLARE time double precision   ;
 DECLARE lastCity    bigint   ;
 DECLARE cities_c_id     bigint   ;
 DECLARE _fetchFailed    integer DEFAULT 0 ;

 DECLARE cities CURSOR FOR select city2_.id as col_0_0_ from route
   route0_ inner join routeCities cities1_ on
   route0_.id= cities1_.routeId inner join city city2_
```

```
on  cities1_ . cityId = city2_ . id  where route0_ . id=route
order by  cities1_ . _index ;
```

DECLARE CONTINUE HANDLER FOR **NOT** FOUND **SET** _fetchFailed = 1;

```
SET km = ( 0 ) ;
SET time = ( 0 ) ;
SET lastCity = ( NULL ) ;
OPEN cities ;
FETCH cities INTO cities_c_id ;
WHILE ( NOT _fetchFailed )
 DO
 IF   ( lastCity IS NOT NULL ) THEN
  SET km = ( ( km +
       ( select neighbours1_ . distance as col_0_0_ from city city0_
    inner join track neighbours1_ on city0_ . id=neighbours1_ . cityId
    inner join city city2_ on neighbours1_ . targetCityId = city2_ . id
    where city0_ . id= cities_c_id  and city2_ . id= lastCity )));
  SET time = ( time +
    ( select neighbours1_ .expectedTime as col_0_0_ from city city0_
    inner join track neighbours1_ on city0_ . id=neighbours1_ . cityId
    inner join city city2_ on neighbours1_ . targetCityId = city2_ . id
    where city0_ . id= cities_c_id  and city2_ . id= lastCity )) ;
  END IF;
 SET lastCity = ( cities_c_id ) ;
 FETCH cities INTO cities_c_id ;
END WHILE;
CLOSE cities ;
SELECT km result1, time result2 ;
END
```

Finally, a question has to be raised how the generated application code conforms to the requirements assumed. The generated procedure's code can be compared with an algorithm described in the previous section.

5.5 Applying Generated Code

The generated source code for data model and associated object-relational mapping files provide an easy and comfortable way to manage the persistence layer. Entity adding, editing or deleting is done simply by creating Java objects and invoking adequate Hibernate session methods. For example, the addition of a new city to a database is as simple as City class object creation and Hibernate session save method invocation. Summing up, Hibernate engine gives an access to persistence layer management functions in very flexible way. Additionally, the generated stored procedures make using application business logic possible, which in a consequence gives an entire well-defined interface to the layer implementing all requirements for a given system.

A Java code snippet is presented below, which applies generated by UML2SQL and complied against database stored procedure to sort a list of all existing routes retrieved from the database:

```
SessionFactory  sessionFactory  = new Configuration (). configure (
        " hibernate –config–sql. xml"). buildSessionFactory ();
final  Session  session  = sessionFactory .openSession ();
List <Route> routes  = ( List <Route>) session . createQuery ("from_Route"). list ();

final  Map<Route, Double> lengths = new HashMap<Route, Double>();
Collections . sort ( routes ,  new Comparator<Route>() {

@Override
public int compare(Route route1 , Route route2 ) {
  double d1  = getDistance ( route1 );
  double d2  = getDistance ( route2 );
  return Double.compare(d1, d2 );
}

private double getDistance (Route route ) {
  if (! lengths .containsKey( route )) {
    Query query = session
      . getNamedQuery(ProceduresRepository.SP_GETEXPECTEDTIMEANDDISTANCE)
      . setLong( ProceduresRepository . SP_GETEXPECTEDTIMEANDDISTANCE_ARGUMENT_1,
      route . getId ());
    List <Object> result  = query. list ();
    if ( result . size () == 1 && result. get (0)  instanceof Object []) {
        lengths . put( route ,  (Double) (( Object []) result . get (0))[0]);
        } else {
                lengths . put( route ,  new Double(0));
        }
  }
  return lengths . get ( route );
}
});
for (Route r :  routes ) {
    System.out. println ( r .getName() + ":_" + lengths . get ( r ));
}
session . close ();
```

At first, Hibernate library is configured and initialized with an XML file generated by UML2SQL too. Then a list of all routes is loaded from a database using an HQL query. Finally, a list of routes is sorted using an anonymous `Comparator` object in which the stored procedure is called to compute a total distance of a particular route.

5.6 Developing a System without MDD/MDA Tool Support

How would the application development process look like without the support of the tool such as UML2SQL? The question should be seen from the perspective of the implementation process comfort and the clarity and integrity of the code. Assuming that the system model is the same, the problem of providing a separate physical data model occurs. That model has to be used to create database schema. Since there are many-to-many relations in object model—prepared data model would have to introduce some additional elements (i.e. association tables) which would cover those issues. So, already at the stage of modeling, models become incompatible and introduce additional difficulties associated by the impedance mismatch problem.

As usually, such an incompatibility introduced at the very beginning of the application development process returns with double power in the following stages. To name only a few, the implementation of data access methods would have to find a solution for translating SQL tables' rows into objects. Mixing SQL code with high-level programming language constructs introduces unclarity and make any code refactoring almost impossible. Mentioned issues can be handled by a programmer but the problem would arise if the application model changes. In such a case, all native SQL code has to be at least double checked, and what is more probable—fixed, so that it conforms to new database schema. The lack of automation in refactoring code responsible for data access layer management and the lack of separation of SQL code handling business logic can coerce into spending long hours on analyzing and fixing existing SQL code.

6 Conclusion

"The aim (of MDA) is to build computationally complete PIMs" [31] where the term "computationally complete" means capable to be executed. As presented in the previous section, using UML2SQL it is possible to create the model of an application, which after translation into particular source code can be used as a component of a "production-ready" system. The activity diagrams enriched with eOQL statements processed by UML2SQL code generators produce SQL stored procedures that can be executed completely independently. It can be said that in the domain of developing data access layer UML2SQL deserves the title of an MDD tool.

Once more we can go back to the advantages of using MDD tools. What is the most frustrating activity for developers? From our experience we would point out coding the same thing countless times and getting to know a new technology without any supportive examples as the most irritating. But having an MDD tool one may overcome those issues quite easily. Instead of writing the same thing again and again, one may simply generate it from the application model. In this way the developer's productivity and satisfaction will certainly increase meaningfully. As an MDD tool for building data access layer UML2SQL can be also used in such aspects.

One may argue that MDA as an idealistic concept which has no rights to exist in reality. However, an example of UML2SQL shows, that in a domain of data access layer code generation, tool based on MDA concept can be quite useful, and what is more, its usage does not require any additional, technology-specific, knowledge. The idea of a system modeled once—deployed anywhere gains its full meaning here.

UML2SQL tool is available at http://uml-2-sql.sourceforge.net/.

References

1. AndroMDA, http://andromda.org/
2. AndroMDA BPM4Struts,
   http://galaxy.andromda.org/docs/andromda-cartridges/
   andromda-bpm4struts-cartridge/index.html

3. Apache Ant 1.7.0 Manual, `http://ant.apache.org/manual/index.html`
4. The Apache Velocity Project, `http://velocity.apache.org/engine/releases/velocity-1.5/`
5. Blue Age tools, `http://www.bluage.com/index.php`
6. Hibernate Reference Documentation, `http://www.hibernate.org/hibdocs/v3/reference/en/html/`
7. Hibernate Tools - Reference Guide, `http://www.hibernate.org/hibdocs/tools/reference/en/html/`
8. Ibm InfoSphere Data Architect, `http://www-01.ibm.com/software/data/studio/data-architect/`.
9. Ibm Rational Rose Data Modeler, `http://www-01.ibm.com/software/awdtools/developer/datamodeler/`
10. Ibm rational rose realtime. IBM
11. Linq: NET Language-Integrated Query, `http://msdn.microsoft.com/en-us/library/bb308959.aspx`
12. LLBLGen Pro, `http://www.llblgen.com/defaultgeneric.aspx`
13. Middlegen, Boss, `http://boss.bekk.no/boss/middlegen/index.html`
14. OpenArchitectureWare, `http://www.openarchitectureware.org/`
15. Orinda Build, `http://www.orindasoft.com/public/features.php4`
16. Rhapsody, Telelogic, `http://www.telelogic.com/products/rhapsody/index.cfm`
17. Visual Paradigm, `http://www.visual-paradigm.com/documentation/`
18. Ambler, S.W.: Agile Database Techniques. John Wiley and Sons, Chichester (2003-2007)
19. Andrzejak, A.: Modelling system behaviour by abstract state machines. Technical report, ZIB Berlin (2004)
20. Ashley McNeile, N.S.: Methods of behaviour modelling - a commentary on behaviour modelling techniques for mda
21. Charles, M.B., Krueger, W.: Leveraging the model driven development and software product line engineering synergy for success. Technical report, Telelogic (2008)
22. Dave Minter, J.L.: Pro Hibernate 3. Apress (2005)
23. Erich Gamma, R.J., Helm, R., Vlissides, J.: Design Patterns: Elements of Reusable Object-Oriented Software. Addison-Wesley, Reading (1994)
24. Engels, S.S.G., Hucking, R., Wagner, A.: Uml collaboration diagrams and their transformation to java (1999)
25. Harrington, J.L.: Object-Oriented Database Design. Academic Press, London (2000)
26. Warmer, J., Kleppe, A.: The Object Constraint Language: Getting Your Models Ready for MDA. Addison-Wesley, Reading (2003)
27. James Rumbaugh, I.J., Nooch, G.: The Unified Modeling Language Reference Manual. Addison-Wesley, Reading (1999)
28. Miller, J., Mukerji, J.: MDA Guide Version 1.0.1. OMG (2003)
29. OMG. Object Constraint Language. 2.0 edn. OMG (2006)
30. OMG. Unified Modeling Language: Superstructure, 2.1.1 edn. OMG (2007)
31. Sims, O.: Mda: The real value (2002), `http://www.omg.org/mda/presentations.htm`
32. Balker, M.J., Mellor, S.J.: Executable UML. Addison-Wesley, Reading (2002)
33. Cormen, T.H., Leiserson, C.E., Rivest, R.L., Stein, C.: Introduction to Algorithms, 2nd edn. MIT Press, Cambridge (2001)

Fuzzy Motivations in Behavior Based Agents

Tomás V. Arredondo

Universidad Técnica Federico Santa María, Valparaíso, Chile,
Departamento de Electrónica,
Casilla 110 V, Valparaíso, Chile
tarredondo@elo.utfsm.cl

Abstract. In this chapter we describe a fuzzy logic based approach for
providing biologically based motivations to be used by agents in evolutionary behavior learning. In this approach, fuzzy logic provides a fitness
measure used in the generation of agents with complex behaviors which
respond to user expectations of previously specified motivations. Our approach is implemented in behavior based navigation, route planning and
action sequence based environment recognition tasks in a Khepera mobile
robot simulator. Our fuzzy logic based motivation technique is shown as
a simple and powerful method for agents to acquire a diverse set of fit behaviors as well as providing an intuitive user interface framework.

Keywords: Agents, fuzzy logic, motivations, evolutionary, mobile robot.

1 Introduction

Having more natural and user friendly interfaces between people and artificial
systems is clearly a desirable and beneficial objective. Through the use of familiar
interfaces, users of technology can become more comfortable and may be better
able to utilize these systems to their full potential. One manner for this to happen
is to develop interfaces which are simple for average users to understand and
which hide some of the underlying technical complexity which enable artificial
systems to function in their environment. This allows people to utilize complex
applications and systems without having to suffer with a deep understanding
of all the details that form a necessary part of modern technology. This is an
ongoing effort and there are many examples of automated systems which aim to
free their user from technical details while providing a comfortable and intuitive
experience [1–5].

Motivation are viewed by psychologists as an internal state or condition (e.g.
a need, desire or want) that serves to influence the intensity and direction of
behavior. Motivation is generally accepted as involved in the performance of
learned behaviors. That is a learned behavior may not occur unless its driven
by a motivation. There are many sources for motivations including: behavioral,
social, biological, cognitive, affective, and spiritual [20]. In nature, differences
in motivations are key drivers in helping to produce a diversity of behaviors
which enables finding behaviors with a high degree of benefit (or fitness) for the

E. Szczerbicki & N.T. Nguyen (Eds.): Smart Infor. & Knowledge Management, SCI 260, pp. 247–272.
springerlink.com   © Springer-Verlag Berlin Heidelberg 2010

organism. For example, in an animal these motivations might represent drives such as hunger or curiosity, while an autonomous vehicle navigation agent will have other motivations concerned with things like refueling, fuel conservation, tire conservation, or parking safety.

Toward the goal of providing simpler interfaces to the users of technology, we have developed a motivation based agent approach that aims to help provide a more natural interface for end-users in a variety of applications and systems. We have implemented it in the form of fuzzy logic based motivations which can be used by behavior based agent systems to obtain fitness scores for a variety of possible behaviors.

To evaluate the validity of our proposal, we test the fuzzy motivation based fitness on behavior based agent navigation and environment recognition tasks within a Khepera robot simulator. The results show that the motivation based approach has the potential for improving human understanding of complex robotic behaviors as well as provide a means for generating a great diversity of robotic behaviors in an intuitive and user friendly manner.

In Section 2, we briefly describe a number of agent architectures. Section 3 refers to fuzzy logic in robotic control. Our proposed method and how it was implemented is described in Section 4. In Section 5 we discuss and summarize the experiments that were performed. Finally, in Section 6 some conclusions are drawn.

2   Survey of Agent Architectures

In this section, in order to illustrate and be able to contrast with other methods, we briefly review a number of popular agent architectures. The architectures reviewed fall under four general classes: deductive approaches, deliberative toward means-ends approaches, motivation based, and behavior based [1, 5, 8].

2.1   Deductive Architectures

Deductive architectures require that a symbolic representation of the environment be implemented. The system manipulates these symbols using logical operations in a manner that corresponds to logical deduction. In the view of agents as theorem provers an executable specification through a set of deduction rules is applied when certain conditions are satisfied. There are several difficulties with these type of methods [1, 5, 8], the main ones being:

- Transduction - the difficulty in translating the real world into an accurate and valid symbolic representation.
- Representation/reasoning - the issue of getting agents to manipulate the symbolic information correctly and in time for these results to be useful, deliberate reasoning requires that the knowledge upon which reasoning is based be consistent, reliable and certain.
- Dynamic and noisy environments - if the information which the agent uses is inaccurate or has changed then reasoning may err seriously (e.g. a robot

in a world with arbitrarily moving objects like a crowded hallway), it is potentially dangerous to rely on old information that may no longer be valid.

- Highly hierarchical - where communication and control flow up and down the agent structure hierarchy. If a change has occurred in the environment the new information must flow completely up the hierarchy in order for the agent to deliberate and decide on a new course of action. Then the order must flow down the chain of command for the action to be executed. Because of the time lag during this processing, hierarchical control systems are best suited for structured and highly predictable applications (e.g. a manufacturing cell).
- Difficulty in engineering a complete system - as incremental competency proves difficult to achieve, the entire system needs to be built before system testing can be done. This is contrary to current engineering methods (e.g. agile, iterative or incremental engineering processes [32, 33]).

2.2 Deliberative towards Means-Ends Architectures

Deliberation toward means-ends reasoning is in general reasoning that is directed towards actions, the process of figuring out what to do next (i.e. *practical reasoning* [1]). The Belief Desire Intention (BDI) paradigm embodies this type of reasoning and contains explicitly represented data structures that loosely correspond to these three mental states.

The earliest and most prevalent implementation of the BDI agent architecture is the Procedural Reasoning System (PRS) which has been applied in a variety of agent based applications [1, 29–31]. As seen in Fig. 1, PRS consists of: (i) a database (i.e. beliefs) containing facts about the environment and internal states, (ii) desires which represent the agent's goals, (iii) an intention structure containing those plans that the agent has committed to achieve, and (iv) a set of plans describing how sequences of actions that have their preconditions satisfied are to be executed to fulfill certain desires. Plans in PRS each have the following components:

- a goal - the postcondition of the plan,
- a context - the precondition of the plan, and
- a body - the recipe part of the plan - the set of actions to execute.

An interpreter runs the entire system by manipulating the components of PRS, selecting the appropriate plans based on current beliefs and desires, putting those that are selected on the intention structure so that they may be executed in order. The body of a plan in PRS may include not only a sequence of actions but may also have desires as components. The idea being that when a plan includes a desire at a particular point means that the desire must be achieved before the remainder of the plan can be executed. This may require that further plans be pushed onto the intention stack which may require finding further plans to execute those plans and so on. Desires are conditions over some interval of time and are described using temporal operators. Desires can include achievement of a certain state, goals of maintenance, or testing of a condition. An action succeeds

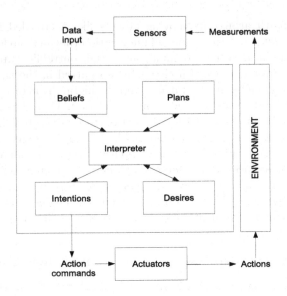

Fig. 1. PRS Architecture

in achieving a given desire if its execution results in the satisfaction of the desires description. Desires can be associated with the external environment or the state of the system [1, 31].

BDI Basic Model. In BDI systems [1], the environment (E) starts in a particular state out of a finite set of discrete states:

$$E = \{e, e', ...\}. \tag{1}$$

The agent (Ag) chooses a particular action out of the set of possible actions available (A) based on the environment and on previous action(s) performed:

$$A = \{\alpha, \alpha', ...\}. \tag{2}$$

A run (R) is a sequence of interleaved states and actions. The environment starts in a state and the agent chooses an action to perform on that state. As a result of that action, the environment can respond with a number of possible states and so on:

$$R : e_0, \xrightarrow{\alpha_0} e_1, \xrightarrow{\alpha_1}, ..., \xrightarrow{\alpha_{u-1}}, e_u \tag{3}$$

The effect of the agent on the environment is mapped as a state transformer function [1] τ which maps a run to a set of possible environment states:

$$\tau : R \rightarrow \wp(E). \tag{4}$$

The Ag function maps runs to actions, the agent decides on the next action based on the history of the system it has seen so far:

$$Ag : R \rightarrow A. \tag{5}$$

We will use $(\rho, \rho', ...)$ to represent percepts and Per as the set of all percepts. The agent uses the $see(...)$ function to map the environment to percepts:

$$see : E \rightarrow Per. \tag{6}$$

Finally, the agent tries to find and execute the best *plan* (π) which is a series of actions to execute in order:

$$\pi = (\alpha_1, \alpha_2, ..., \alpha_n). \tag{7}$$

BDI Process Flow. The BDI process flow [1, 23] is shown in Fig. 2. In this process the agent performs the following functions:

1. An agent belief revision function (i.e. brf) updates the agent's current beliefs based on the set of all beliefs (Bel) and current percepts:

$$brf : \wp(Bel) \times Per \rightarrow \wp(Bel). \tag{8}$$

2. During deliberation, the desire to be achieved is read from the *intention* stack. This stack contains all the *desires* (i.e. *goals*) that the agent has committed to and are pending achievement. The agent searches through its plan library to find the *plans* that have the desire (i.e. goal) that is on top of the *intentions* stack as their postcondition. From those plans, only those that have their precondition satisfied (based on its current *beliefs*) become the possible *options* or *desires* for the agent.

    Formally, the agent begins to *deliberate* using two functional components: the *options* function takes the agent's current set of beliefs and intentions (Int) and produces a set of possible desires (Des) or options. In order to select between competing options, the agent uses the *filter* function to find the best option(s) to commit to:

$$options : \wp(Bel) \times \wp(Int) \rightarrow \wp(Des), \tag{9}$$

$$filter : \wp(Bel) \times \wp(Des) \times \wp(Int) \rightarrow \wp(Bel). \tag{10}$$

Eventually, through deliberation one option will be chosen to be executed. This may involve pushing further desires onto the intention stack, which may then in turn involve finding more plans to achieve these goals, and so on. The process ends with individual actions that may be directly computed. If a particular plan to achieve a goal fails, then the agent is able to select another plan to achieve the desire from all candidate plans.

There are several mechanisms for deliberating between competing options in PRS-like architectures. In the traditional PRS system, deliberation is implemented via *meta-level plans* which are plans about plans and are able to modify

Algorithm BDI Agent
Input:
  B_0; /* B_0 are initial beliefs */
  I_0; /* I_0 are initial intentions */
Variables:
  $B := B_0$; /* B are current beliefs */
  $I := I_0$; /* I are current intentions */
  D : current desires;
  π : current plan;
begin
  while true do
    get next percept ρ;
    $B := \mathbf{brf}(B, \rho)$;
    $D := \mathbf{options}(B, I)$;
    $I := \mathbf{filter}(B, D, I)$;
    $\pi := \mathbf{plan}(B, I)$;
      while not (**empty**(π) **or succeeded**(I,B) **or impossible**(I,B)) **do**
        $\alpha := \mathbf{head}(\pi)$;
        execute(α);
        $\pi := \mathbf{tail}(\pi)$;
        get next percept ρ;
        $B := \mathbf{brf}(B, \rho)$;
        if reconsider(I,B) **then**
          $D := \mathbf{options}(B, I)$;
          $I := \mathbf{filter}(B, D, I)$;
        end-if
        if not sound(π,I,B) **then**
          $\pi := \mathbf{plan}(B, I)$;
        end-if
      end-while
  end-while
end;

Fig. 2. BDI Agent Algorithm

an agent's intention structures at runtime (e.g. the **reconsider**(I,B) function). The BDI algorithm may also use a plan soundness function (**sound**(π,I,B)) to evaluate whether π is a sound plan for achieving current intentions (I) given current beliefs (B). Simpler methods of deliberation are to use *utilities* for options which are scoring mechanisms (i.e. numerical values) so that the agent can choose the option with the best score.

2.3   Motivation Based Architectures

Motivations have been modeled in deliberative schemes to represent an agent's basic needs and desires. In deliberative motivation based schemes, associated with each motivation is a numeric measure of strength or motivational value. Motivations may also assist in modeling the context of an agent, this is due to the fact that the state of an agent's environment is an essential driver in its plan-generation capabilities. What is contextual depends upon an agent's sensors, actuators, size, internal states and so on. The context of an autonomous planning agent is captured partly by representing and reasoning about the motivations of the agent and is important in several ways: it constrains the *desires* that the agent can generate, it enables the agent to prioritize its goals and to perform plan selection [24].

Cognitive Overload. A traditional deliberative (e.g. BDI based) agent can generate goals on the fly in response to a changing domain but there are resource constraint issues once a multiplicity of goals have been generated. Computing if a propositional planning instance has solutions is a PSPACE-complete problem which may be reduced to an NP-complete once severe restrictions are placed on the operators and the formulas used [27].

With this in mind, an agent may be physically able of achieving all its goals given ample time but there are restrictions as to the amount of processing time that can be given to planning, goal management and other deliberative tasks. The agent will become overloaded by such tasks and they will exclude all other activity. This is known as cognitive overload and a possible solution is to focus agent attention on a number of relevant goals so that cognitive overload is avoided [28].

As shown in Fig. 2, filtering can be used to reduce the amount of reasoning that the resource constrained BDI agent has to perform. In the IRMA approach [29], options are filtered on the basis of how similar they are to current intentions. Two component filters are described: a compatibility filter and an override filter operating in parallel. The compatibility filter passes options that are compatible with the agents current intentions, and the override filter passes options that may be incompatible but are sufficiently relevant to allow consideration.

Unfortunately, the filtering mechanisms previously mentioned may not be very successful in reducing cognitive overload because as time goes by a number of valid alternatives may be generated which are all compatible with the agent's current intentions. Also if the override filter is too sensitive to alternative options, the agent will tend to be easily distracted from its present intentions. But, if it is too insensitive to other options, the agent will behave in a too single minded fashion. The options that pass these filters are considered a deliberative process and, if selected, are added to the agents intention structure. As the size of the intention structure gets larger, the computational requirements of scheduling, improving and evaluating these intentions may also cause cognitive overload [23].

Using motives is an alternative that has been investigated in BDI based agents. Motives may be used to observe the environment and the internal state of the agent and ensure that a particular interest is served according to the specified motives. Specific goals are thus generated which have the potential of supporting this interest. Hence motives cause an agent to act and have no other purpose. The generation of goals may involve making decisions on what goals to generate and under what conditions they should be acted on [23].

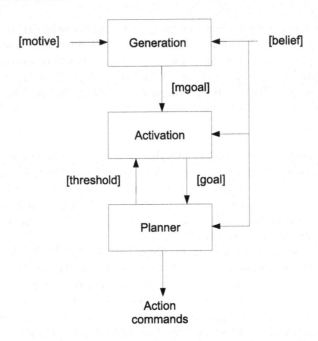

Fig. 3. Deliberative motivated agent architecture

In typical schemes the numeric value associated with a motivation (e.g. *intensity*) will vary [24] with changing external or internal environmental circumstances. The changes represent the driving force that directs action to satisfy the motivation. For example, in a vehicle driving agent the motivational value associated with parking safety would be very low if not zero if the vehicle is moving at a high speed but would increase when the vehicle is stopped or moving very slow (in case it might be getting ready to park).

Deliberative Motivated Agents. In the deliberative *motivated agent* approach (Fig. 3) [23], a *motive* is a mapping from the current *beliefs* of the agent about the state of its domain to a, possibly empty, list of motivated goals (*mgoal*). Such an agent will have two functions: goal generation and goal activation both of which have the capacity to generate *motivated goals* in response to detected changes in its current *beliefs*:

$$motive : \wp(belief) \rightarrow \wp(mgoal). \tag{11}$$

In this approach, the motivation value associated with a goal changes over time as the agent's beliefs change and reflects the current relevance of that goal to the agent:

$$mgoal = (motivation, goal). \tag{12}$$

This type of agent is driven, not by a conjunction of top level goals (like BDI or IRMA based agents), but by a number of motives that can generate motivated goals in response to detected environment or domain changes. This is referred to as goal *generation*, motivated goals that are generated are added to the agents list of motivated goals.

$$generation : \wp(motive) \times \wp(belief) \rightarrow \wp(mgoal). \qquad (13)$$

The second main function of a motivated agent is the *activation* of goals. Only if a goal is active will the agent act on that goal. A goal will be activated if:

- the intensity of the motivation to acting on the goal is sufficient to exceed a threshold, and
- the agent decides that it is relevant to act on the goal now.

Triggering of goals occurs when the intensity of the motivation associated with a goal exceeds a *threshold*. Thresholds are derived from the state of the planner and are manipulated to control the *triggering* of goals. Typically, if a threshold is high, goals will be less likely to trigger, and less likely to be considered for *activation*.

$$triggering : \wp(mgoal) \times \wp(belief) \times \wp(threshold) \rightarrow \wp(goal). \qquad (14)$$

If a goal triggers, it is considered by the *considergoal* deliberative process, and if it is selected then it is added to the agents focus of planning attention. This process checks the conditions under which the goal was generated. If these conditions still hold and the goal should still be acted on then it is passed on to the planner.

$$considergoal : \wp(goal) \times \wp(belief) \rightarrow \wp(goal). \qquad (15)$$

So, the *activation* process acts as a two-stage filter where only goals with an associated motivation that exceeds the threshold and are still considered relevant for action will pass.

$$activation : \wp(mgoal) \times \wp(belief) \times \wp(threshold) \rightarrow \wp(goal). \qquad (16)$$

Only activated goals will influence the planning processes of the agent. If the level of the threshold depends on the load on the agents planning processes, the goals that are active should be small in number so that the planner is not overloaded and restricted to the goals that the agent is most motivated to achieve. Once activated, a goal is added to the set of goals which constitutes that agents focus of planning attention. The function of the planner is then to direct action in the pursuit of these active goals. However, its specific capabilities are not prescribed by the architecture [23].

If the motivation of one goal is greater than that of another then the agent will act on the former rather than the latter. Consider a warehouse agent that has two *mgoals*, one is to prepare an order in a few hours time and another to remove some old inventory from an area in the warehouse. Cleaning the warehouse is less important than preparing the order, but it is more relevant for the agent to clean the room now than to prepare the order for a client that will not arrive for several hours. So, the motivation to clean the area will be greater than the motivation to prepare the order.

In the motivated agent architecture, goals are filtered on the basis of their importance, temporal relevance and the current load on the cognitive resources of the agent, and so the feedback from the planner is some measure of its load. In generating a value for a motivation it is important to find a balance between the influence of subjective importance versus temporal relevance. Otherwise, the agent will tend to act in one of two undesirable ways: considering nothing but the most important goals regardless of when these goals are to be achieved or if the temporal relevance of the goal has too great an influence on motivation intensity, the agent will consider only its most urgent goals regardless of their importance [23].

The process of selecting between alternative goals (e.g. *considergoal*) is not combined with the process of selecting between possible refinements of adopted goals (i.e *intentions*). If the motivation associated with a particular goal which depends on both the importance and temporal relevance of the goal exceeds the current level of a threshold, the goal is considered for activation. These thresholds depend on measures of the load on the agents computational resources and the activation of goals is limited by the computational resources available.

The motivated deliberative agent architecture uses external mechanisms by which goals are generated (e.g. alarms) and presented as options for adoption. A system of alarms and associated guidelines could trigger goals automatically and cause the intensity of motivations to vary (generally to increase but also could be mitigated) in time [23].

2.4   Behavior Based Architectures

Behavior based systems were first introduced in the realm of robotics (e.g. subsumption architecture), but these concepts [5, 8] have extended into general agent based systems [1]. Behavior based (i.e. reactive) agent architectures in general do not use world models, representative or symbolic knowledge and there is a tight coupling between sensing and action. This design philosophy promotes the idea that agent based systems should be inexpensive, robust to sensor and other noise, uncalibrated and without complex computers and communication systems.

Two key concepts of behavior based robotic systems (e.g. subsumption) that can be extended into agents are situatedness and embodiment. Situatedness refers to the ability of an agent to sense and readily react to its current surrounding without the use of abstract representations (i.e. maps, symbolic representations) and embodiment means that the agent(s) (e.g. robot, control system, web search engine, etc) should experience the world directly. Behavior based learning systems may typically include softcomputing based methods such as reinforcement learning, neural networks, genetic algorithms, fuzzy systems, case and memory based learning and others [1, 5, 8].

Planning actions based on complex internal world representations in not seen as something beneficial because of its inherent error and associated temporal costs. In general, a purely reactive agent (Ag_r) will produce a response out of all possible responses (R) based on current stimulus (S) which can be represented by the following map:

$$Ag_r : \beta(S) \to R \qquad (17)$$

Each individual stimulus or percept s_i (where $s_i \in S$) is a tuple consisting of a particular type or perceptual class (p) and a strength value λ:

$$s_i = (p, \lambda) \qquad (18)$$

The agent (Ag_r) chooses a current response (r_i) out of the set of possible responses available (R) whenever λ is greater than a threshold τ:

$$R = \{r_1, r_2, ...\}. \qquad (19)$$

Behavior Function. The behavior functions β_i (where $\beta_i \in \beta$) responsible for mapping stimuli into responses can be discrete or continuous [5]:

- Discrete: the stimulus produces a response out of an enumerable set of choices (e.g. go-straight, turn-right, turn-left). R consists of a bounded set of responses to stimuli S as specified by β. Typical discrete behavior function implementations may include: fuzzy rules, production rules, subsumption encoded rules using the Behavior Language [8] and others (e.g. neural networks). There is no general restriction that precludes other methods as long as they produce bounded discrete outputs to stimuli.
- Continuous: the stimulus produces a response that is continuous over the range R. Continuous response allow an infinite space of responses to its stimuli, a common technique is the potential field method [5].

Several methods are available for representing behavior based systems [5, 8]. These include: stimulus-response (SR) diagrams, functional notation and finite state acceptor (FSA) diagrams. SR diagrams provide one of the most intuitive means of representing behavior based robotic systems. As can be seen in Fig. 4, a stimulus-response (SR) diagram is a mapping between a stimulus, a behavior and a response.

Behavior Layering and Coordination. Behavior based systems tend to be developed in an incremental manner in which multiple behavior functions are layered on top of one another in an iterative process and without changing previous layers. In behavior based systems layering is done in a different dimension

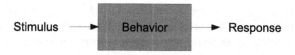

Fig. 4. SR diagram

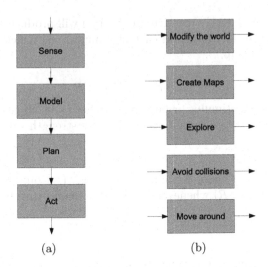

(a)   (b)

Fig. 5. Layering models (a) vertical (b) horizontal

than what traditionally had been pursued in the AI community, the conventional layering model being vertical (Fig. 5(a)) and the behavior based model being horizontal (Fig. 5(b)).

Layered behavior based systems require a mediation mechanism (i.e. a *coordinator*) that produces an appropriate overall output for the agent at any point in time given current stimuli (Fig. 6). There are competitive and cooperative types of coordination mechanisms for behavior based systems. In competitive methods when conflicts arise as a result of two or more behaviors being active, the coordinator typically will select a winner take all response in which the winning response is chosen to be executed by the robot. This can take the form of strict hierarchical methods through the use of suppression of lower layers (e.g. subsumption) [8], action-selection methods which arbitrarily select the output of a single layer based on activation levels determined by the agent's goals and current stimulus, and voting based methods in which behaviors generate votes and the action with the most votes is chosen. Cooperative methods require that behaviors be fused or somehow added, this could take the form of vector summation based on individual behavior gains (similar to how a neural network aggregates it's inputs) but in general requires an output representation that is amenable to such fusion [5].

Mediator Issue in Horizontally Layered Systems. One advantage of horizontally layered architectures is their apparent simplicity as well as their inherent parallelism. In many behavior based architectures (e.g. subsumption) higher layers may manage lower layer outputs by inhibiting or suppressing them [5, 8]. These parallel layers could also act more independently but as previously

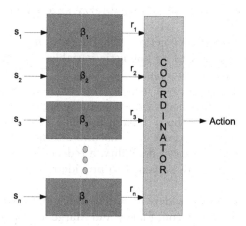

Fig. 6. SR diagram showing layered behaviors with a coordinator

mentioned, this approach requires a mediator to ensure consistency of behavior and also to determine which layer is in charge of the system (i.e. producing the output). As indicated in [1]:

> If there are n layers in the architecture and each layer is capable of suggesting m possible actions, then this means that there are m^n such interactions to be considered.

The value of m^n interactions is the upper limit in terms of the number of interactions or possible actions that a horizontal layered agent system would have to consider. If a mediator was to consider each possible action in turn in order to pick the best one it simply would shift all deliberation from the n horizontal layers to a single layer (the mediator). This is another form of cognitive overload as this approach is clearly not practical since problems requiring exponential time are impossible to solve in constrained systems except for small values of n.

One possible way to get around this problem is to utilize a less granular partitioning of the possible action space. For example, this could take the form of a much smaller deliberation which selects one of the m layers to take control of the system based on a number of internal and external variables (e.g. significant environment, system state and sensor inputs). While this may seem like a less than optimal approach, it is a typical compromise found in softcomputing based methods which do not in general try to find the best possible solution to intractable problems but by using inexact methods try to find useful solutions [11, 34].

3   Fuzzy Logic in Agent Behavioral Control

Behavior based agents implicitly demand embodiment and a tight interaction with the environment [5]. Due to this, we have focused our method on robotic applications as a natural means of physical agent evaluation but clearly there are multiple other possible applications of fuzzy logic based agents in regard to intelligent systems [11, 25, 26, 34].

Fuzzy logic systems produce actions using a set of fuzzy rules and variables, these variables are referred to as linguistic variables and indicate a degree of membership of the variable to a particular fuzzy set. The degree of membership (between 0 and 1) is defined by a membership function which maps a crisp input to a fuzzy output (fuzzifier). Typically, fuzzy logic control systems consist of: fuzzifier, fuzzy rule base, fuzzy inference engine, and a defuzzifier. In many applications fuzzy logic provides a more natural interface with a large tolerance for imprecision that gives greater flexibility, a reduced processing cost than traditional logic and which may be tuned to provide useful results in a variety of applications [11, 34].

Fuzzy logic has been used widely in various behavior based agent applications such as robotic behavioral fusion. Flakey and Marge are two examples of robots that used fuzzy based implementations to blend different behaviors for things such as collision avoidance, goal seeking, docking, and wall following [12, 13]. As can be expected with the exponential growth in the fuzzy rule base, these systems had issues with scalability which they attempted to manage by: a reduction of input spaces and using contexts with a limited world model [12], or by using independent distributed fuzzy agents and weighted vector summation via fuzzy multiplexers for producing the final command signals for its drive and steer mechanism [13].

More recently other fuzzy logic strategies have been used in mobile robotics including: neuro-fuzzy controllers for behavior design (based on direction, distance and steering) [14], fuzzy based modular motion planning [15], fuzzy integration of groups of behaviors [16], multiple fuzzy agents used in behavior fusion [17], GA based neuro fuzzy reinforcement learning agents used in training a walking robot [18], behavior based fuzzy logic integration for robotic navigation in challenging terrain [19] and motivations as fuzzy fitness functions for robotic behaviors [6].

4   Fuzzy Motivations for Behavior Based Agent Learning

As indicated previously, different motivations are key drivers in helping to produce a variety of behaviors which have a high degree of benefit (or fitness) for the organism. Our approach is a behavior-based extension of goal autonomous motivation based agents in which motivations drive the planning of agents and their corresponding actions [23]. In this behavior-based extension, motivations are not used to reduce the search space of behaviors as is the case with the deliberative motivation implementations but rather to provide a scoring mechanism so that the agent(s) responsible for said behaviors can improve over time. As a behavior based extension of the deliberative method, this method does not use a separate goal activation system and associated triggers or guidelines. Towards evaluating the feasibility of this approach we implement and test fuzzy motivations in behavior based mobile robotics. The benefit of this approach versus

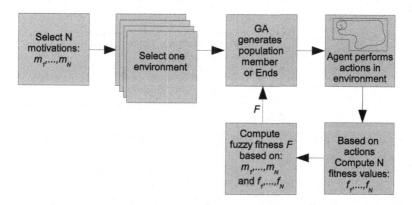

Fig. 7. Motivation Implementation Overview

traditional behavior based implementations is that user expectations of agent behavior are are mapped to a motivation profile which can then be compared to the agent's behavior. Traditional behavior based agents focus specifically on the agent and the behaviors that are required in order for him to complete his tasks but without a tweaking mechanism that the end-user can manipulate in a simple and intuitive fashion.

4.1 Motivation Implementation

In our behavior based motivation implementation [6], we use motivation settings in order to determine the Takagi-Sugeno-Kang (TSK) based fuzzy fitness of an agent's behavior in various environments (Fig. 7). The motivation set (M) that we consider in this implementation includes: curiosity, homing, orientation, and energy (the opposite of laziness).

In our sample implementation we utilized a feed-forward neural network as the agent in order to map stimuli S to responses R with its outputs discretized by using Action-based Environmental Modeling (AEM). The choice for this was based on the simplicity of having a single neural network providing the β mapping function as this choice did not require a coordination mechanism. The lack of a behavior coordinator provided simpler implementation and more obvious exposure to the effects of fuzzy fitness (and motivations) on agent behavior.

In this implementation, the following general steps are considered:

- Select N motivations: the user selects the motivation values (i.e. $m_1, ..., m_N$) that are to be used by the fuzzy fitness calculation (i.e. the motivations that help determine the best agent out of the population of agents).
- Select one environment: the user selects one environment out of many in order to train the agent.
- GA generates population: a genetic algorithm generates a population of different behavior based agents.
- Agent performs actions: each agent of the population performs actions in the environment in turn.

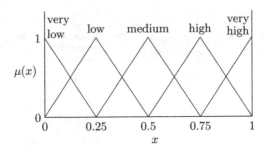

Fig. 8. Fuzzy membership functions

- Based on actions compute N fitness values: based on each agent's behaviors a set of N fitness values is generated.
- Compute fuzzy fitness F: fuzzy fitness is calculated based on the motivations and fitness values.
- GA Ends: The optimization ends when the number of maximum iterations of the GA is reached. The best agent out of the population is kept as the final solution.

Other possible alternatives to the implementation presented include: using other types of fuzzy logic such as Mamdani and using a different discrete agent implementation (β mapping function) such as fuzzy rules, production rules, neuro-fuzzy, or subsumption encoded rules.

Fuzzy Fitness. In order to determine a fuzzy fitness score for a particular behavior we use a TSK fuzzy logic model, TSK fuzzy logic does not require deffuzification as each rule has a crisp output that is aggregated as a weighted average [21]. We have used triangular fuzzy functions toward determining robotic fitness. The membership functions used for each of the four motivations used in our experiments are shown in Fig. 8. As shown in Fig. 9, four fuzzy variables with five membership functions each are used ($5^4 = 625$ different fuzzy rules) and for the coefficient array C we used an increasing linear function [6].

The motivation set (M) considered includes: homing (m_1), curiosity (m_2), energy (m_3), and orientation (m_4). These motivations are used as input settings (between 0 and 1) prior to running each experiment.

The set of fitness criteria and the fuzzy variables that correspond to them are: proper action termination and escape from original neighborhood area (f_1), amount of area explored (f_2), percent of battery usage (f_3) and environment recognition (f_4). The values for these criteria are normalized (range from 0 to 1) and are calculated after the robot completes each run:

- f_1: a normalized final distance to home.
- f_2: percentage area explored relative to the optimum.

Algorithm FuzzyFitness
Input:
  N : number of fuzzy motivations;
  M : number of membership functions per motivation;
  $X[N]$: array of motivation values preset;
  $Y[N]$: array of fitness values;
  $C[N]$: array of coefficients;
  $\mu[N][M]$: matrix of membership values for each motivation;
Variables:
  $w[n]$: the weight for each fuzzy rule being evaluated;
  $f[n]$: the estimated fitness;
  $n, x_0, x_1, \ldots, x_N$: integers;
Output:
  F : the fuzzy fitness value calculated;
begin
  $n := 1$;
  for each $x_1, x_2, \ldots, x_N := 1$ **step** 1 **until** M **do**
  begin
    $w[n] := \min\{\mu[1][x_1], \mu[2][x_2], \ldots, \mu[N][x_N]\}$;
    $f[n] := \sum_{i=1}^{N} X[i]Y[i]C[x_i]$;
    $n := n + 1$;
  end;
  $F := (\sum_{i=1}^{N^M} w[i]f[i])/(\sum_{i=1}^{N^M} w[i])$;
end;

Fig. 9. Fuzzy Fitness Algorithm

- f_3: estimated percent total energy consumption considering all steps taken.
- f_4: binary value determined by having the robot establish which room he is in (output node from a previously trained SOM network) versus the correct one.

During training, a run environment (room) is selected and the GA initial population is randomly initialized. After this, each individual in the population performs its task (navigation and optionally environment recognition) and a set of fitness values corresponding to the performed task are obtained. Finally, robotic fuzzy fitness is calculated using the fitness criteria information obtained and the different motivations at the time of exploration.

Action-based Environmental Modeling. Action-based environmental modeling (AEM) also follows the less is more philosophy of behavior based systems by using a small set of set actions which are combined to produce complex behaviors. AEM based robots use this small action set (e.g. go straight, turn left, turn right, turn around) in order to perform a sequence of actions based on sensed states in a specific environment. Only local sensors are used in order to navigate and perform environment recognition in various scenarios (rooms). The search space of suitable behaviors is huge and designing suitable behaviors by

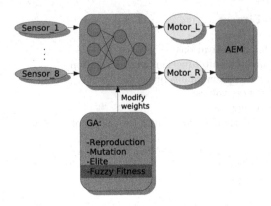

Fig. 10. Fuzzy fitness and AEM implementation

hand is very difficult therefore Yamada [10] has used a genetic algorithm within a Khepera simulator to find suitable behaviors for AEM.

In the training phase of the AEM procedure and for each room of the set of rooms being recognized the robot executes an action sequence which is converted into an environment vector. These vectors are repeatedly fed into a SOM [22] network in order for the neural network to learn without supervision which is the output node (r-node) that corresponds to each room. The environment vector used for each room and the winning output node is also stored as a room instance. The next step is a test phase in which the robot executes an action sequence in one of the rooms previously used and the r-node for the test room is determined using the previously trained SOM network. The robot then determines which room used during training has the minimum distance to the current test room by using 1-Nearest Neighbor with Euclidean distance [10].

Fig. 10 shows how fuzzy fitness, GA and AEM are used toward training an agent.

As seen in Fig. 11, to implement AEM, we select a highly fit agent (corresponding to the neural network in Fig. 10) and make him navigate in all environments (rooms). This navigation produces actions which are saved as action sequences. These action sequences are converted using chain coding into an environment vector [10]. These vectors are fed into the SOM network for unsupervised learning. After learning the SOM network associates a vector with one of its output nodes (r-nodes).

Implementation Environment. We used the simulator YAKS (Yet Another Khepera Simulator) for our agent implementation. YAKS is a simple open source behavior based simulator [9] that uses neural networks and genetic algorithms in order to provide a navigation environment for a Khepera robot. Sensor inputs are directly provided into a multilayer neural network in order to drive left and right wheel motors. A simple genetic algorithm is used with 200 members, 200 generations, mutation of 1%, and elite reproduction. Random noise (6%) is injected into motor actions and sensors to improve realism. The GA provides

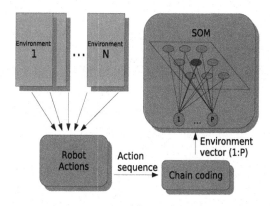

Fig. 11. AEM Overview

a mechanism for updating neural network weights used by each robot in the population that is being optimized [6].

For our implementation of SOM we select an input of 400 actions (steps) and a linear output layer of 128 nodes. The actions are obtained by converting each motor command (Motor_L and Motor_R) which is a real value between 0 and 1 into an action (left 30°, right 30°, turn 180°, go straight).

After training SOM we evaluate it by using the same robot and at random select a room for him to navigate in. The SOM network is evaluated by the ability of the robot to recognize which room (neural output node or r-node) it navigates in.

5   Experimental Evaluation

In order to evaluate motivation based behavior learning we performed several experiments. These experiments included: navigation, environment recognition, and the effects of various motivations on robot behavior. All these experiments were also a test of whether the motivation based interface could aid end-user understanding of agent behavior. As previously mentioned, the motivation set (M) considered included: homing (m_1), curiosity (m_2), energy (m_3), and orientation (m_4). The set of fitness criteria and the fuzzy variables were: proper action termination and escape from original neighborhood area (f_1), amount of area explored (f_2), percent of battery usage (f_3) and environment recognition (f_4) [6].

5.1   Scenario 1: Navigation

In this scenario, the agent was trained and then made to navigate in each of five rooms in order. The rooms have the following shapes: H, T, L, Square and Rectangle. In order to contrast the effect of curiosity we selected two sets of motivation criteria, the first motivation criterion was $(0.1, 0.8, 0.1, 0)$ and the second was $(0.3, 0.5, 0.2, 0)$. We stopped each experiment after 400 steps. As seen in the examples of Fig. 12, during the simulations performed average fuzzy

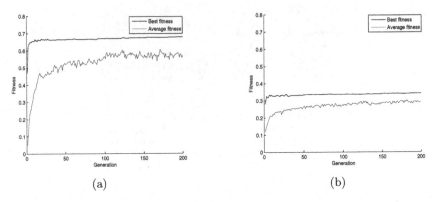

Fig. 12. Fitness Evolution Examples: (a) high curiosity (b) low curiosity

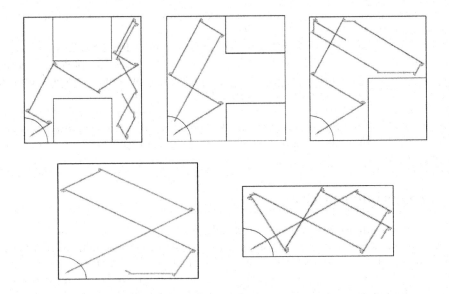

Fig. 13. Navigation with M as $(0.1, 0.8, 0.1, 0)$

fitness in the first generation was 0.1 and gradually increased. Figs. 13 and 14 show representative results of navigation by various robots.

By changing the different motivation factors the characteristics of robot navigation were changed. With lower curiosity and higher homing the robot conserved its battery (battery usage was around 50%) and the robot managed to explore only around 50% of the region. With higher curiosity the agent explored more and used a higher amount of battery but it did not return home very often. Agent learning was very rapid in both cases and in general by the 10th generation good solutions were found to the navigation problem.

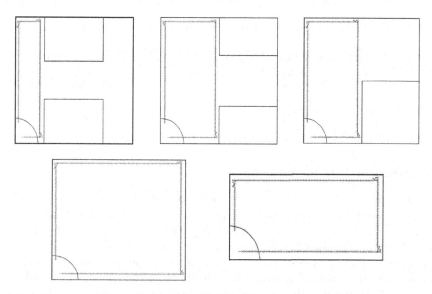

Fig. 14. Navigation with M as $(0.3, 0.5, 0.2, 0)$

Table 1. Navigation and Environment Recognition M as $(0.1, 0.8, 0.1, 0)$

Room shape	Fuzzy Fitness	% of Optimal Exploration	% Recognition
H	0.6421	93.84	80.00
T	0.6686	91.77	100.00
L	0.6507	97.04	100.00
Square	0.6699	99.88	100.00
Rectangle	0.6377	95.29	100.00

5.2 Scenario 2: Environment Recognition

The agent was trained and then made to navigate and recognize in the same five rooms as the previous scenario. For comparison we selected two sets of fuzzy motivation criteria (M), the first criterion used was $(0.1, 0.8, 0.1, 0)$ and the second was $(0.3, 0.5, 0.2, 0)$. We stopped each experiment after 400 steps.

The procedure was repeated 10 times for each room. The threshold for r-node recognition was set at 10 nearest neighbor nodes in the SOM output layer. Environment recognition is the % recognition of which room the agent actually visited. Tables 1 and 2 show our results.

Different motivation factors affected the capacity of the robot to recognize its environment. With lower curiosity the robot managed to recognize its environment much less than with higher curiosity values. With higher curiosity (m_2), the environment vector generated was more representative of the room shape which would explain this result.

Table 2. Navigation and Environment Recognition with M as $(0.3, 0.5, 0.2, 0)$

Room shape	Fuzzy fitness	% of Optimal Exploration	% Recognition
H	0.2465	54.66	20.00
T	0.2694	57.84	0.00
L	0.2707	57.93	0.00
$Square$	0.3274	80.63	100.00
$Rectangle$	0.267	57.70	0.00

5.3 Scenario 3: Behavioral Effects

For this experiment, in order to see the behavioral effects of a variety of motivations, we used four sets of motivations. In order to have an objective comparison, indicated score values are a weighted combination (according to M) of the obtained behavior fitnesses $\{f_1, f_2, f_3, f_4\}$.

Training consisted of consecutively training the population in three different randomly selected rooms (based on our own office layouts) $r_a - r_c$. The population was trained for 30 generations in each of the three rooms in order. During testing the best individual of the 90 generations was tested in six rooms $r_a - r_f$.

Ten complete runs were performed of each experiment (each run consisting of one training and 10 test executions) and average values are reported in our results.

Table 3. Exp. 2: Behaviors for M as $(0, 1, 0, 0)$.

Room	Score	Homing	% Exploration	% Battery usage
r_a	0.92	0.44	91.48	79.18
r_b	0.93	0.67	92.92	79.32
r_c	0.91	0.56	91.23	79.60
r_d	0.92	0.48	91.82	78.73
r_e	0.92	0.49	91.68	79.14
r_f	0.89	0.48	88.82	79.47
Avg.	0.91	0.52	91.33	79.24

Table 4. Exp. 2: Behaviors for M as $(0.45, 0.55, 0, 0)$.

Room	Score	Homing	% Exploration	% Battery usage
r_a	0.86	0.91	82.50	80.13
r_b	0.89	0.93	85.03	81.49
r_c	0.89	0.95	84.28	81.31
r_d	0.86	0.92	81.50	81.46
r_e	0.90	0.93	86.75	81.44
r_f	0.87	0.94	80.94	81.30
Avg.	0.88	0.93	83.50	81.19

Table 5. Exp. 2: Summary of Behaviors for (M) as $(0.25, 0.5, 0.25, 0)$.

Room	Score	Homing	% Exploration	% Battery usage
r_a	0.83	0.83	84.85	79.49
r_b	0.88	0.96	88.14	80.88
r_c	0.88	0.94	87.65	81.41
r_d	0.85	0.86	85.92	80.67
r_e	0.85	0.83	88.08	81.14
r_f	0.86	0.91	84.94	81.42
Avg.	0.86	0.89	86.60	80.84

Table 6. Exp. 2: Behaviors for (M) as $(0.09, 0.41, 0.09, 0.41)$.

Room	Score	Homing	% Exploration	% Battery usage	% Recognition
r_a	0.86	0.43	81.38	80.44	100
r_b	0.50	0.68	88.26	80.77	0
r_c	0.48	0.61	85.20	80.75	0
r_d	0.63	0.46	76.52	75.59	50
r_e	0.90	0.62	87.22	80.89	100
r_f	0.46	0.57	82.08	80.39	0
Avg.	0.64	0.56	83.44	79.81	41.67

In tables 3 - 6 we show the results of the testing phase after applying the respective training method specified for these experiments.

Greater diversity in the specified motivations seem to cause lower average score values. This could be attributed to the many conflicting requirements of these different motivations upon a constrained agent brain (i.e. neural network). Even though obtained fitness was generally lower with more diverse motivations, the obtained behaviors demonstrated very good capability (e.g. room exploration, battery usage, etc) and were in close agreement with the specified motivations. In our experiments, environment recognition (e.g. SOM) was generally successful in determining the robots location.

6   Conclusions

This method was shown as an effective mechanism for generating a variety of agent behaviors within a framework that is simple and intuitive to understand. Using motivations in the context of behavior based systems seems to have been validated by our experimental results. As seen by our results, motivations provide a mechanism by which to set testable end-user expectations of behavior based agents. There is a large potential area of applications as these lessons would apply to both physical (e.g. robots) and non physical implementations (e.g. web and video game agents).

Clearly there is a benefit in having intuitive agent interfaces so that end-users can configure and understand (at a very basic level) what their agent is doing. No one wants an agent that has to be told exactly how to act, but we would also like an agent that roughly behaves in an expected manner. This is a delicate balance and the user should have a means of managing it. If these behaviors were to not match end-user expectations then there would be a need for root cause analysis (as with any such situation) but the linkage (or lack thereof) between observed behaviors fitnesses and previously set motivations could provide some clues. In this case, the fuzzy logic rules could be changed in order to make behaviors better match user specified motivations or the user could simply tweak his choice of motivations.

In our experiments, behavior based agents rapidly learned behaviors that fulfilled required tasks but failed to improve their objective scores when a more complex set of motivations was given. This is probably caused by limits on the inference engine being used vis-á-vis problem complexity and is an area of potential further research.

This method should be extensible into multiple-agent systems by giving agents different roles (e.g. mapper, reactive navigator, strategic planner/navigator) in the overall system. As in any hybrid (i.e. deliberative and behavior based) system, representation/reasoning issues would have to be carefully considered in order to deliver a good implementation.

Future work includes the integration of other motivations such as urgency for completing specific missions, integration into multiple-agent systems (MAS), the construction of a robot specifically designed to test this method, using a variable set of motivations (e.g. personality profiles), and possible real-time extensions to this method.

Acknowledgements

This research was partially funded by the research authority (DGIP) of UTFSM (230808). The author would also acknowledge the contributions of Wolfgang Freund and Ethel Vidal Hintze in the review of this chapter.

References

1. Woolridge, M.: An Introduction to MultiAgent Systems. Wiley, England (2002)
2. Park, H., Kim, E., Kim, H.: Robot Competition Using Gesture Based Interface. In: Moonis, A., Esposito, F. (eds.) Innovations in Applied Artificial Intelligence. LNCS (LNAI), vol. 3353, pp. 131–133. Springer, Berlin (2005)
3. Tasaki, T., Matsumoto, S., Ohba, H., Toda, M., Komatani, K., Ogata, T., Okuno, H.: Distance-Based Dynamic Interaction of Humanoid Robot with Multiple People. In: Moonis, A., Esposito, F. (eds.) Innovations in Applied Artificial Intelligence. LNCS (LNAI), vol. 3353, pp. 111–120. Springer, Berlin (2005)
4. Jensen, B., Tomatis, N., Mayor, L., Drygajlo, A., Siegwart, R.: Robots Meet Humans - Interacion in Public Spaces. IEEE Transactions on Industrial Electronics 52(6), 1530–1546 (2006)
5. Arkin, R.: Behavior-Based Robotics. MIT Press, Cambridge (1998)

6. Arredondo, T., Freund, W., Muñoz, C., Navarro, N., Quirós, F.: Fuzzy Motivations for Evolutionary Behavior Learning by a Mobile Robot. In: Ali, M., Dapoigny, R. (eds.) IEA/AIE 2006. LNCS (LNAI), vol. 4031, pp. 462–471. Springer, Heidelberg (2006)
7. Pezzulo, G., Calvi, G.: Modulatory Influence of Motivations on a Schema-Based Architecture: A Simulative Study. In: Paiva, A.C.R., Prada, R., Picard, R.W. (eds.) ACII 2007. LNCS, vol. 4738, pp. 374–385. Springer, Heidelberg (2007)
8. Brooks, R.: A Robust Layered Control System for a Mobile Robot. IEEE Journal of Robotics and Automation RA-2(1), 14–23 (1986)
9. YAKS simulator website, http://freshmeat.net/projects/yaks/
10. Yamada, S.: Evolutionary behavior learning for action-based environment modeling by a mobile robot. Applied Soft Computing 5, 245–257 (2005)
11. Jang, J., Chuen-Tsai, S., Mitzutani, E.: Neuro-Fuzzy and Soft Computing. Prentice-Hall, Englewood Cliffs (1997)
12. Konolige, K., Meyers, K., Saffiotti, A.: FLAKEY, an Autonomous Mobile Robot. SRI technical document, July 20 (1992)
13. Goodrige, S., Kay, M., Luo, R.: Multi-Layered Fuzzy Behavior Fusion for Reactive Control of an Autonomous Mobile Robot. In: Proceedings of the Sixth IEEE International Conference on Fuzzy Systems, July 1997, pp. 573–578 (1997)
14. Hoffman, F.: Soft computing techniques for the design of mobile robot behaviors. Information Sciences 122, 241–258 (2000)
15. Al-Khatib, M., Saade, J.: An efficient data-driven fuzzy approach to the motion planning problem of a mobile robot. Fuzzy Sets and Systems 134, 65–82 (2003)
16. Izumi, K., Watanabe, K.: Fuzzy behavior-based control trained by module learning to acquire the adaptive behaviors of mobile robots. Mathematics and Computers in Simulation 51, 233–243 (2000)
17. Martnez Barber, H., Gmez Skarmeta, A.: A Framework for Defining and Learning Fuzzy Behaviours for Autonomous Mobile Robots. International Journal of Intelligent Systems 17(1), 1–20 (2002)
18. Zhou, C.: Robot learning with GA-based fuzzy reinforcement learning agents. Information Sciences 145, 45–68 (2002)
19. Seraji, H., Howard, A.: Behavior-Based Robot Navigation on Challenging Terrain: A Fuzzy Logic Approach. IEEE Trans. on Robotics and Automation 18(3), 308–321 (2002)
20. Huitt, W.: Motivation to learn: An overview. In: Educational Psychology Interactive, Valdosta State University (2001),
http://chiron.valdosta.edu/whuitt/col/motivation/motivate.html
21. Jang, J.-S., Sun, C.-T., Sun, M.E.: Neuro-Fuzzy and Soft Computing: a computational approach to learning and machine intelligence. Prentice-Hall, Englewood Cliffs (1997)
22. Teuvo, K.: The self-organizing map. Proceedings of the IEEE 79(9), 1464–1480 (1990)
23. Norman, T.J.: Motivation-based direction of planning attention in agents with goal autonomy, PhD Thesis, Department of Computer Science, University College of London, UK (1997)
24. Coddington, A.M., Luck, M.: A Motivation Based Planning and Execution Framework. International Journal on Artificial Intelligence Tools 13(1), 5–25 (2004)
25. Karray, F.O., De Silva, C.: Soft Computing and Intelligent Systems Design. Addison Wesley, England (2004)
26. Passino, K.: Biomimicry for Optimization, Control and Automation. Springer, London (2005)

27. Bylander, T.: The computational complexity of propositional STRIPS planning. Artificial Intelligence 69, 165–204 (1994)
28. Cherniak, C.: Minimal Rationality. MIT Press, Boston (1986)
29. Bratman, M.: Plans and Resource-Bounded Practical Reasoning. Computational Intelligence 4(4), 349–355 (1988)
30. Rao, A.S., Georgeff, M.P.: BDI Agents From Theory to Practice. In: Proceedings of the First International Conference on Multi Agent Systems (1995)
31. Georgeff, M.P., Ingrand, F.F.: Decision-Making in an Embedded Reasoning System. In: Proceedings of the Eleventh International Joint Conference on Artificial Intelligence (1989)
32. Agile Development at the Wikipedia,
    http://en.wikipedia.org/wiki/Agile_software_development
33. Iterative Development at the Wikipedia,
    http://en.wikipedia.org/wiki/Iterative_and_incremental_development
34. Zadeh, L.A.: Fuzzy Logic, Neural Networks, and Soft Computing. Communications of the ACM 37(3), 77–84 (1994)

Designing Optimal Operational-Point Trajectories Using an Intelligent Sub-strategy Agent-Based Approach

Zdzislaw Kowalczuk and Krzysztof E. Olinski

Gdansk University of Technology, Department of Decision Systems,
Gdansk, Poland
{kova,kolin}@eti.pg.gda.pl

Abstract. This paper presents a method intended for designing optimal and safe control for nonlinear dynamical processes. The sought control signal results from elementary control strategies induced by different agents implementing their (partial) task of minimizing a common control cost measure (index). The issue of designing optimal control is therefore treated as a decision process, where the decisions are made in particular regions of the state space of the dynamical process under consideration. The regions thus constitute local decision spaces being searched by a group of agents in a multistage searching procedure. At each stage, every agent can increment its cost index only by a limited value. This guarantees that at the end of each stage all the agents represent control strategies which are cost equivalent (approximately). The algorithm starts off by generating an initial population of agents (each for one of the previously defined elementary control strategies). Each of these agents realizes a different kind of possible elementary control strategies, which determine predefined agent behaviors. When an agent reaches one of the decision regions, it generates a new/local population of the seeking/hunting agents (they are, again, of different kinds of the elementary control strategies). After getting explored, such a decision region turns to a forbidden zone for all agents but those belonging to the newly created population. In such a way the successive populations of the agents allow to complete the path to a prearranged destination point in a competitive way. The first agent which reaches the destination area in the state space determines an optimal solution in the sense of the above assumptions.

1 Introduction

The notion of 'agent system' refers, in general, to an algorithmic architecture in which autonomous entities (referred to as agents) observe and act upon an environment and directs towards achieving goals [1,2]. The agents are popular in the fields of telecommunication, computer science, crowd simulation, etc. Different forms of multi-agent systems can also be applied in control theory problems. In particular, such techniques can be found in team-oriented formations of robots [4]

E. Szczerbicki & N.T. Nguyen (Eds.): Smart Infor. & Knowledge Management, SCI 260, pp. 273–282.
springerlink.com

and fault-tolerant strategies of keeping robotic formations [3]. The idea which stands behind the presented method slightly differs from the general definition of the multi-agent system. While, for instance, in [3,4] the authors consider a group of agents which search through the state space, in our approach both the interaction and 'intelligence' of the agents are very limited. The agent realizes a prescribed control strategy without acting upon the environment, which in our case will be represented by a known distribution of the autonomous dynamics of the process under consideration. The mechanism of interacting between agents is reduced to consecutively excluding some neighborhoods of certain decision points from the further exploration process. Nevertheless, the parallel and distributed type of processing is still utilized in the proposed search procedure.

The principal aim of this paper is thus to present a method for designing trajectories of the system operational points (and the corresponding control signals) using an 'intelligent' sub-strategy agent-based approach. The operational point trajectory sought is a simple combined sequence of the control signals being the elements of a set $\mathcal{U}$ which represent predefined elementary (partial) control strategies. The agent encapsulates all the necessary quantities needed to describe the partial control strategy, along with control cost. The algorithm starts off by locating some N_u agents at a given initial point of the state-space (each for one of the previously defined elementary control strategies). The searching procedure is divided into N_s stages (where N_s is not a predefined fixed value but results from the course of the searching process). At each stage, every agent can increment its own cost index only by a certain fixed value. This guarantees that at the end of each stage all the agents present the control strategies which are treated as cost-equivalent. There are N_d decision points defined in the state-space. If an agent is the first to visit a defined neighborhood of a given decision point, it becomes a **predecessor** for a new local population by generating its agents, one for each elementary control strategy. These new agents memorize the control strategy of their predecessor, along with its control cost value. The decision point being visited is marked as inactive. An inactive decision point and its neighborhood become a forbidden zone for all agents but those belonging to the newly created population (which means that the agents are forbidden from entering it). This rule makes allowances for controlling the size of the agent population. When one of the agents enters the defined terminal region, the search procedure stops. This agent and the history of the control strategies of its predecessors represent the sought solution.

2   Formal Description

This section comprises a terminology, definitions and a raw algorithm description along with all the rules necessary for an implementation.

Let us start the general problem formulation by defining the <u>autonomous</u> and <u>forced</u> components of the dynamics of a given process. Let us assume that the function which describes the process dynamics has the following form:

$$\dot{\mathbf{x}}(t) = \mathbf{f}(\mathbf{x}(t), \mathbf{u}(t)) \tag{1}$$

and belongs to the class

$$\mathbf{f} : \mathcal{R}^{N_x} \times \mathcal{R}^{N_u} \to \mathcal{R}^{N_x} \tag{2}$$

Definition 1. *Consider the following decomposition of (1) into two parts:*

$$\dot{\mathbf{x}} = \dot{\mathbf{x}}_x + \dot{\mathbf{x}}_u = \mathbf{f}_x(\mathbf{x}) + \mathbf{f}_u(\mathbf{x}, \mathbf{u}) \tag{3}$$

where $\mathbf{f}_u(\mathbf{x}, 0) = 0$. With the above let us introduce the following nomenclature:
$\dot{\mathbf{x}}_x = \mathbf{f}_x(\mathbf{x})$ represents the <u>autonomous dynamics</u>,
$\dot{\mathbf{x}}_u = \mathbf{f}_u(\mathbf{x}, \mathbf{u})$ portrays the <u>forced dynamics</u>,
$\mathbf{x}_x, \mathbf{x}_u$ are their two corresponding <u>state contributors</u>.

Definition 2. *The sequence $E(\mathbf{u})$ of consecutive states of a given dynamical system for a feasible control input $\mathbf{u}$ is said to be the **system's trajectory**, or the **trajectory of operational points** in the state space of the system.*

Definition 3. *A bounded subset P of the state-space in the form of a hypercube in $\mathcal{R}^{N_x}$, which is taken into account while seeking an optimal trajectory, is said to be the **operational workspace**.*

Definition 4. *A subset Z of P which is prohibited for operational points is referred to as the **forbidden zone**. This means that the sought optimal trajectory cannot enter it.*

Definition 5. *Let $\mathcal{A}$ be the set of all trajectories: $E \in \mathcal{A} \subset P$. Any real function of the class $\mathcal{A} \to \mathcal{R}^+$ can be considered as a **cost function** $J(E(\mathbf{u}))$ of these trajectories.*

Definition 6. *Let $\Xi \subset \mathcal{A}$ be the subset of all possible trajectories which start at a given point x_0 and terminate at x_k. We say that a trajectory E^* is **optimal** if it satisfies the following conditions:*

$$\forall_{E \in \Xi} J(E) \geq J(E^*) \tag{4}$$

$$\forall_{\mathbf{x} \in E^*} \mathbf{x} \in P \setminus Z \tag{5}$$

Definition 7. *Any function which describes feasible control as a function of the operational point in the state-space can be assumed an **elementary control strategy**.*

$$\mathbf{u}(t) = \mathbf{f}_u(\mathbf{x}) \tag{6}$$

In a complete set $\mathcal{U}$ of N_u elementary strategies:

$$\mathcal{U} = \{\mathbf{f}_{u1}, \dots, \mathbf{f}_{uN_u}\} \tag{7}$$

each elementary strategy $\mathbf{f}_{ui}$ is identified by its index $i = 1 \dots N_u$.

Definition 8. *A **combined control strategy** spanned on a time interval $\langle t_0, t_k \rangle$ is composed as a sequence of N_e elementary control strategies taken from the set U:*

$$\mathbf{U}(t_0, t_k) = \{\mathbf{f}_{ui_1}(t_0, t_1), \mathbf{f}_{ui_2}(t_1, t_2), \ldots, \mathbf{f}_{ui_k}(t_{k-1}, t_k)\} \qquad (8)$$

where $i_1, i_2, \ldots, i_k \in \{1, 2, \ldots, N_e\}$.

Definition 9. *The **decision point and its neighborhood**, determined in the state space of a given process, defines a region where one agent (the first reaching it) is permitted to generate a new (local) population of agents. Once visited, such a region becomes a forbidden zone for all agents but those belonging to the newly created population.*

Definition 10. *The **agent** is a representation of an elementary control strategy. Each agent is characterized by the following quantities:*

- *the realized elementary control strategy*
- *the coordinates ($\mathbf{x}$) of the operational point, which is subject to the elementary control strategy being realized*
- *the control cost index*
- *the history of the elementary control strategies (previously) utilized by the agent's predecessors.*

2.1   Algorithm Description

The optimization problem to be solved shall be stated according to the following points:

- the description of the process dynamics in the form of the function $\dot{\mathbf{x}} = \mathbf{f}(\mathbf{x}, \mathbf{u})$ (we assume that initial conditions and the control trajectory define unambiguously the trajectory of the process states),
- the operational subspace P subject to the search process,
- the initial and the terminal points (or regions) of the sough trajectory in the state space,
- the set U of the elementary control strategies (components of the sought control),
- the forbidden zones Z (optional),
- the cost function,
- the value $\Delta_{max}J$ by which, each of the agents can increment its own cost index at a single stage.

The proposed algorithm can be described as follows:

- <u>Initialization.</u> The algorithm starts off by locating N_u agents in the initial point of the trajectory in the process state space. Each of the N_u agents realizes a unique control strategy (from the set U).

- Search procedure. The search procedure is divided into N_s stages. At each stage, every agent implements its own control strategy (defined by a respective function from the set $\mathcal{U}$). Agents (with their operational points) entering forbidden zones are removed. The stage ends if none of the agents can further perform its own control strategy. An agent cannot perform its control strategy (i.e. disappears) if, at least, one of the following conditions is met:
  - the agent's operational point get out of the operational subspace P,
  - the agent's operational point enters the forbidden zone,
  - the agent enters the neighborhood of one of the allowable decision points and generates a new (local) population.

  The above search procedure ends if all the agents disappear (die out), or if one of them enters the pre-defined terminal region.
- Reconstruction of the sought trajectory. If one of the agents enters the terminal region, reconstruction of the sought trajectory takes place. The agents store the information about their predecessor control strategies. Based on these data the sought control signal trajectory is compiled.

3 Examples

Consider the system described by the following equations:

$$\dot{x}_1 = \begin{cases} \frac{4x_2+4.2}{2} + u_1 & \text{for } x_1 \in (-1.5, -0.5) \\ -x_2 + u_1 & \text{for } x_2 \in (-\infty, -1.5) \cup (-0.5, +\infty) \end{cases}$$

$$\dot{x}_2 = \begin{cases} \frac{\sin(4x_1)}{2} + u_2 & \text{for } x_1 \in (-1.5, -0.5) \\ -x_1 + x_1^3 + 0.4x_2 - x_1^2 x_2 + u_2 & \text{for } x_2 \in (-\infty, -1.5) \cup (-0.5, +\infty) \end{cases} \quad (9)$$

Assume that the operational workspace P is represented by a rectangle defined by the two vertex pairs :$\{(-2,-2),(2,2)\}$. The corresponding autonomous dynamics map is presented in Fig. 1. In the following problems we consider the task of finding the optimal operational-point trajectory. Let us assume that the set of elementary control strategies has the following form:

$$\mathcal{U} = \begin{cases} \mathbf{f}_{u1}(\mathbf{x},\mathbf{u}) = [u_1 = u_{max}, & u_2 = 0], \\ \mathbf{f}_{u2}(\mathbf{x},\mathbf{u}) = [u_1 = u_{max}, & u_2 = -u_{max}], \\ \mathbf{f}_{u3}(\mathbf{x},\mathbf{u}) = [u_1 = 0, & u_2 = -u_{max}], \\ \mathbf{f}_{u4}(\mathbf{x},\mathbf{u}) = [u_1 = -u_{max}, & u_2 = -u_{max}], \\ \mathbf{f}_{u5}(\mathbf{x},\mathbf{u}) = [u_1 = -u_{max}, & u_2 = 0], \\ \mathbf{f}_{u6}(\mathbf{x},\mathbf{u}) = [u_1 = -u_{max}, & u_2 = u_{max}], \\ \mathbf{f}_{u7}(\mathbf{x},\mathbf{u}) = [u_1 = 0, & u_2 = u_{max}], \\ \mathbf{f}_{u8}(\mathbf{x},\mathbf{u}) = [u_1 = u_{max}, & u_2 = u_{max}], \\ \mathbf{f}_{u9}(\mathbf{x},\mathbf{u}) = [u_1 = 0, & u_2 = 0] \end{cases} \quad (10)$$

where $u_{max} = 10.0$ is the largest absolute value of the control signal.

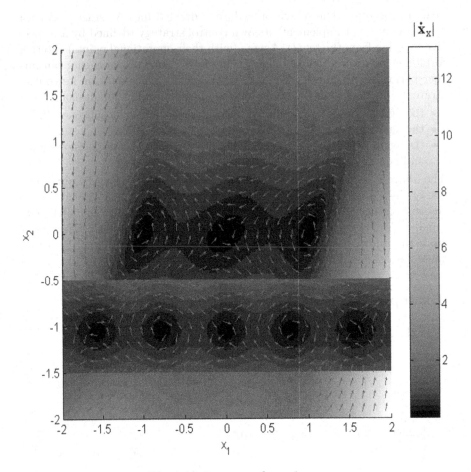

Fig. 1. Autonomous dynamics map.

The cost criterion is given in the following form:

$$J(E(\mathbf{u})) = \int_0^T (\sum_{i=1}^{N_x} |u_i(t)|)dt \qquad (11)$$

where T is a transition time interval resulting from the applied optimal control procedure.

The set of the decision points takes shape of a regular grid of the size 0.4×0.4. The neighborhood around each of the decision points forms a rectangle region of the size 0.3×0.3.

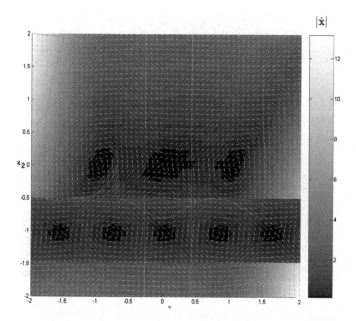

Fig. 2. Computed optimal trajectory for Problem 1.

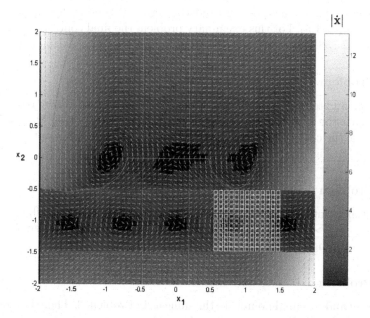

Fig. 3. Resulting optimal trajectory for Problem 2. The forbidden zone is marked as a rectangular region.

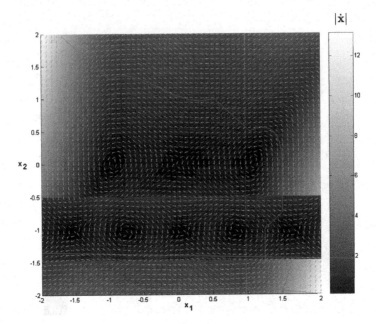

Fig. 4. Optimal operational-point trajectory for Problem 3.

3.1   Problem 1

The initial point is $\mathbf{x}_0 = (-2, 2)$ and the terminal zone is described by $\mathbf{x}_T = (x_1 > 1.5, x_2 > 1.5)$. For this exemplary problem the final control cost index has been $J_1 = 2.738$, and the corresponding sought trajectory is depicted in Fig. 2.

3.2   Problem 2

The initial and terminal points are the same as in Problem 1. There is a forbidden zone given in the form of a rectangle defined by the vertices $(0.5, -0.5), (-1.5, -1.5)$. In this case the control cost indicator has been $J_2 = 3.4$, and the optimal solution is depicted in Fig. 3.

3.3   Problem 3

The initial point is $\mathbf{x}_0 = (2, -2)$ and the terminal zone is described by $\mathbf{x}_T = (x_1 < -0.5, x_2 > 1.5)$. This time, the control cost indicator has been $J_3 = 3.782$, and the resulting optimal trajectory is shown in Fig. 4.

3.4   Problem 4

The initial and terminal points are the same as in Problem 3. There is a forbidden zone in the shape of a rectangle defined by the vertices $(0.5, 0.5), (1.5, 1.5)$. The control cost for this example has been $J_4 = 3.79$, while the results of optimization are illustrated in Fig. 5.

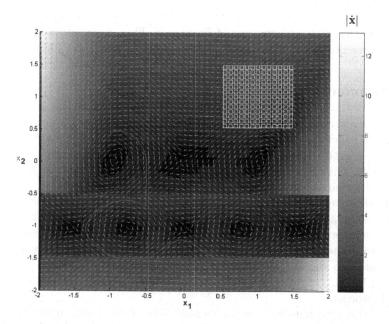

Fig. 5. Optimal trajectory for Problem 4. The forbidden zone is marked as a rectangle.

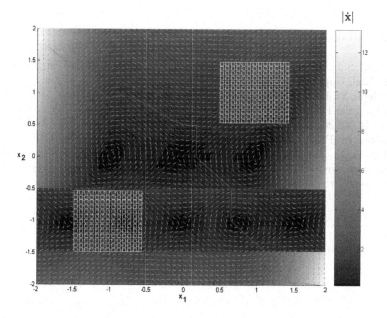

Fig. 6. Optimal trajectory for Problem 5. The forbidden zone is denoted as a rectangle.

3.5   Problem 5

The last example is an extension of Problem 4. There is an additional forbidden zone represented by the rectangle of the vertices $(-1.5, -1.5), (0.5, 0.5)$.

4   Summary

This paper presents an intelligent sub-strategy agent-based approach to optimization, in which the search for the optimal solution is performed simultaneously by a group of agents. A compilation of the partial solutions delivered by the agents, which results in shifting the operational point from the initial to the designed terminal point, forms the final solution being sought.

The effectiveness of the presented method in the case of a high order dynamical processes can as well turn out to be unsatisfactory. The worst scenario in terms of a total number of the agents present in the operational state space can be easily calculated. The resulting trajectory (if exists) represents a sub-optimal solution, what refers to the fact that the set of control strategies with the control cost varying by $\Delta_{max}J$ are treated as 'cost equivalent' during a single stage of the search process.

In further research steps one can consider sensitivity of this approach to the set specification of the elementary control strategies and to the applied arrangement of the decision points in the state space. Other developments in this area can deliberate more sophisticated agent behaviors and interacting mechanisms.

References

1. Russell, S.J., Norvig, P.: Artificial Intelligence: A Modern Approach, 2nd edn. Prentice Hall, Upper Saddle River (2003)
2. Shoham, Y., Leyton-Brown, K.: Multiagent Systems: Algorithmic, Game-Theoretic, and Logical Foundations. Cambridge University Press, Cambridge (2008)
3. Sun, B., Chen, W., Xi, Y.: Team-Oriented Formation Control for Multiple Mobile Robots. In: Proc. of the IFAC Congress, Prague (Czech Republic), IFAC (2005)
4. Abel, R.O., Dasgupta, S., Kuhl, J.G.: Coordinated fault-tolerant control of autonomous agents: Geometry and communications architecture. In: Proc. of the IFAC Congress, Prague (Czech Republic), IFAC (2005)

An Ontology-Based System for Knowledge Management and Learning in Neuropediatric Physiotherapy

Luciana V. Castilho and Heitor S. Lopes

Federal University of Technology - Paraná,
Bioinformatics Laboratory / CPGEI
Av. 7 de setembro, 3165, Curitiba, Brazil
luciana.neurologia@gmail.com, hslopes@utfpr.edu.br

Abstract. This chapter first presents an extensive review of the current state of art in knowledge management and ontologies. Next, we propose a methodology for modeling and building an ontology-based system for knowledge management in the domain of Neuropediatric Physiotherapy and its application to supporting learning. This area of Physiotherapy includes diagnosis, treatment and evaluation of patients with neurological injuries. The domain knowledge in Physiotherapy is, by nature, complex, ambiguous and non-standardized. In this work knowledge was elicited from domain experts and complemented with information from reference textbooks. The acquired knowledge was represented as an ontology. The formal procedures allowed the development of a knowledge-base for further use in an educational tool. The completeness and consistency of formal model was verified. Overall, the main contribution of the work are a domain ontology based on consensus vocabulary for an important area of health sciences, and the possibility of using it as a tool for supporting the learning of undergraduate students. In particular, the application of the ontology for learning in Physiotherapy is of great importance, since it includes multimedia resources as well as active learning concepts, together with traditional instructional methods.

Keywords: ontology, knowledge management, neuropediatric physiotherapy, learning.

1 Introduction

Similarly to Medicine, Physiotherapy also has different areas of specialization. One of them is the Neuropediatric Physiotherapy that includes diagnosis, technical procedures and continuous evaluation of patients that have motor or postural diseases due to lesions in the central nervous system [1].

There are many reference publications focusing all aspects of diagnosis and clinical treatment in Neuropediatric Physiotherapy. However, not all physiotherapists have extensive knowledge of such domain [1].

Recent developments of information technology and the widespread availability of the internet have lead to huge amounts of data in all segments of human knowledge, including those related with health sciences [2]. Physiotherapy in general, and, more

E. Szczerbicki & N.T. Nguyen (Eds.): Smart Infor. & Knowledge Management, SCI 260, pp. 283–307.

specifically, Neuropediatric Physiotherapy, is a domain where knowledge is subjective by nature and concepts are poorly systematisized. This has been the main drawback for creating a consensus vocabulary and, consequently, sharing and reuse of data, information and knowledge. Efforts towards this issue would allow efficient management of technical knowledge in this area, by organizing, validating, maintaining and spreading the available expert knowledge. As side effect, both teaching and learning could be enhanced, by introducing formalized concepts and vocabulary.

Modeling and developing a formal structure for representing knowledge in the domain of Neuropediatric Physiotherapy can be of great interest and an important contribution not only for Physiotherapy, but also, for other health-related areas. Such areas are frequently characterized by subjectiveness and the use of non-standardized information. Therefore, they could benefit from the use of knowledge management methodologies.

An ontology is a formal description of a given knowledge domain based on concepts and relationships. Recent literature has demonstrated that such approach is an efficient way to structure knowledge in many areas. This is the formal approach used in this work, which methodology can be extended to other similar areas.

Besides the importance of ontologies for knowledge management, we will show that a developed ontology can be also useful for the teaching-learning process in the related area. Ontologies are frequently used for the development of consensus vocabulary. However, the use of ontologies for learning is poorly explored, especially in the health-related areas.

The objectives of this work are: (1) apply formal procedures for knowledge management in the specific domain of Neuropediatric Physiotherapy; (2) develop a reusable and extensible ontology for representing knowledge in that domain; (3) propose a methodology for using the developed ontology as an educational tool in Neuropediatric Physiotherapy.

2 Knowledge Acquisition and Representation

In the Artificial Intelligence (AI) area, the word "knowledge" means the information that a computer program needs to solve problems in such a way considered intelligent [3].

Knowledge is made up of data and information [4] Data are raw, isolated facts. Information is a set of organized facts. The term information is defined in a more generic sense as knowledge obtained from investigation, study, or instruction. Finally, knowledge is information within a context [5]. Knowledge leverages experience and interpretation to make sense out of information and data. In other words, knowledge is a set (information) of facts (data) and relationships (context) used or needed to obtain insight or to solve a problem [6].

Knowledge can be of two types: explicit and tacit (or implicit) [7]. Explicit knowledge is the one that is available in concrete media (such as books or CD-ROMs) and can be easily shared among people. The tacit knowledge refers to the individual knowledge that aggregates the experience and intuition of each one. Tacit knowledge is implied or understood from the context without being actually stated. It is accepted that people knows much more than they can speak about or transmit [8]. Therefore, the knowledge acquisition process from a given expert domain may be quite difficult, if tacit knowledge is wanted to acquire.

Overall, the knowledge acquisition and representations can be considered as a linear and hierarchical progression, in which data are converted into information, and information is converted into knowledge [4].

2.1 Knowledge Acquisition

The Knowledge Acquisition (KA) process includes elicitation, transformation and transfer of information from a knowledge source to a computer program. The objective of KA is to obtain specialized knowledge from an expert to solve problems [9].

The KA process is usually divided into two stages: initial analysis, when it is decided which knowledge is necessary; and knowledge elicitation and interpretation, when the knowledge itself is acquired from the expert [3].

The main potential knowledge sources are the human experts. Also, other sources of explicit knowledge are considered as complimentary, such as textbooks, data bases, experimental reports, as well the personal experience of the knowledge engineer [10].

There are several techniques for KA, such as text analysis, behavioral analysis, analysis of scenarios and interviews.

In the text analysis, knowledge is extracted by means of a careful analysis of textbooks accepted as reference in the corresponding area. This is an indirect way by which the knowledge engineer tries to assimilate knowledge from the expert (who wrote the textbook). This method has the advantage of being possible without the need of a human expert. However, this is also its main drawback, since the direct contact with the expert is much more efficient for explaining terminology and clarifying possible doubts.

Behavioral analysis is a technique that consists in a systematic observation of the tasks that an expert executes during his/her professional activity. The observer, although passive most time, is allowed to interrupt the expert requesting further explanations of specific points not understood. Obviously, questioning has to be done with parsimony so as to avoid excessive disturbance.

In the analysis of scenarios, the knowledge engineer submits selected cases (tasks), either real or hypothetic, to the expert and observes their resolution. The selection of cases should be based on the premise that they reflect relevant problems that cover a considerable portion of the domain, as well as problems that include different levels of uncertainty. This technique emphasizes the case-based reasoning, where a solution of the problem is based on the adaptation of a known solution for a similar problem.

Interview is an interactive activity between the knowledge engineer and expert. It is based on an answer-reply strategy and, usually, several sessions are necessary according to the depth and complexity of the knowledge to be elicited. Interviews can be directed, structured and semi-structured, as follows:

- A direct interview is similar to a habitual conversation in which the expert talks with the knowledge engineer about specific subjects of his/her domain. The interview usually follows a predefined agenda, focusing selected topics of the domain. Such agenda is previously sent to the expert to allow the familiarization with the subjects. The main objective of the interview is to acquire a broad overview of the area of expertise as well as the tasks involved.

- The semi-structured interview is similar to a questioning. The information required is more specific and at a deeper level than that focused in the directed interview. The objective here is to acquire a better understanding of the issues involved in the solution of a given problem. The strategy is to divide the most general tasks into subtasks. The order of questioning is changeable so as to allow the knowledge engineer to adopt a terminology according to the progress of the interview and the appropriation of knowledge. This kind of interview combines open and closed questions.

- Structured interviews have some characteristics that make them useful in KA. They require a careful previous planning of the questions to be done and the order of questioning, besides the actions expected from the knowledge engineer. This kind of interview should take place after the interviewer has already acquired enough knowledge about the domain, so as to explore specific issues. The interview is based on closed questions, previously elaborated with the objective of extracting information that was missing in previous semi-structured interviews.

There are many obstacles to be considered during the KA process, for instance: experts have extensive and specialized knowledge, usually tacit (that is, they are not aware of all they know, but use such knowledge to solve problems); frequently, experts are very busy and difficult to approach; due to the level of specialization, experts do not know everything about the domain. Consequently, to achieve success in the KA process, it is necessary to devise ways to circumvent the obstacles previously mentioned.

To illustrate in a general sense the KA process, it is presented an example proposed by Milton [11]. This method starts with a simple approach and then proceeds with more elaborated techniques, as follows:

- The first step is to conduct an initial interview with the expert, aiming at establishing the objectives and the scope of the knowledge to be acquired. Also, it is important to make clear how and for what purposes the knowledge will be used. Establishing a communication channel with the expert, allows the basic terminology of the area to be acquired, as well as facilitate further approaches. This interview (as well as all remaining ones) should be recorded to preserve information.

- Next, the initial interview should be transcribed and the resulting document analyzed. From this analysis, a hierarchy of concepts about the knowledge is constructed, thus obtaining a general overview of the domain. The hierarchy can be further used for producing a set of questions about the main topics of the domain, as well as serving as guide for the KA process.

- In the third step, a semi-structured interview with the expert is conducted, using the questions previously planned. The objective here is to enhance structure and improve focus.

- The expected result of the previous step is a documented protocol with the main concepts of the domain, their attributes, typical and limit values, relationships and explicit rules.
- In the fifth step it is suggested the representation of the knowledge acquired to date using appropriate analytical models (rules, diagrams, hypertexts and others).
- Based on the previous models, a questionnaire is elaborated for a structured interview, so as to complement and extend the information modeled.

The steps described above should be repeated until the formal model generated meets the expectations of both expert and knowledge engineer. After finishing the KA process with an expert, it is desirable to validate the knowledge with other experts, who may require changes. In this stage, the knowledge engineer must to have a strategy for managing possible conflicts.

2.2 Knowledge Representation

One of the main concerns of AI researchers is how to represent knowledge. The question is how to capture, in a formal language suitable for being processed by a computer, knowledge in its full extension, so as to enable its use to simulate intelligent behavior [9].

Therefore, Knowledge Representation – KR, is the method used by the knowledge engineer to model expert knowledge in a given domain. The representation should be efficient enough for using by a computer, thus, it may include a combination of data structures and interpretative procedures. KR is always related with the ways by which humans express information. Although there is much research towards the development of general languages and systems for KR, still different types of knowledge require different representation methods.

Sowa [3] points that KR is the application of logics in the task of constructing computational models in a given domain. Frequently, KR is referred as "knowledge representation and reasoning" because KR formalisms are useless without the possibility of reasoning and inference with them.

KR is closely related to the KA process. Actually, as the knowledge engineer conducts the KA process he/she has to record the acquired knowledge using formalism and so, KR takes place. This is the way by which real-world facts and events, human convictions and expertise are computationally modeled and used [10].

Amongst the many methods for KR proposed in the literature, possibly the most frequently used are logics, rules, semantic nets and frames [9]. Fig. 1 shows the methods for knowledge representation.

- Logics: the logical representations are based on Mathematics and Philosophy, trying to characterize the principles of correct reasoning. It concerns about the development of formal representation languages with consistent and complete deduction inference rules (deduction).
- Rules: production rules consist of propositions, usually in the form "IF A THEN C". The antecedent (A) is a logical conjunction of conditions, and the consequent (C) is a given class. The conditions of the antecedent are t-uplets

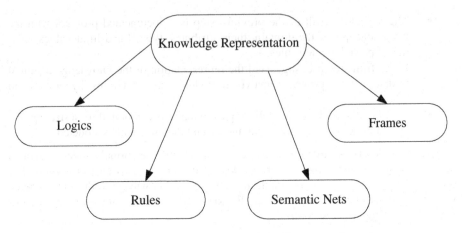

Fig. 1. Main methods for Knowledge Representation

in the form $<A_i\ Op\ V_{ij}>$, where A_i is the i-th attribute, Op is a relational operator, and V_{ij} is the j-th possible value of the corresponding i-th attribute. The combination of several conditions in the antecedent is accomplished by means of the logical operators. The consequent of the rule consists of a simple condition in the form $<M_i = V_{ij}>$, where M_i is one of the possible target attributes and V_{ij} is the same as above.

- Semantic nets: represents knowledge explicitly as a graph, where vertices correspond to facts or concepts and edges correspond to relationships or associations between concepts.
- Frames: are data structures that group elements into classes, subclasses, down to instances. Each frame is composed by slots that contain features and properties of a class or instance. Frames connect each other to build a compete idea.

Therefore, a knowledge base can be defined as a mapping between objects and relationships of a given domain, and the computational objects and structures within a computer program. Results of inferences in the knowledge base should correspond to results of actions or observations of real-world facts. The objects, the relationships and the inferences are all mediated by the knowledge representation language.

3   Knowledge Management

Knowledge Management (KM) includes the procedures for creating, maintaining, applying, sharing and updating of knowledge, aiming at to increase the organizational performance and aggregate value to the established knowledge [12]. According to Keeling [13], the main objective of KM is to use the experience and comprehension of people in an organized way so as to enrich the intellectual property. A more significant definition of KM is an innovative practice that allows collaboration and communication between knowledge developers of the same or different domains [5].

Smith and Farquhar [14] summarize KM as a procedure that improves the organizational performance, because allows capture, sharing and application of the collective knowledge to take correct decisions. To accomplish this, organizational knowledge has to be constantly updated and reviewed.

KM, in its basic form, exists since long ago, and can be identified in many professions and areas, such as, philosophy, religion, education, and politics. However, the concept of KM, as a subject or specific branch of knowledge, has developed only from a decade ago. KM has become more technical and formal as the necessity and value of knowledge has increased in large organizations, to be competitive with the growing technological advancements.

3.1 Knowledge Life Cycle

According to the Merriam-Webster dictionary [15], life cycle is a series of stages by which something (such as individual, a culture or a manufactured product) undergoes during its lifetime. Many researchers describe the life cycle of knowledge. For instance, Birkinshaw and Sheehan [16] described four stages: creation, mobilization, diffusion and commoditization. Staab et al [17] described the knowledge life cycle as a circular process that includes: creation and/or importation, capture, access and use. Also, Bhatt [18] described a cycle composed by four stages: creation, revision, distribution and adoption. However, independently of the terms describing the life cycle of knowledge, attention should be paid to each stage of Knowledge Management, otherwise, knowledge can become invalid, outdated and unreliable.

3.2 Knowledge Management in Healthcare

The widespread use of informatics in health areas has fostered the need for information systems, diagnosis support systems, and teaching/learning support systems. Consequently, an underlying problem that emerges is the acquisition, representation and management of knowledge in such systems [13]. Knowledge Management (KM), in particular, is essential for supporting and improving the efficiency of health professionals in their daily activities [7].

Davenport and Glaser [19] report that KM helps health professionals to avoid errors, to learn with other colleague's experience, and to access updated and specialized information, when necessary. There are, also, other circumstances that contribute to popularize KM in this area: the health professional can give support to the system so as to create, extend or improve the knowledge-base and, more importantly, he/she still will have control over the situation, being the only responsible for the final decision about a diagnosis or treatment.

During decades, health professionals have seen the exponential growth of knowledge in their areas of expertise, and the growing difficulty in accessing, manipulating and sharing information. Nowadays, the access to information is essential to provide a satisfactory clinical and therapeutic support to patients. It is a matter of fact that, in the near future, health professionals will need complimentary education to deal with the ubiquitous information technology and manage knowledge in the respective area [20]. Information technology is a critical issue that establishes a clear division between past and future for health professionals and the way they manage patients.

4 Ontologies

Several data structures can be used for organizing and formalizing knowledge, as mentioned before. Recently, an emerging approach that has drawn the attention of researchers is the ontologies. Based on a set of concepts and their relationships, an ontology establishes concisely a formal descriptions of a given knowledge domain.

The origin of the word "ontology" relies in the Philosophy, and was introduced by Aristotle. In this context, philosophers try to answer the questions: "What is a being?" and "What are the common characteristics of all beings?" [21]. More recently, both the AI and KM communities have adopted this term to express concepts that can be used to describe a given area of knowledge or, else, to construct a representation of it.

A frequently used definition of ontology is provided by Gruber [22], who asserts that it is a formal and explicit specification of a conceptualization. Such definition requests further explanation of the meaning of words used [23]:

- Conceptualization is referred to an abstract model of a given phenomenon that identifies relevant concepts of such phenomenon;
- Explicit means that the type of concepts used and the limitations of their use must be clearly defined;
- Formal indicates that the ontology must be capable of being processed by a machine.

In fact, the literature about ontologies presents several different definitions, some of them are complimentary each other. Fig. 2 shows an schema with the main components of an ontology.

For instance, Guarino [24] presents a extensive discussion about the meaning of the term within the scope of Computer Science, as follows:

- In AI, an ontology is a theory about which entities can exist in the mind of a knowledge agent [25];
- From the point of view of a knowledge about a particular task or a domain, an ontology describes a taxonomy of concepts that define the semantic interpretation of the knowledge [26];
- Ontologies are consensus about shared conceptualizations. Shared reflects the notion that the ontology captures consensual knowledge. That is, this knowledge should not be restricted to a few number of individuals but, instead, accepted by a group of experts in the domain of the ontology [27];
- An ontology is an explicit, but partial conceptualization, a logic theory that restricts models into a logic language [28];
- An ontology is an explicit and partial specification of a conceptualization that is expressible, from the meta-level point of view, in a set of possible domain theories, with the objective of modular design, redesign and reuse of intensive knowledge [29];
- Ontology is an explicit specification of knowledge level of a conceptualization, which can be affected by a particular domain or task, for which it has been created [30].

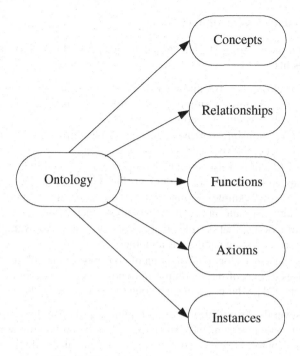

Fig. 2. Main components of an ontology

Another complimentary definition of ontology is proposed by Gómez-Pérez [31] who includes information about its structure: an ontology is a set of hierarchically ordered terms aimed at describing a domain that can be used as skeleton for a knowledge base. According to such description, an ontology groups a set of terms organized with a hierarchy or associated taxonomy. An important detail of this description is to present one of the main utilities of an ontology, which is to serve as the starting point of a knowledge base.

This definition makes an important distinction between ontology and knowledge base. Ontology creates the structure over which it is possible to construct a knowledge base. It provides a set of concepts and terms to describe knowledge in a given domain. On the other hand, the knowledge base uses those terms and concepts to describe a given reality. If this reality is modified, the knowledge base will be modified as well to reflect it, but, even so, the ontology remains unchanged, provided the domain is the same.

In general, there are some important benefits in using ontologies, as follows. Ontologies can provide a common vocabulary for representing knowledge among a group of professionals, thus decreasing ambiguities and interpretation errors. By using ontologies a formal representation of knowledge can be constructed, thus allowing information sharing. Differently from natural language, where words are subject to contextual semantics, ontologies offer an exact description of knowledge. Finally, the same conceptualization represented in an ontology can be expressed in several different languages and its reuse may extend a generic ontology to be suitable for specific domains.

4.1 Structure and Classification of Ontologies

According to Gómes-Pérez [31] and Maedche [21], ontologies are structured over several components:

- A set of concepts (also known as classes) and a hierarchy among concepts, that is, a taxonomy. A simple example of taxonomy is the concept of "man" being a sub-concept of "people";
- A set of relationships between concepts. An example of relationship between concepts "people" and "car" is "the owner of";
- A set of functions (also known as properties). A function is a special case of relationship in which a set of elements has a unique relationship with another element. An example of function is "to be parent", where the concepts "man" and "woman" are related to another concept "people";
- A set of axioms, that is, rules that are always valid. An example of axiom is the assertion "every people has a mother";
- A set of instances, or specialization of concepts. Gómez-Pérez [31] considers instances as part of the ontology, in opposition to the definition proposed by Maedche [21], where instances belong to the knowledge base.

There are several classifications of ontologies, provided by different authors. Mizoguchi, Vanwelkenhuysen and Ikeda [32] classify ontologies according to the function: domain ontologies, task ontologies and general ontologies. Uschold and Gruninger [33] classify ontologies according to their degree of formalism: highly informal, semi-formal and rigorously formal ontologies. Jasper and Uschold [34] classify ontologies according to their application: neutral authorship, as specification and common access to information. Haav and Lubi [35] classify ontologies according to the structure: high-level, domain, and task ontologies. Van-Heijist, Schreiber & Wielinga [30] classify ontologies according to their contents: terminological, information-based, knowledge modeling, application, domain, generic and representation ontologies.

Guarino [36], on the other hand, classifies ontologies in a simple and intuitive way, according to their level of generality, therefore having some overlapping with other previously mentioned classifications:

- High-level ontologies: they describe general concepts, such as space, time, event, and other. These concepts are usually independent of a given problem or domain;
- Domain ontologies: they describe a particular vocabulary related to a given domain, and can be a specialization of a high-level ontology;
- Task ontologies: they describe the vocabulary for a given task or generic activity;
- Application ontologies: they are more specific and particularize concepts from both the domain and the task ontologies.

In general, the high-level ontologies are those that have the largest capacity of reuse, and application ontologies, the smallest one. This is because high-level ontologies define generic concepts, and application ontologies define concepts regarding a specific application.

4.2 Applicability of Ontologies

Currently, there are many areas in which ontologies have been successfully applied, for instance: knowledge management, electronic commerce, natural language processing, web information retrieval, education, and other.

There are KM-related projects that include acquisition, representation, maintenance and access to knowledge within the scope of an organization. Ontologies can help to provide the basic structure over which enterprise knowledge bases are constructed.

In projects related to electronic commerce, it is possible to develop automated transaction systems. They require a formal description of products, beyond syntactic exchange formats. An ontology can provide a common description and understanding of terms, thus allowing interoperability and ways to accomplish an intelligent integration of information [21].

In natural language processing, domain knowledge is essential for a coherent comprehension of the text. Ontologies can play an important role for elucidating the ambiguities inherent to text interpretation, and to establish a dictionary of concepts within the text domain.

Due to the exponential expansion of the information available in the internet, much attention has been given to web information retrieval (or semantic web). The search engines available are not able to improve search and obtain precise results without discovering the precise meaning of the web pages searched. To circumvent this problem, Tim Berners-Lee [37] proposed the semantic web that includes semantics to the web pages by using three technologies: Extensible Markup Language (XML), Resource Description Framework (RDF), and ontologies. Basically, the role of ontologies is to provide a semantic structure in the annotations of web pages.

Ontologies are essential for the development of knowledge-based systems. Every knowledge-based model is, explicitly or implicitly, committed to some kind of conceptualization, which, in turn, is the basis for ontological models [38].

The most important KR projects are based on ontologies, such as CYC [39] and TOVE [33]. Specifically in the health sciences, there are important research projects that include ontologies and knowledge-base construction, for instance, SNOMED-CT [40] and GO [41].

In education-related projects, ontologies can become learning environments that describe a physical domain with rich details and standardization of terminology. Consequently, the formal representation of knowledge is accomplished with educational purposes [42].

4.3 Methodologies for the Development of Ontologies

It is important to adopt a methodology for modeling an ontology in order to avoid jumping from the KA process directly to the implementation phase. Such procedure may cause problems such as: difficulty or impossibility of reuse, since the ontology is implicit in the code, and difficulty in communication, because the domain expert usually does not understand computer languages in which the ontology was implemented.

Some methodologies for the systematic development and manipulation of ontologies are available [43]. Currently, the most widely known and cited in the literature are:

- Methodology of Uschold and King [44]: it is based on the construction of the Enterprise Ontology and comprehends four development stages: identification of the purpose of the ontology, construction, evaluation and documentation. However, this methodology does not describe in details the techniques for executing those activities. Data for the construction of the ontology are obtained by means of interviews with the domain experts, and also reusing existing ontologies;

- Methodology of Grüninger and Fox [45]: it is based on the experience of the authors in developing ontologies for small enterprises. The methodology has a formal procedure for identifying scenarios for using the ontology and includes questioning in natural language for establishing the scope of the ontology and for extracting the main concepts, properties, relationships and axioms. The methodology comprehends six steps: definition of motivational scenarios (problems demanding a new ontology and a set of possible solutions); informal definition of competencies (set of questions that require an ontology to being answered); specification of the terminology of the domain(using first-order logic); verification of completeness (matching of the ontology with the competence issues previously defined). Differently from the previous methodology, this one provides more than general principles. After KA, at the second step, a formal language is immediately required in the subsequent steps;

- Methodology of Fernández, Gómez-Péres and Jurino [46]: it is also known as Methontology and describes more deeply the steps to be followed and the artifacts to be generated for creating the conceptual model. It also proposes a life cycle based on the evolution of prototypes. The development process is divided in ten steps, as follows:

  ✓ Identify the tasks of the ontology and plan the use of available resources;
  ✓ Specify the purpose of the development and their potential users;
  ✓ Acquire knowledge about the domain of the ontology;
  ✓ Create a conceptual model that describes both the problem and the solution;
  ✓ Create a formalization for transforming the conceptual model into a formal model;
  ✓ Integrate, as far as possible, other existing ontologies to the new ontology;
  ✓ Implement the ontology in a formal and computable language;
  ✓ Evaluate the ontology;
  ✓ Document the ontology so as to facilitate its reuse and maintenance;
  ✓ Update the ontology, whenever necessary.

- Methodology of Noy and McGuiness [47]: it includes an interactive development through successive refinements. The development process is divided into six steps: define the domain and scope of the ontology; reuse existing ontologies; list terminology; define classes (concepts) and their hierarchy; define the priorities of classes or concepts, create instances of the concepts within the hierarchy;

- Methodology of Sure and Studer [48]: it is also known as On-To-Knowledge Methodology and is useful for the management of knowledge in organizations. This methodology is divided into five steps: Kick-off (identification of requirements and competence isssues); refinement (from the scratch to an application-oriented mature ontology); evaluation (focused on the technology, the user and the ontology); and maintenance (evolution and corrections, if necessary).

4.4 Software Tools for Developing Ontologies

In recent years, the number of computational tools for constructing ontologies has grown significantly. These tools aim at helping the knowledge engineer not only in building an ontology itself, but also, in reusing knowledge. Possibly, the most relevant tools available to date are [49]:

- Ontolingua Server, Ontosaurus and WebOnto: they were the first editors for ontologies.
- Protégé, WebODE and OntoEdit: they represent a new generation of development environments for ontologies.
- OILEd and DUET: tools especially suited for developing ontologies for semantic web.

In particular, Protégé was developed by the Medical Informatics group at Stanford University (USA) and is constantly updated. Its core is an ontology editor and has a large library of plug-ins that adds more functionality to the environment. Currently, there are plug-ins that allow to import/export contents in the format of ontology languages (such as FLogic, Jess, OIL, XML and Prolog), flexible access and manipulation of data bases, creation of restrictions and fusion of ontologies [38]. Besides, Protégé is open source and has a Graphic User Interface (GUI) that allows easy access to its resources. With Protégé it is possible to make explicit consensual knowledge, separate the knowledge domain from the operational knowledge, and analyze the domain at a high level [47]. All these features contribute to make Protégé an outstanding tool for KM, widely used by knowledge engineers, facilitating the development, sharing of structure and information, and reuse of knowledge.

5 The Use of Ontologies for Learning

Frequently, "seeing the big picture" is a key element in learning. Ontologies could play an important role in showing the big picture of a subject, allowing students to view knowledge in any sequence they wish and taking the time they need. Using ontologies, students are not forced to follow the order of the instructor; they may start at any location and follow the relationships in any order that is most beneficial to each individual student [50].

Ontologies have been used in colleges and universities for teaching. Milam [51] has described some uses: marketing to future students, describing academic disciplines, documenting data, providing metadata about learning management systems, describing the nature of higher education enterprise, and delineating online resources.

Wilson [52] provided a list of reasons why ontologies might be useful in a learning environment:

1. Students are provided with advanced browsing and searching support in their quest for relevant material on the Web. Especially where their understanding of a topic is low, students can be directed intelligently towards resources of relevance.
2. Syntactically different but semantically similar resources can more easily be located.
3. Information can be shared across educational applications, enabling reuse.
4. Distance learners can be provided with the intelligent and personalized support.

However, ontologies have been sparsely used for learning, although in the current applications very promising results can be observed. Examples are provided by Macris and Georgakellos [53] who developed an ontology for learning environmental education. Also, Hausmanns [54] created an ontology for illustrating contents of dynamical systems. Finally, a relevant reference is Wilkinson [50], who proposed an ontology for Physiotherapy undergraduate students learn anatomy. The work reported here is also focused on Physiotherapy.

6   A Case Study in Neuropediatric Physiotherapy

As mentioned before, Neuropediatric Physiotherapy is an area that includes diagnosis, treatment and evaluation of patients. Such patients, usually babies or young children, have to be frequently evaluated by the physiotherapist in order to observe the progress of treatment [1] [55].

When a child with neurological lesion is under diagnosis by a physiotherapist, its motricity and movement functionality is evaluated, regarding to the normal motor development. For instance, a normal child of 8 months old of *chronological age* is expected to have also 8 months old of *motor age*. On the other hand, a child affected by a neurological lesion can have 8 months old of chronological age, but 2 months old of motor age. This discrepancy is considered as a motor delay or abnormal condition. Starting from this presupposition, the physiotherapist is in charge of analyzing all the complex components of the normal motor development to be stimulated during the treatment of the patient. The objective is to foster motor development in such a way to make motor and chronological ages to match.

To treat children with neurological lesions, the physiotherapist must know the normal motor development (NMD) of a child, with all its peculiarities, so as to be able to recognize what would be abnormal. Therefore, the several steps of NMD are used as reference in the diagnosis procedure, as well as during treatment [56].

Understanding the underlying complexity, the extension, and non-standardization of terms in Neuropediatric Physiotherapy, it becomes clear the importance of correctly content learning and diagnosing to be able to carry out an effective treatment. It is in this scenery where the building an ontology takes place, establishing clear and definite concepts and relationships.

6.1 Knowledge Acquisition Procedure

Building an ontology is a labor-intensive activity and it becomes even more complex due to the absence of a standard vocabulary in the Neuropediatric Physiotherapy domain.

Uschold [57] emphasizes that there is no unified methodology capable of fulfilling all requirements for modeling any domain. In this work we followed the two steps associated with the development of an ontology, as proposed by Zhou et al [2]: (i) knowledge acquisition and management of the concepts between different sources of information (management of conflicting opinions), and (ii) implementation of the ontology itself using the represented knowledge.

The classical artificial intelligence suggests that the knowledge engineer should use a single knowledge source (expert) [10]. However, in this work we use an ontology for representing knowledge. The main authors in this area recommend that ontologies should be based by a consensus of a group of experts [22] [36]. Therefore, to cope with such contradiction, we decided to engage three expert physiotherapists. All of them had extensive expertise in Neuropediatric Physiotherapy, including educational (theoretical) and therapeutic (practical) experience.

Experts took part of several individual interviews. First, previously planned semi-structured interviews were used, and then, structured interviews for deepening specific subjects. To meet the requirements of the domain, on those interviews we adopted a six-phase questioning system proposed by LaFrance [58]:

1. Broad overview: a semi-structured interview was applied to the experts aiming at to understand the reasoning used during both diagnosis and therapy.
2. Categories cataloguing: all the classes (concepts) and subclasses relative to the domain were clearly defined.
3. Attribute detailing: structured interviews were carried out for analyzing how frequent was the use of each concept for different types of diagnostic outcomes.
4. Weight determination: weighting factors for each diagnostic class and subclass were obtained
5. Cross correlation: a consistency check was done after experts have exanimate all the information stored necessary for creating the ontology for Neuropediatric Physiotherapy.

Another important issue in the knowledge acquisition process is managing conflicts and divergence of opinions between experts. We used the methodology known as IBIS (Issue-Based Information System) [59] to manage conflicts between experts. This methodology helps to evolve a divergence of opinions to a convergence, thus emerging a consensus. When the knowledge engineer comes upon a question with different answers from the experts, he/she decides in favor of the one with better arguments. That is, the answer that is better supported by approval or justification. When two answers have justifications, one should choose the one with the large number of supporting arguments.

When finished the knowledge acquisition process with the experts, all information collected was checked against the main textbooks in Neuropediatric Physiotherapy [1] [55] [60].

As result of the knowledge acquisition process, the relevant information for diagnosis and learning was grouped into five main classes: reflexes, reactions, movement plans, movement patterns and motor skills. The divisions of these classes were also defined, as well as all relationships between the classes of the ontology.

6.2 Knowledge Representation in the Ontology

Acquired knowledge was represented in a hierarchical structure of an ontology. First, a taxonomy of terms was created with the main concepts (classes): *MotorAge* (corresponding to the diagnosis), *NormalMotorDevelopment* (NMD – set of characteristics belonging to a given diagnosis) and *Patients* (representing specific cases). This hierarchy was refined by creating subclasses from derived concepts: *MotorAge* included the 12 first months of life; *NormalMotorDevelopment* included the main components analyzed by the physiotherapist (reflexes, reactions, movement plans, movement patterns, motor skills and values); and *Patients* included some case-studies of real patients. Subclasses of NormalMotorDevelopment were later refined.

Next, the properties pertaining to each motor age (diagnosis) were represented, including their respective components of the NMD. An example is the property *has-Reflex* that connects individuals of the *Reflex* class with individuals of *MotorAge* class. For the full description of the domain, the definition axioms of each subclass of MotorAge were declared, thus fulfilling the components of NMD necessary to accomplish the diagnosis.

The tool chosen for knowledge representation was an ontology because it allows the formal representation of tacit knowledge (kept in mind of the experts, but not concretely expressed) usually found in the domain area.

During the development of the ontology, two methodologies were used: Methontology [46] and On-To-Knowledge Methodology [48]. To model the ontology, the following steps of the life cycle of Methontology were done: development, managing and support. In the development process the following activities were done: specification, conceptualization, formalization, implementation and maintenance. In the management process, the control and quality assurance activities were done. The support process was done in parallel to the previous mentioned processes, accomplishing knowledge acquisition, evaluation (analysis of competencies issues and coherence of the taxonomy) and documentation activities. It is important to note that in the specification activity, the principles of On-To-Knowledge Methodology were extensively used.

The implementation of the ontology was done using a computational tool for editing, Protégé[1], version 3.3.1. This tool has extensible architecture, allows good level of details, and its interface is user-friendly. The formal language for representation chosen was OWL-DL (Web Ontology Language – Description Logic), which is recommended by World Wide Web Consortium (W3C). Fig. 3 shows the high-level class hierarchy of the developed ontology.

The classes mentioned in the figure are those defined above. Notice that class *NormalMotorDevelopment* includes all components of the NMD (not expanded in the

---

[1] http://protege.stanford.edu/

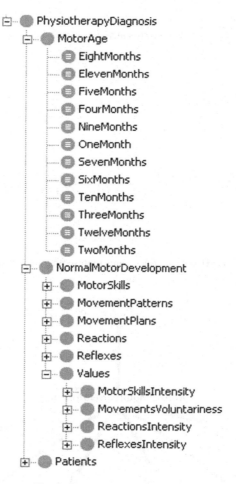

Fig. 3. High-level class hierarchy.

figure) necessary for the diagnosis of the patient in each class of *MotorAge*. Class Values includes the (relative) intensities of each component of the NMD.

6.3 Ontology Instantiation as a Learning Activity

The next step is the use of the ontology in the learning environment. This work explores the use of a computational tool for knowledge management for the education of Physiotherapy undergraduate students. These students are expected to use the ontology for developing and improving their own learning abilities. Therefore, using the ontology for studying includes the creation of specific instances, as an active learning process. Students use the preexisting class hierarchies to add contents to the ontology. They are instructed by the teacher to add a given patient profile and their associated features: reflexes, reactions, movement plans, movement patterns and motor skills.

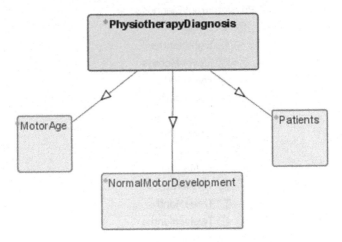

Fig. 4. SHriMP interface - arcs representing relationships

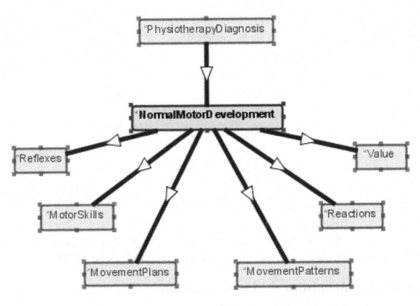

Fig. 5. SHriMP interface and navigation – class hierarchy notions.

The ontology for learning is presented to the student by means of a software known as Simple Hierarchical Multi-Perspective (SHriMP)[2]. Shrimp is both an application and a technique, designed for visualizing and exploring any information space. SHriMP is a domain-independent visualization technique designed to enhance how students browse, explore and understand complex knowledge-bases.

---

[2] http://www.thechiselgroup.com/shrimp

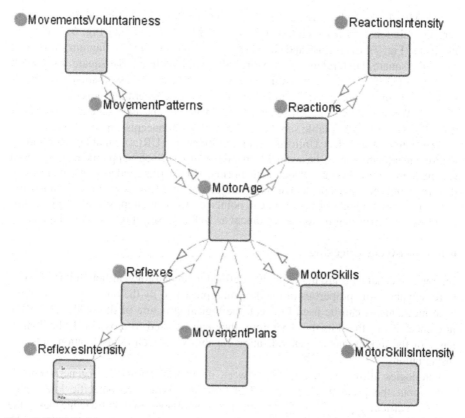

Fig. 6. SHriMP learning environment - edges represent domain and range between existing classes.

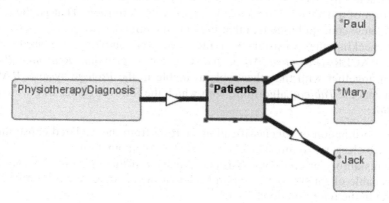

Fig. 7. SHriMP showing individuals (patients).

SHriMP allows a visual representation of the ontology, including edges that represent the relationships between existing classes (Fig. 4, Fig. 5 and Fig. 6) and given instances (Fig. 7). Each class and each instance are presented by a diagram shown in a separated square. Hierarchies, in turn, are fully included in a large square. Its content is represented by smaller squares. For instance, the generic class "patients" is represented by a large square that includes several squares concerning individual patients (Fig. 7). When clicking in each square, the student can visualize several other useful information, such as available subclasses, definition of concepts, and properties.

The ontology also has Uniform Resource Locators (URLs) capable of providing additional information to be available in the Internet. Such supplementary material can be web pages, Portable Document Format (PDF) files, video clips, pictures or drawings. Internet pages present specific subjects about the area of study. Pictures and drawings help to highlight anatomical points of interest or positions. Video clips demonstrate the normal motor development as well as cases of real-world patients.

6.4 Consistency Checking

Inference mechanisms are not explicitly defined in an ontology, although it is possible to reason about the properties of the domain represented by the ontology. Such inference mechanisms can be used to check the logical structure of the model and make inferences about the domain. Therefore, they can be used to crosscheck the consistency of the model and its generalization capability, as well as its relationships and instantiations.

Ontologies allow the distinction between intentional knowledge (general knowledge about the problem domain) and extensional knowledge (specific knowledge about a particular problem). Typically, in an ontology-based knowledge base, the Description Logic (DL) is composed by two components: a *TBox* and an *ABox* [61]. The *TBox* contains the intentional knowledge in the form of a terminology and it is constructed by declarations that describe general properties of concepts. The basic form of a declaration in a *TBox* is a concept definition. That is, the definition of a new concept based on other previously defined.

For checking the consistency of the developed ontology, we used a tool, named RACER (*Renamed ABox and Concept Expression Reasoner Profissional*[3]), together with the other tools available in the Protégé system. RACER implements the *Tableau* algorithm, with which the following checking were done in a *TBox*:

- Subordination or subclassification: starting from the declared constraints in each class, try to infer if a class is subclass of another one;
- Satisfability or concept consistence: analyze if there is some interpretation capable of satisfying the axiom such that the concept denotes a non-empty set in the interpretation;
- Equivalence: verify if two concepts are equivalent;
- Disjunction: determine if two disjoint concepts share the same instance;.

---

[3] http://www.racer-systems.com/

6.5   Results of the Case Study and Discussion

This section presents the main results and acquired experience during the development of the ontology for the Physiotherapy domain.

In the knowledge acquisition phase, during the structured interviews with the three domain experts, 12 questionnaires were requested to be filled in by them. These questionnaires had 49 items each, making up a total of 588 items evaluated.

It is important to note that, in Neuropediatric Physiotherapy, as well as in many health sciences, there are different schools of though that directs the professional practice, giving different approaches to the diagnosis problem. Due to the difference of approaches between schools of though, it could be quite difficult to establish consensual knowledge, thus making impracticable to build an ontology. As consequence of the lack of consensus, the created knowledge base could be inconsistent, thus making it useless for decision-support. Therefore, this work is directed towards the most widely spread school, created by Karel and Bertha Bobath [1] [55] [60], usually referred to as Neurodevelopment Treatment. As mentioned in section 4.1, knowledge acquisition was carried out with three expert physiotherapists. All of them belonged to the same school of though, thus taking more consistency and reliability to the resulting ontology and the knowledge-base. Even so, considering the large number of items to be evaluated by the experts, some divergences of opinions occurred. The occurrence of conflicts was relatively low, corresponding to only 7% of the items (that is, 41 out of 588). Such level of divergence between experts of the same school is promptly manageable and the IBIS methodology was adequate and efficient for this task.

Knowledge representation was carried out using Protégé. This hierarchical structure gives as result the full organization and formalization of diagnostic knowledge in Neuropediatric Physiotherapy. The current version of the developed ontology is composed by 100 classes and subclasses, 30 properties and 200 axioms. This ontology allowed the creation of vast consensus vocabulary for the domain, including concepts with full definitions through their relationships and axioms.

We believe that the application of the created ontology for supporting learning in Physiotherapy is of great importance, since it includes multimedia resources as well as active learning concepts, together with traditional instructional methods. Consequently, with this complimentary and illustrated resource, the learning of students can be more effective. Also, it promotes the approximation of health sciences with informatics.

7   Conclusions

In this work knowledge was elicited from domain experts and complemented from reference textbooks. Knowledge was formally represented as an ontology, using well-defined methodological procedures, thus enabling efficient management of knowledge during the whole process.

The formalism inherent to the methodology allowed the development of a knowledge-base which completeness and consistency were verified. Such ontology represents a consensus vocabulary in the domain of Neuropediatric Physiotherapy diagnosis, allowing knowledge reuse, sharing and maintenance, accomplishing the Knowledge Management life cycle.

It is important to recall the integration of different artificial intelligence-based methodologies, such as the LaFrance's questioning technique, the IBIS methodology for managing opinion conflicts, the Methontology and On-To-Knowledge Methodology for developing the ontology.

The use of an ontology for structuring knowledge was helpful not only for categorizing the collected information into hierarchies of concepts, but also, to comprehend the relationships between concepts, and, mainly, allowed full definition of concepts using axioms.

The use of this ontology for learning, by means of SHriMP, makes concepts more clearly defined to the student. Also, it facilitates the understanding of the hierarchy of concepts in Neuropediatric Physiotherapy, mainly the dependency relationship between them. Overall, the proposed approach gives the necessary broad view to the students, giving them a solid starting point to deepen the study.

Overall, the main contribution of this work is establishing a complete and effective methodology for knowledge management in the area of Neuropediatric Physiotherapy, an area with many unstructured and non-standardized information that lacks computational approaches for support. Also, the proposed methodology can be extended to other areas of health sciences.

Acknowledgements

Authors would like to thanks the Brazilian National Research Council (CNPq) for the research grant to H.S.Lopes; as well as to CAPES for the PhD scholarship to L.V. Castilho.

References

1. Bly, L.: Motor Skills Acquisitions in the First Year. Therapy Skill Builders, USA (1994)
2. Zhou, X., Wu, Z., Yin, A., Wu, L., Fan, W., Zhang, R.: Ontology Development for Unified Traditional Chinese Medical Language System. Artif. Intell. Med. 32, 15–27 (2004)
3. Sowa, J.F.: Knowledge Representation: Logical, Philosophical, and Computational Foundations Pacific Grove. Brooks/Cole, California (2000)
4. Georgiou, A.: Data Information and Knowledge: The Health Informatics Model and its Role in Evidence-based Medicine. J. Eval. Clin. Pract. 8, 127–130 (2002)
5. Fischer, G., Ostwald, J.: Knowledge Management: Problems, Promises, Realities, and Challenges. IEEE Intell. Syst. 16, 60–72 (2001)
6. Van Bemmel, J.H., Musen, M.A.: Handbook of Medical Informatics, http://www.mieur.nl/mihandbook/r_3_3/handbook/home.htm
7. Stefanelli, M.: Knowledge Management to Support Performance-based Medicine. Methods Inf. Med. 41, 36–43 (2002)
8. Polanyi, M.: The Tacit Dimension. Doubleday & Co, Massachusetts (1983)
9. Luger, G.F.: Artificial Intelligence: Structures and Strategies for Complex Problem Solving. Addison-Wesley, Pearson Education, Boston (2009)

10. Russel, S.J., Norvig, P.: Artificial Intelligence: a Modern Approach. Prentice-Hall, New Jersey (2003)
11. Milton, N.: Knowledge Acquisition,
    http://www.epistemics.co.uk/Notes/63-0-0.htm
12. Bates, D.W., Evans, R.S., Murff, H., Stetson, P.D., Pizziferri, L., Hripcsak, G.: Detecting Adverse Events Using Information Technology. J. Am. Med. Inform. Assoc. 10, 115–128 (2003)
13. Keeling, C., Lambert, S.: Knowledge Management in the NHS: Positioning the Healthcare Librarian at the Knowledge Intersection. Health Libr. Rev. 17, 136–143 (2000)
14. Smith, R.G., Farquhar, A.: The Road Ahead for Knowledge Management. AI Magazine 1, 17–40 (2000)
15. Merriam-Webster's Collegiate Dictionary. Merriam Company, Massachusetts (2002)
16. Staab, S., Studer, R., Schnurr, H., Sure, Y.: Knowledge Processes and Ontologies. IEEE Intell. Syst. 1, 26–34 (2001)
17. Birkinshaw, J., Sheehan, T.: Managing the Knowledge Life Cycle. MIT SMR 44, 75–83 (2002)
18. Bhatt, G.D.: Organizing Knowledge in the Knowledge Development Cycle. J. Knowl. Manag. 4, 15–26 (2000)
19. Davenport, T.H., Glaser, J.: Just-in-time Delivery Comes to Knowledge Management. Harv. Bus. Rev. 80, 107–111 (2002)
20. Ash, J.S., Bates, D.W.: Factors and Forces Affecting EHR System Adoption: Report of a 2004 ACMI discussion. J. Am. Med. Inform. Assoc. 12, 8–12 (2005)
21. Maedche, A.: Ontology Learning for the Semantic Web. Kluwer Academic, Massachusetts (2002)
22. Gruber, T.R.: A Translation Approach to Portable Ontology Specifications. Knowl. Acquis. 5, 199–220 (1993)
23. Fensel, D.: The Semantic Web and its Languages. IEEE Intell. Syst. 15, 67–73 (2000)
24. Guarino, N.: Understanding, Building, and Using Ontologies. Int. J. Hum. Comp. Stud. 46, 293–310 (1997)
25. Wielinga, B.J., Schreiber, A.T.: Reusable and Sharable Knowledge Bases: a European Perspective. In: Proceedings of First International Conference on Building and Sharing of Very Large-Scaled Knowledge Bases, Tokyo, pp. 103–115 (1993)
26. Alberts, L.K.: YMIR: An Ontology for Engineering Design. PhD Thesis. University of Twente, Enschede (1993)
27. Gruber, T.R.: Toward Principles for the Design of Ontologies Used for Knowledge Sharing. Int. J. Hum. Comp. Stud. 43, 907–928 (1995)
28. Guarino, N., Giaretta, P.: Ontologies and Knowledge Bases: Towards a Terminological Clarification. In: Mars, N.J.I. (ed.) Towards Very Large Knowledge Bases: Knowledge Building and Knowledge Sharing, pp. 25–32. IOS Press, Amsterdam (1995)
29. Schreiber, G., Wielinga, B., Jansweijer, W.: The Kactus View on the 'o' Word. In: Workshop on Basic Ontological Issues in Knowledge Sharing, AAAI Press, Montreal (1995)
30. Van Heijist, G., Schreiber, A.T., Wielinga, B.J.: Using Explicit Ontologies in KBS Development. Int. J. Hum. Comp. Stud. 46, 183–192 (1997)
31. Gómez-Pérez, A.: Ontological Engineering: a State of the Art. Exp. Upd. 2, 33–43 (1999)
32. Mizoguchi, R., Vanwelkenhuysen, J., Ikeda, M.: Task Ontology for Reuse of Problem Solving Knowledge. In: Proceedings of ECAI 1994 Towards Very Large Knowledge Bases, Amsterdam, pp. 46–59 (1995)
33. Uschold, M., Gruninger, M.: Ontologies: Principles, Methods and Applications. Knowl. Eng. Rev. 11, 93–155 (1996)

34. Jasper, R., Uschold, M.: A Framework for Understanding and Classifying Ontology Applications. In: Proceedings of IJCAI 1999 Ontology Workshop, Stockholm (1999)
35. Haav, H.M., Lubi, T.L.: A Survey of Concept-based Information Retrieval Tools on the Web. In: Proceedings of East-European Conference ADBIS, Vilnius (2001)
36. Guarino, N.: Formal Ontology and Information Systems. IOS Press, Amsterdam (1998)
37. Berners-Lee, T., Hendler, J., Lassila, O.: The Semantic Web. Scient. Amer. 5, 34–43 (2001)
38. Noy, N.F., Hafner, C.D.: The State of the Art in Ontology Design: a Survey and Comparative Review. AI Mag. 18, 53–74 (1997)
39. Lenat, D.B.: CYC: a Large-scale Investment in Knowledge Infrastructure. Commun. ACM. 38, 33–38 (1995)
40. Spackman, K.A., Campbell, K.E., Cote, R.A.: SNOMED-RT: A Reference Terminology for Health Care. In: Proceedings of American Medical Informatics Association Fall Symposium, pp. 640–644 (1997)
41. Consortium, T.O.: Gene Ontology: Tool for the Unification of Biology. Nat. Gen. 1, 25–29 (2000)
42. Zdrahal, Z., Mulholland, P., Domingue, J., Hatala, M.: Sharing Engineering Design Knowledge in a Distributed Environment. Behav. Inf. Tech. 19, 189–200 (2000)
43. Fernández, M., Gómez-Pérez, A., Jurino, N.: Methontology: From Ontological art Towards Ontological Engineering. In: Proceedings of AAAI-Spring Symposium on Ontological Engineering, AAAI Press, Stanford (1997)
44. Uschold, M., King, M.: Building Ontologies: Towards a Unified Methodology. In: Proceedings of 16th Annual Confeerence of the British Computer Society Specialist Group on Expert Systems, Cambridge (1995)
45. Grüninger, M., Fox, M.S.: Methodology for the Design and Evaluation of Ontologies. In: Proceedings of Workshop on Basic Ontological Issues in Knowledge Sharing, Montreal (1995)
46. Fernández, M., Gómez-Pérez, A., Jurino, N.: Methontology: From Ontological Art Towards Ontological Engineering. In: Proceedings of AAAi-Spring Symposium on Ontological Engineering, California (1997)
47. Noy, N.F., Mcguinness, D.L.: Ontology Development 101: A Guide to Creating Your First Ontology. Stanford University, Stanford (2000)
48. Sure, Y., Studer, R.: On-To-Knowledge Methodology: Final Version. In: Institute of Applied Informatics and Formal Description Methods, Karlsruhe (2002)
49. Corcho, O., Férnandez-López, M., Gómez-Pérez, A.: Methodologies, Tools and Languages for Building Ontologies: Where is their Meeting Point. Data Knowl. Eng. 46, 41–64 (2003)
50. Wilkinson, S.G.: Computerized Ontology Methods for Teaching Musculoskeletal Topics to Physical Therapy Students. PhD Thesis. The University of Utah, Utah (2007)
51. Milam, J.: Ontologies in Higher Education, http://highered.org/docs/milam-ontology.pdf
52. Wilson, R.: The Role of Ontologies in Teaching and Learning, http://www.jisc.ac.uk/media/documents/techwatch/acf11ac.pdf
53. Macris, A.M., Georgakellos, D.A.: A New Teaching Tool in Education for Sustainable Development: Ontology-based Knowledge Networks for Environmental Training. J. Clean Prod. 14, 855–867 (2006)
54. Hausmanns, C., Zerry, R., Goers, B., Urbas, L., Gauss, B., Wozny, G.: Multimedia-Supported Teaching of Process System Dynamics Using an Ontology-Based Semantic Network. Comput. Aided Chem. Eng. 15, 1453–1459 (2003)

55. Levitt, S.: Treatment of Cerebral Palsy and Motor Delay. Blackwell Science, Oxford (1995)
56. Torre, C.A.: Follow up and Purpose of Physiotherapy Treatment for Teenagers and Young Adults With Cerebral Palsy. Brain Dev 23, 170–178 (2001)
57. Uschold, M.: Building Ontologies: Towards a Unified Methodology. In: 16th Annual Conference of the British Computer Society Specialist Group on Expert Systems, University of Edinburgh (1996)
58. LaFrance, M.: The Knowledge-acquisition Grid: A Method for Training Knowledge Engineers. Int. J. Man Mach. Stud. 26, 245–255 (1987)
59. Rittel, H.W.J., Webber, M.: Dilemmas in a General Theory of Planning. Pol. Sci. 4, 155–169 (1973)
60. Flehmig, I.: Normal Infant Development And Borderline Deviations, Early Diagnosis And Therapy. Thieme Medical Pub., Switzerland (1992)
61. Baader, F., Calvanese, D., McGuinness, D., Nardi, D., Patel-Schneider, P.: The Description Logic Handbook: Theory, Implementation and Applications. Cambridge University Press, Cambridge (2007)

Mining Causal Relationships in Multidimensional Time Series

Yasser Mohammad and Toyoaki Nishida

Graduate School of Informatics,
Kyoto University

Abstract. Time series are ubiquitous in all domains of human endeavor. They are generated, stored, and manipulated during any kind of activity. The goal of this chapter is to introduce a novel approach to mine multi-dimensional time-series data for causal relationships. The main feature of the proposed system is supporting discovery of causal relations based on automatically discovered recurring patterns in the input time series. This is achieved by integrating a variety of data mining techniques.

The main insight of the proposed system is that causal relations can be found more easily and robustly by analyzing meaningful events in the time series rather than by analyzing the time series numerical values directly. The RSST (Robust Singular Spectrum Transform) algorithm is used to find interesting points in every time series that is further analyzed by a constrained motif discovery algorithm (if needed) to learn basic events of the time series. The Granger-causality test is extended and applied to the multidimensional time-series describing the occurrences of these basic events rather than to the raw time-series data.

The combined algorithm is evaluated using both synthetic and real world data. The real world application is to mine records of activities during a human-robot interaction experiment in which a human subject is guiding a robot to navigate using free hand gesture. The results show that the combined system can provide causality graphs representing the underlying relations between the human's actions and robot behavior that cannot be recovered using standard causal graph learning procedures.

Keywords: Mining Time Series, Robust Singular Spectrum Transform, Granger-Causality, Mining Causal Relations.

1   Introduction

In today's information societies, effective knowledge management became an important field of study. With advancements of sensing technologies, data are now collected from various kinds of natural, industrial and other human activities in various forms and stored in large databases. The need for discovering useful knowledge from such huge databases was behind the recent interest of many scholars and industrial entities in data mining.

Time series data is one of the most ubiquitous data formats that captures activities and information. Moreover many other complex types of data formats

E. Szczerbicki & N.T. Nguyen (Eds.): Smart Infor. & Knowledge Management, SCI 260, pp. 309–338.

can be converted to time series data which adds more importance for algorithms that can discover useful knowledge in time series databases. For example Wang et al. [1] developed an algorithm to convert shapes to time series while Inokuchi and Washio [2] proposed a graph-edit mechanism to convert series of graphs to multidimensional time series.

A variety of techniques have been proposed in literature to deal with multidimensional time series analysis including algorithms for efficient clustering [3], searching [4] and discovery of recurrent patterns [5]. We can distinguish two types of time series analysis algorithms. The first type is pattern-based algorithms including motif discovery [6], motif detection, pattern search [5] and some types of compression and discretization algorithms. These algorithms do not deal with the whole time series but with subcomponents of it called patterns, subsequences, or motifs. On the other hand there are holistic algorithms that deal with whole time series, including regression, and clustering[3]. Most causality inference algorithms available belong to the holistic algorithms and so they can discover relations like *smoking cause cancer*. In some cases we are more interested in discovering relations between specific patterns in the input time series rather than the relations between the time series themselves. For example we may like to check if a specific pattern of change in oil prices causes a specific kind of change in the stock market. Holistic approaches cannot answer this question and they can only tell whether or not there is a causal relation between oil prices and stock market in general. In this chapter we focus on unsupervised discovery of causal relations between recurring patterns in multidimensional time series. Given a set of n time series that are *roughly* synchronized, our goal is to first discover the recurring patterns on these time series and then to discover *causal* relations between these recurring patterns.

There are many possible industrial, engineering, and social applications of the proposed techniques. For example behavioral data – in the form of motion captured signals from humans during face to face interactions – can be used to discover the effects of one partner's verbal or nonverbal behavior patterns on the behavior of the partner(s).

The main insight of the proposed techniques in this chapter is to base the analysis of causality on recurring events in the input signals (time series) rather than on the instantaneous values of these series.

The rest of this chapter is organized as follows: Section 2 details the state of the art in causal relation discovery from time series. Section 3 gives a bird's eye view of the proposed technique which has two major phases described in sections 4 and 5. Section 6 reports preliminary evaluation of the system on synthetic data (section 6.1) and real world human-robot interaction data (section 6.2). The paper is then concluded.

2   Related Work

Causality was – and is still – a subject of great disagreement between philosophers. The study of causality can be traced back to Aristotle who defined four types of causal relations [7]:

1. Material Causality: This is the relation between a statue and the bronze it is made of.

2. Formal Causality: This is the relation between the statue and its shape.

3. Efficient Causality: This is the relation between what makes to what is made or what changes to what is changed.

4. Final Causality: This is the relation between the goal and whatever exists (or happens) to produce this goal.

From these four types, only efficient causality has survived as *causality* until today. In his *Treatise of Human Nature* (1739-1740), David Hume described causality in terms of regular succession. For Hume, causality is a regular succession of event-types: one thing invariably following another. His famous first definition of causality runs as follows: *We may define a CAUSE to be 'An object precedent and contiguous to another, and where all the objects resembling the former are placed in like relations of precedence and contiguity to those objects, that resemble the latter'.*

Some philosophers support the positivistic claims that causality is a metaphysical concept that has been replaced in science by nonmetaphysical concepts such as functional interdependence (Mach, 1883/1960), association or correlation (Pearson 1911), or functional relation (Schlick, 1932/1959). The aim of these positivists' efforts was to move science away from the ideas of necessary connection and determinism that they identified with their contemporaries' ideas of causality. On the other hand other philosophers [8] take causality for granted and even try to build mathematical models of causal inference. The only ingredient that *most* definitions of causality agree upon is that causation involves time. Simply causes *precede* effects. In fact this is the basis of the definition of Causal Linear Systems as used in electrical engineering literature.

Some views of causation do not highlight this time asymmetry between causes and effects explicitly even though in most cases the concept implicitly exists. For example, some theorists have equated causality with manipulability [9]. Under these theories, x causes y just in case one can change x in order to change y. Even though the assumption that a cause must precede its effect is not explicit in this case, it does exist implicitly in the definition because manipulation can only cause changes in the future (or present if infinite effect speed is allowed) but no manipulation can *change* the past values of y. Under this assumption about manipulation – which is consistent with both common sense and physical knowledge – the definition of causality through manipulation involves time and assumes that causes *precede* effects.

Another example is the definition of causality as a counterfactual which means that the statement "x causes y" is equivalent to "y would have happened if x". This kind of definition of causality is the basis of Pearl's formalization of the Structure Equation Model [10]. In this definition also time asymmetry between causes and effects enters the definition through the connective *would have happened* which assumes that x should happen in the past or at least the present for y to happen.

In this chapter, we will not attempt answering these centuries old questions about the meaning of causation or its implications or even contribute in any way to the philosophical discussions about it. Our goal in this chapter is to propose a practical way to discover specific kinds of relations that can be treated as causal ones by simply rejecting relations that cannot satisfy the condition *causes precede effects in time*. This condition is a necessary but not a sufficient condition which means that the proposed technique may generate output relations that are not actually causal depending on the definition of *causality* used. Even though this is not satisfying from the theoretical point of view, it is the only practical solution given that an exact definition of causality is far from being agreed upon. The produced relations can be filtered using a stricter sufficient condition but this will invariably lead to rejection of authentic *causal* relations depending on some definitions of causality.

Discovering relations between variables from data is a subject of great importance to science and there are two main techniques currently available.

The first approach is Rule Discovery using algorithms like DBMiner [11]. In this case the input is a set of n variables $X_{1:n}$ and the output is a set of association rules in the form:

$$X_i = x_i \wedge X_j = x_j \wedge \dots \wedge X_k = x_k \Rightarrow X_l = x_l \tag{1}$$

The most frequently mentioned example is the supermarket scenario. $X_i(c)$ is a binary variable representing the existence of product i in the basket of customer c at the check-out. Determining which products customers are likely to buy together is considered valuable for marketing applications such as mailing or catalogue design, as well as store layout and customer segmentation. Rule Discovery can be classified to Classification Rule Discovery which tries to find only rules with high prediction value and Association Rule Discovery which tries to discover all rules covering the input data set. The former is more important for efficient decision making while the later is more useful for expanding our understanding of the domain. The strength of the discovered rule is described using two concepts: support and confidence. Support indicates the frequency that the association rule is true for the population being examined and can be defined as:

$$Support\,(A \Rightarrow B) = \frac{|\text{coverset}\,(A \cup B)|}{|X|} \tag{2}$$

where coverset (A) is the set of all records in X that satisfy A.

Confidence indicates the prediction strength of the rule and can be defined as:

$$Confidence\,(A \Rightarrow B) = \frac{|Support\,(A \cup B)|}{|\text{coverset}\,(A)|} \tag{3}$$

Examination of the definitions of *Support* and *Confidence* indicates that Rule Discovery algorithms – in this form – do not take into account the time at which A and B happens. Given the importance of time in the causality relation, Rule Discovery – either classification or association – can be dismissed as a way for discovering underlying causality models. In fact the same supermarket example clearly indicates that Rule Discovery algorithms do not even try to discover causal relations. For example if we found with 100% confidence that:

$$NeckTie \Rightarrow FullSuit \tag{4}$$

We can only infer that there is a hidden variable (SocialRule) that causes the people who buy neck ties to buy full suits (e.g. because neck ties are only used with full suits). This is exactly the kind of rule that an authentic causal relation discovery algorithm should reject because buying a neck tie was not the *cause* of buying a full suit.

To overcome this difficulty, Das, Lin and Mannila [12] proposed an algorithm that can discover association rules in temporal time series that have this extended form:

$$A \xrightarrow{T} B \tag{5}$$

which means that when A happens at time step t, B happens at time step $t + T$. The method is based on discretizing the sequence $X(t)$ by methods resembling vector quantization. First subsequences are formed by sliding a window through the time series, and then these subsequences are clustered using a suitable measure of time-series similarity. The discretized version of the time series is obtained by concatenating the cluster identifiers corresponding to the subsequence. Once the time-series is discretized, simple rule finding methods are used to obtain rules from the sequence. Many researchers suggested improvements to the basic algorithm provided in this work to increase its speed [13], to utilize constraints in the search for rules [14] and to use parallel implementations rather than serial processing of the time series [15]. One common problem of all of these algorithms is that what the rule $A \xrightarrow{T} B$ actually tells us is a kind of association (this is why it is called Association Rule Mining) but not causal relation because A and B can still be caused by a hidden factor with different time delays T_A and T_B so that $T_B - T_A = T$.

The second approach to causal relation discovery can be called Bayesian Causal Network Induction approach and this approach can overcome the limitation of Rule Discovery concerning causal relation discovery. The goal of these algorithms is to build a Directed Graph (DG) describing the causal relations between variables. As with the Association Rule Discovery case, the input is a set of n variables $X_{1:n}$. If Feedback (e.g. X causes Y and Y causes X) is not supported then the graph is a DAG (Directed Acyclic Graph). There are three categories of Bayesian Causal Network Induction algorithms that utilize different aspects of causality to discover causal rules.

The first of these techniques can be called Causation from Perturbation Algorithms under this category use some form of change or perturbation of the generating process to distinguish causal and non-causal correlations. This technique is rooted in the work of the economist Kevin Hoover [16] who inferred the direction of causation among correlated economical variables (e.g. employment and money supply) by observing changes that sudden modifications of the economy (e.g. tax reformation) induce on the statistics of these variables.

The main assumption was that the conditional probability of an effect given its causes should not vary due to economy modifications while the conditional probability of a cause given its effect can vary with such modifications. Formally, if E is an effect, C is its cause, $P_a(x)$ is the probability distribution of x after the disturbance/modification of the generating process, $P_b(x)$ is the probability distribution of x before the disturbance, and $\Psi(P)$ is some statistic of P, then:

$$\Psi(P_a(E|C)) = \Psi(P_b(E|C)) \tag{6}$$

$$\Psi(P_a(C|E)) \neq \Psi(P_b(C|E)) \tag{7}$$

This technique is usually used for confirming or rejecting specific causal models rather than discovering them from scratch. One problem of this approach is the need for some system wide perturbation or modification but if this modification is already available it can augment available causal relation discovery algorithms.

Tian and Pearl [17] proposed an algorithm based on this approach and applied it to simulated data showing that it can be superior to other techniques that rely only on static conditions. It should be noted that fixing variables (or variable statistics) during experimental procedures corresponds to a perturbation of the underlying generation processes and so experimental scientific discovery can be related to this approach.

The second Bayesian Causal Network Induction approach can be called Causality Model Assessment. In this category of algorithms, first a set of hypothesized causal structures (*models*) is generated by the modeler and then the algorithm is used to find which of them has the highest fit to the data.

Structural Equation Modeling (SEM) [10] technique is an example of this approach. In SEM, the modeler first specifies the model and specifies the exogenous and endogenous variables. Exogenous variables are the variables that cause other variables while endogenous variables are ones that are caused by other variables. The same variable can be both an endogenous and exogenous variable in the same model. The model is specified as a set of structure equations where a structure equation takes the form:

$$E = \sum_{causes} a_i C_i + D \tag{8}$$

where E is an effect (endogenous variable), and C_i are its causes (exogenous variables). The coefficients a_i are called structural coefficients, while D represents unmodeled disturbance or noise in the system. Once the model is specified, its best parameters are estimated from the data using either multiple regression or a more sophisticated SEM specific estimation procedure like SPSS's AMOS program. The estimated model parameters are used to predict the correlations or covariances between measured variables and the predicted correlations or covariances are compared to the observed correlations or covariances. There are many model fitting techniques that can be used including χ^2 fitting, Root

Mean Square Error of Approximation, Comparative Fit Index etc. The main disadvantage of this approach is that it does not specify a way to construct (specify) the set of models to be tested in the first place. Most scientific discovery using the standard scientific procedure of generating hypotheses and testing them using experiments can be considered a combination of Causality Model Assessment and Causation By Perturbation.

The third approach to Bayesian Causal Network Induction can be called Causality from Data. In this category, the algorithm does not require a predefined causal model to test but it constructs its own model from the data directly. Usually more data points are required to generate robust models. Granger-causality tests fall under this category. Granger causality [18] is a technique for determining whether one time series is useful in forecasting another.

A *Ganger-Causes B* ($A_{GC} \rightarrow B$) *iff*, it can be shown, usually through a series of F-tests on lagged values of A, that those A values provide statistically significant information about future values of B.

There are many ways in which to implement a test of Granger causality. One particularly simple approach uses the autoregressive specification of a bivariate vector autoregression. First we assume a particular autoregressive lag length ρ, and estimate the following unrestricted equation by ordinary least squares (OLS):

$$\hat{A}(t) = \epsilon_1 + u(t) + \sum_{i=1}^{\rho} \alpha_i \hat{A}(t-i) + \sum_{i=1}^{\rho} \beta_i \hat{B}(t-i) \tag{9}$$

Second, we estimate the following restricted equation also by OLS:

$$\hat{A}(t) = \epsilon_2 + e(t) + \sum_{i=1}^{\rho} \lambda_i \hat{A}(t-i) \tag{10}$$

We then calculate the sum of squared residuals (SSR) in both cases:

$$\begin{aligned} SSR_1 &= \sum_{i=1}^{T} u^2(t) \\ SSR_0 &= \sum_{i=1}^{T} e^2(t) \end{aligned} \tag{11}$$

We then calculate the test statistic S_ρ as:

$$S_\rho = \frac{(SSR_0 - SSR_1)/\rho}{SSR_1/(T-2\rho-1)} \tag{12}$$

If S_ρ is larger than the specified critical value then reject the null hypothesis that A does not granger cause B. The p-value in this case is $1 - F_{\rho, T-2\rho-1}(S_\rho)$. As presented here Ganger Causality can be used to deduce causal relations between two variables assuming a linear regression model. The test was extended to multiple variables and nonlinear relations (using radial basis functions) [19].

Another algorithm that falls under this category is the algorithm proposed by Kamiri and Hamilton [20] to distinguish causal and accusal temporal relations called Temporal Investigation Method for Enregistered Record Sequences (TIMERS). The main idea is to run time forward and backward while mining the relations from system data. If the reversed run provides better model fit then the relation is estimated as an accusal relation. If the best fit was in the forward direction, this means that it is a causal relation. A relation that relates variables at the same point of time (same time step) is called an instantaneous relation. The authors also designed TimeSleuth software that can be used to mine causal relations in time series data. This is the first approach that explicitly uses the direction of time to test for causation and in this it is similar to the approach presented in this paper. The main difference between this approach and our proposed approach is in the kind of knowledge it reveals about the system. While TIMERS only reveals the direction of causation, the proposed system discovers the exact patterns in every time series that cause changes in other time series.

Another more ambitious example of systems that fall under this category is the TETRAD II and III programs developed by Spirtes, Glymour and Scheines [21]. These programs aim at automatically discovering the causal structure of input data. The program has two inference engines PC and FCI. The choice of the engine depends on whether the data examined is causally sufficient for the population, that is, whether there exist unmeasured hidden or latent causal variables outside of X that explain spurious associations between variables in X. If data is causally sufficient, the PC algorithm is used. Otherwise, the Fast Causal Inference (FCI) algorithm is used. Instead of logical rules, the PC and FCI algorithms find different kinds of causal relationships between variables X_i and X_j in X, graphically represented as follows:

- $X_i - X_j$, meaning in PC that either X_i causes X_j or X_j causes X_i, but the direction is indeterminate. In FCI it means that there is a correlation between X_i and X_j but causation cannot be inferred
- $X_i \rightarrow X_j$, meaning that X_i causes X_j.
- $X_i \leftrightarrow X_j$, indicating a common hidden cause. In this case PC cannot be used.
- $X_i \bullet \rightarrow X_j$, indicating that common cause cannot be ruled out but cannot be confirmed (potentially causal relation).

The resulting graph represents a set of Markov equivalent Bayes networks. This causal inference algorithm makes strict assumptions about the data: Variables in X must satisfy the Markov Condition; i.e., variables can be organized into a directed acyclic graph so that any variable X_i in X conditioned on X_i's parents is independent of all sets of variables that do not include X_i or its descendants.

The following factors affect whether or not the Markov Condition is satisfied by data:

1. Values of one unit of the population must be independent of values in other units of the population (e.g. i.i.d. data).

2. Mixtures in populations that result in contradictory causal connections between two variables violate the Markov Condition.
3. Cyclical processes that reach equilibrium also violate the Markov condition.
Consider the sequence: in time t the value of X_i affects the value of X_j.
Then in time t+1 the value of X_j affects the value of X_i. In time t+2 the
value of X_i affects the value of X_j. This circular relationship violates the
Markov condition. For example supply and demand variables in economics
frequently violate this condition.
4. The sample must be representative of the population.

One limitation of TETRAD is that it does not allow the user to incorporate
temporal ordering so a time series is actually treated as a list of unordered values.
Another limitation of this system is that for continuous variables it supports
only normally distributed variables. A third limitation is that it cannot allow
Feedback which will result in a cycle in the causal graph.

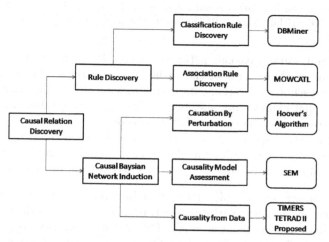

Fig. 1. A taxonomy of causal relation discovery/assessment techniques

Fig. 1 shows a summary of the reviewed approaches to causal relation discovery with examples of each approach. The algorithm proposed in this paper falls
under the causality from data category of causal Bayesian induction algorithms.
The main contributions of the proposed algorithm are:

– It does not search for causal relations in the instantaneous values of the
input but in the time series generated by finding interesting events in the
input. This allows the proposed approach to handle inputs that cannot be
handled using any of the reviewed algorithms. Fig. 2 shows an example of
such system.
– It combines Granger-Causality's statistical testing with TIMERS-like reversed time check of causal connections
– It supports Feedback (finding causal relations when there is a cycle in the
causal graph).

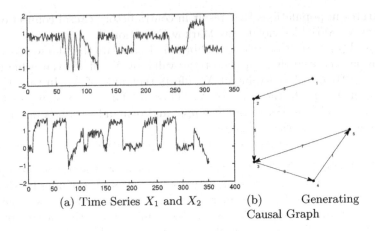

(a) Time Series X_1 and X_2 (b)   Generating
Causal Graph

Fig. 2. Example of two variables X_1 and X_2 where patterns in X_1 cause patterns in X_2. This kind of causal relation is not discoverable by any of the reviewed techniques. The proposed algorithm can handle this kind of signals

- It can find subdimensional patterns and so does not require patterns to appear in all dimensions of the variable. For example if one of the variables represents acceleration data from a moving object, the system can find patterns that exist in any one, two or the three spatial dimensions even if *distractive* noise exists in the other dimensions.
- It can find the response time (or dead time) between the activation of the cause and the beginning of the effect. This value is found for every causal relationship separately and is represented as a weight on the causal graph.

Fig. 2(a) shows an example of two variables X_1 and X_2. In this example both X_1 and X_2 has five patterns. Activation of some pattern in X_1 causes an activation of another pattern in X_2. The causal graph used to generate this data is shown in Fig. 2(b). The reviewed techniques cannot discover this causal structure because the instantaneous values of the signals are not what matter – and they do not contain any information useful for causality discovery – but the pattern in which these values happen.

There are many real world examples of systems that generate time series of the form shown in X_1 and X_2. For example X_1 could represent movements of a robotic head while X_2 represents the respiration pattern of its human partner during a face to face interaction and we would like to know if the head motion of the robot has some effect on the respiration pattern of the human which in turn is correlated with the stress level of him/her. In this case the exact value of robot head angles is not of any importance in most cases but the pattern of movement (e.g. looking to the subject, looking to unexpected location etc) is what can affect the respiration rate. Another example is analyzing the relation between two frequency modulated signals to discover if one of them is causing the other which may inform us about the internal workings of an unknown device. A third example is analyzing the relation between a bee's movement pattern and the movement patterns of other bees to determine if there is some form of communication happening (communication entails a causal relation between the message sent and the message received).

3   Proposed System Architecture

Fig. 3 summarizes the steps of the proposed approach to discover causal relations in time series. The inputs to the system are a set of n synchronized time series $X_{1:n}$ and a parameter specifying if causality should be discovered in the instantaneous values of the time series or in unknown patterns inside the time series. The algorithm proceeds in two phases. The aim of the first phase is to discover a set of important events that are useful for discovering the underlying causal structure using the algorithm described in section 4. The second phase takes the occurrence times of these events as input and estimates the causal structure used to generate the original time series in the form of a weighted directed graph (WDG) where the weights correspond to the response time/dead time between the activation of the cause and the start of the effect. This phase is described in section 5.

In the remaining of this paper we use the following operational definition of causality:

Definition: $A \xrightarrow{T} B$ – which means A causes B after T time units – if and only if:

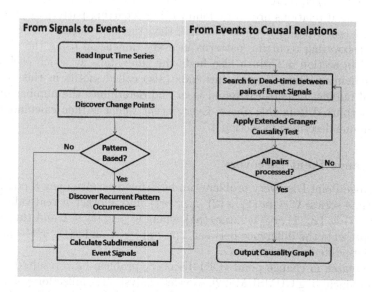

Fig. 3. Overview of the proposed causal relation mining technique. The system proceeds in two main steps: discovering interesting events in the time series followed by discovery of the underlying causal structure.

1. $A(t)_{EGC} \rightarrow B(t+T)$. The condition means that A causes B based on the extended granger causality test described in section 5.
2. There is no variable X in the set of tested variables so that $A(t)_{EGC} \rightarrow X(t+T_A)$ and $X(t)_{EGC} \rightarrow B(t+T_B)$ and $T = T_A + T_B$.

4   From Signals to Events

The first phase of the proposed algorithm aims at finding interesting events in the time series that can be utilized to reliably discover the underlying causal structure. The input to this phase is the set of input time series $X_{1:n}(t)$ as well as a Boolean parameter (*pattern-based*) that takes the value $FALSE$ if causal relations are to be found between input dimensions $1 : n$ using instantaneous values of the input $(X_{1:n})$.The output of this phase is a set of m time series $E_{1:m}(t)$ where $0 \le E_{1:m}(t) \le 1$ called *event signals*. The way to calculate *event signals* depends on the value of *pattern-based* parameter.

In all cases we start by finding the change points of the input time series (applied to every dimension of the input separately) using a Change Point Discovery algorithm. The result of this step is another set of n time series $\hat{X}_{1:n}$ where $0 \le \hat{X}(t) \le 1$. Semantically $\hat{X}(t)$ is an estimation of the amount of change in the generating dynamics at time step t. Section 4.1 details the change point discovery algorithm used in this research.

If the causal relations are to be found using the instantaneous values of the input time series (*pattern-based* equals $FALSE$); then the *event signals* are equal to $\hat{X}_{1:n}$ and m is equal to n.

If the causal relations are to be found using recurring patterns in the input time series (*pattern-based* equals $TRUE$); then the *event signals* are calculated by first discovering recurring patterns in every dimension using the algorithm described in section 4.2 which also finds the time steps at which each one of these recurring patterns begins and ends (also called motifs in this chapter). The total number of patterns found is m and determines the number of *event signals* outputted from this phase. Every *event signal* is then calculated using the technique described in section 4.

4.1   Change Point Discovery

The Change Point Discovery problem can be defined as: given any n points real-valued time series $X = \{x(1), x(2), ..., x(n)\}$ find another n real valued time series $X_s(t) = \{x_s(1), x_s(1), ..., x_s(n)\}$ where $0 \le x_s(t) \le 1$ and the value of $x_s(t)$ represents the difference between $x(t-p) : x(t)$ and $x(t) : x(t+f)$ for an appropriate difference measure D.

The research in change point (CP) discovery problem have resulted in many techniques including CUMSUM [22], wavelet analysis [23], inflection point search [24], autoregressive modeling [25], Discrete Cosine Transform, and Singular Spectrum Analysis SST [26].

Most of these methods with the exception of SST either discover a single kind of change (e.g. CUMSUM discovers only mean shifts), require ad-hoc tuning for every time series (e.g. wavelet analysis), or assumes a restricted generation process (e.g. Gaussian mixtures). The main disadvantages of SST though are the sensitivity to noise and the need to specify five different parameters.

The main idea of SST is to use PCA to discover the degree of 'difference' between the past and future signal pattern around every point in the time series and use the difference as the change score for this point.

Many researchers suggested improvements to traditional SST even though most of these suggestions targeted increasing the speed of the algorithm not its accuracy. [27] introduced online SVD and [28] proposed Krylov Subspace Learning. [28] also proposed using the angle between the subspaces associated with the major PCA components of the past and the future to calculate the change score at every point. The main problem of this proposal is the assumption that all Eigen vectors are equal in importance which can be an inaccurate assumption if the distribution of the top Eigen values is not nearly uniform (a condition that happens most of the time in our experience with real world time series). In this paper we propose a different approach to utilize the information of the Eigen values as well as the Eigen vectors for finding the change score.

The authors designed the Robust Singular Spectrum Transform (RSST) [29] for discovering change points in time series which reduces the number of parameters required into two parameters rather than five and dramatically increases the specificity of the traditional SST without decreasing its sensitivity. RSST is linear in time and space requirements as SST and adds very small constant increase in the processing time. Moreover speedup techniques like the use of Krylov Subspace Learning suggested in [28] can be directly utilized with RSST. Extensive comparisons between SST and RSST on synthetic data support the superiority of RSST in both synthetic and real world data [29]. In this work we use RSST for the discovery of change points in all dimensions of the input time series. A brief explanation of the algorithm is given in section 4.1

Robust Singular Spectrum Transform. The essence of the $RSST$ transform is to find for every point $x(i)$ the difference between a representation of the dynamics of the few points before it (i.e. $x(i-p) : x(i)$) and the few points after it (i.e. $x(i+g) : x(i+f)$). This difference is normalized to have a value between zero and one and named $x_s(i)$.

The dynamics of the points before and after the current point are represented using the Herkel matrix which is calculated as:

$$H(t) = [seq(t-n), ..., seq(t-1)] \tag{13}$$

where $seq(t) = \{x(t-w+1), ..., x(t)\}^T$

Singular Value Decomposition (SVD) is then used to find the singular values and vectors of the Herkel Matrix by solving:

$$H(t) = U(t) S(t) V(t)^T \tag{14}$$

where $S(i+1, i+1) \leq S(i,i) \leq (i-1, i-1)$.

Algorithm 1. Robust Singular Spectrum Transform (RSST)

1: **Inputs:** $X = \{x(1), x(2), ..., x(T)\}$.
2: **parameters:** w, n.
3: **Outputs:** $X_s(t) \leftarrow \{x_s(1), x_s(1), ..., x_s(T)\}$ where $0 \leq x_s(i) \leq 1$ for $i \leftarrow 1 : T$.
4: **procedure** SST   ▷ Calculates the SST Transform
5:       **for** $t \leftarrow w \times n : T - w \times n$ **do**
6:             **1. Extract past pattern:**
7:             1.1. $m \leftarrow n$ and $g \leftarrow 0$
8:             1.2. Calculate Subsequences $SEQ(t)$
9:             1.3. Find the Herkel Matrix $H(t)$ at point t (Eq. 13)
10:            1.4. Find the SVD decomposition of $H(t)$ by solving 14.
11:            1.5. Estimate the optimal value for $l(t)$
12:            1.6. $U_l(t) \leftarrow$ first $l(t)$ left singular vectors
13:            **2. Extract future pattern:**
14:            2.1. Calculate subsequences $R(t)$
15:            2.2. Find the Herkel Matrix of the future $G(t)$ (Eq. 16)
16:            2.3. Find the optimal value for $l_f(t)$
17:            2.4. Find the largest $l_f(t)$ Eigen vector of $G(t)$ $(\beta_{1:f}(t))$
18:            **3. Calculate Change Score Guess:**
19:            calculate $\hat{x}_s(t)$ using 23
20:      **end for**
21:      **4. Remove Noise Effect:**
22:      4.1. Estimate x_s^{min} and x_s^{max}
23:      **for** $t \leftarrow w \times n : T - w \times n$ **do**
24:            4.2. Find the average and variance of the w points subsequences before and
      after point t
25:            4.3. Modify $\hat{x}_s(t)$ using Eq. 28
26:      **end for**
27:      4.4. Keep only local maximums of $x_s(t)$
28:      4.5. Normalize $x_s(t)$
29: **end procedure**

In RSST the value of $l(t)$ is allowed to change from point to point in the time series depending on the complexity of the signal before it. To calculate a sensible value for l we first sort the singular values of $H(t)$ and find the corner of the accumulated sum of them $(l(t))$ [the point at which the tangent to the curve has an angle of $\pi/4$]. The singular vectors with singular values higher than this value are assumed to be caused by the genuine dynamics of the signal while the other directions encode the effect of noise. This dynamic setting of l reduces the effect of noise on the final results as will be shown in the following section.

A similar procedure is used to find the direction of largest change in the dynamics for the future of the signal by concatenating m overlapping windows of size w starting g points after t according to:

$$r(t+g) = \{x(t+g), ..., x(t+g+w-1)\}^T \tag{15}$$

$$G(t) = [r(t+g), ..., r(t+g+m-1)] \tag{16}$$

The Eigen vector $\beta(t)$ corresponding to the direction of maximum change in the future of the signal is found by solving:

$$G(t) G(t)^T u^g = \mu u^g \qquad (17)$$

$$\beta(t) = u_m^g \qquad (18)$$

where $m = \arg\max_i (\mu_i)$

To find a first guess of the change score around every point, RSST uses the $l_f(t)$ Eigen vectors of $G(t) G(t)^T$ with highest corresponding Eigen values $(\lambda_{1:l_f})$ rather than only the first Eigen vector used in SST. The value of $l_f(t)$ is selected using the same algorithm for selecting $l(t)$.

$$G(t) G(t)^T u^g = \mu u^g \qquad (19)$$

$$\beta_i(t) = u_i^g, \, i \le l_f \, and \lambda_{j+1} \le \lambda_j \le \lambda_{j-1} \, for \, 1 \le j \le w \qquad (20)$$

Each one of these l_f directions are then projected onto the hyperplane defined by $U_l(t)$

The projection of $\beta_i(t)$s and the hyperplane defined by $U_l(t)$ is then found using:

$$\alpha_i(t) = \frac{U_l^T \beta_i(t)}{\|U_l^T \beta_i(t)\|}, i \le l_f \qquad (21)$$

The change scores defined by $\beta_i(t)$s and $\alpha_i(t)$s are then calculated as:

$$cs_i(t) = 1 - \alpha_i(t)^T \beta_i(t) \qquad (22)$$

The first guess of the change score at the point t is then calculated as the weighted sum of these change point scores where the Eigen values of the matrix $G(t)$ are used as weights.

$$\hat{x}(t) = \frac{\sum_{i=1}^{l_f} \lambda_i \times cs_i}{\sum_{i=1}^{l_f} \lambda_i} \qquad (23)$$

After applying the aforementioned steps we get a first estimate $\hat{x}(t)$ of the change score at every point t of the time series. RSST then applies a filtering step to attenuate the effect of noise on the final scores. The main insight of this filter is that the reduction of SST specificity in noisy signals happens in the sections in which noise takes over the original signal in the time series. The response of SST at these sections can be modeled by a random walk around a high average for uncorrelated white noise. The filter used by RSST discovers these sections in which the average and the variance of $\hat{x}(t)$ remains nearly constant and attenuates them. The filter first calculates the average and variance of the signal before and after each point using a subwindow of size w:

$$\mu_b(t) = \frac{\sum_{i=0}^{w-1} \hat{x}(t-i)}{w} \tag{24}$$

$$\sigma_b(t) = \frac{\sum_{i=1}^{w}(\hat{x}(t-i) - \mu_b(t))^2}{w-1} \tag{25}$$

$$\mu_a(t) = \frac{\sum_{i=1}^{w} \hat{x}(t+i)}{w} \tag{26}$$

$$\sigma_a(t) = \frac{\sum_{i=1}^{w}(\hat{x}(t+i) - \mu_a(t))^2}{w-1} \tag{27}$$

The guess of the change score at every point is then updated by:

$$\tilde{x}(t) = \hat{x}(t) \times |\mu_a(t) - \mu_b(t)| \times \left|\sqrt{\sigma_a(t)} - \sqrt{\sigma_b(t)}\right| \tag{28}$$

where μ_a and σ_a are the mean and variance of $\hat{x}(t)$ in a subsequence of length w before the point t while μ_b and σ_b are the mean and variance of $\hat{x}(t)$ in a subsequence of length w after the point t

RSST then keeps only the local maxima of $\tilde{x}(t)$ and normalizes the resulting time series by dividing with its maximum. This normalized signal $x(t)$ represents the final change score of RSST.

4.2   Pattern/Motif Discovery

If the causal relations are to be found in recurring patterns in the input streams rather than in the instantaneous values of these streams; a reliable algorithm to extract these patterns and their occurrences is needed. This problem is known in data mining literature as motif discovery in time series. The research in unsupervised motif discovery have led to many techniques including the PROJECTIONS algorithm [30], PERUSE [31], Gemoda [32] among many others ([33]). These algorithms try to find all recurring motifs with lengths between l_{min} and l_{max} time steps where l_{min} and l_{max} are input parameters to the algorithm (or at least one of them).

With the exception of Gemoda which is quadratic in time and space complexities, these algorithms aim to achieve sub-quadratic time complexity by first looking for candidate motif stems using some heuristic method and then doing exhaustive motif detection instead of motif discovery which is linear in time.

The most used method for finding these stems is the PROJECTIONS algorithm [30] which requires discretization of the data using the SAX [5] algorithm.

A major drawback of all methods relying on discretization of the data is the need to specify a word length and a vocabulary size that are difficult to decide in real world situations.

Another problem of most of the previously proposed techniques is the need to specify an exact or at least roughly correct motif length.

A third problem is deciding when to stop searching for new motifs ([33] suggested using density estimation for this purpose). One common problem to all the algorithms based on the PROJECTIONS algorithm [33,30] is the need to construct and keep the collision matrix which is in general quadratic in the length of the SAX word describing the time series. Given that the optimal word size can be very short for short motifs in signals with high frequency components, the size of this matrix can grow quadratic with the length of the time series.

Catalano et al. [34] suggested a very efficient algorithm for locating variable length patterns in data series using random sampling that allows it to run in linear time and constant memory. The main problem of this approach is that it relies on random sampling which can lead to poor performance for long time series with infrequent embedded motifs including long records of human activities.

Most of the methods proposed for motif discovery that we are aware of assume no prior knowledge of the probable locations of the motifs which leads to this explosion in the processing time or space needed. In response to that, the authors designed two variations of Catalano et al.'s algorithm called MCInc and MCFull [6] that can utilize prior information about the locations of motif occurrences to speed up the operation of discovery to linear and in some cases sublinear both in time and space. Moreover, the output of any change point discovery algorithm (e.g. RSST described in section 4.1 can be used as the input constraint to limit the search to the points of the time series around the points of high change score based on the assumption that interesting patterns happen near the points in which there is a change in the underlying dynamics.

Catalano et al.'s algorithm [34] works by processing a fixed sized set of candidate and comparison windows randomly sampled from the time series. The input to the algorithm are minimum and maximum motif lengths ($\hat{w}$,w) as well as the number of candidate windows (n_c) which determines the space requirements of the algorithm. The steps of the algorithm given a single dimension input time series $X(t)$ can be summarized as:

1. Select a subsequence s^w of length $w \geq l_{max}$.
2. Select w values randomly from X and concatenate them to form a noise sequence n^w
3. select a set of n_c comparison subsequences of X ($\{c^w_i\}$) each of length w.
4. Find the set $S^{\hat{w}}$ of subsequences of length $\hat{w}$ where $\hat{w} \leq l_{min}$ for s^w,c^w_i, and n^w. Then normalize all of the resulting subsequences to have unit mean square.
5. For the candidate subsequence of s^w ($s^{\hat{w}}_k$) do the following
   (a) Randomly select $w - \hat{w} - 1$ subsequences from the set of all subsequences of the comparison windows (c^w_i). call this set the comparison set $\hat{c}^{\hat{w}}_j$
   (b) Find the distances $d_{kj} = d\left(s^{\hat{w}}_k, \hat{c}^w_j\right)$.

(c) Group the set $\hat{c}_j^{\hat{w}}$ with their parent subsequence $c^w{}_i$ and for every group select $\hat{c}_j^{\hat{w}}$ that has the minimum distance d_{kj}. This leads to a set of R subsequences $\tilde{c}_r^{\hat{w}}$ where $R \leq n_c$

(d) Keep only the $\hat{R}$ subsequences of $\tilde{c}_r^{\hat{w}}$ with least d_{kr}.

(e) Repeat the previous three steps for subsequences of the noise subsequence n^w. This leads to another set of $\hat{R}$ subsequences called $\tilde{n}_r^{\hat{w}}$

6. Remove all candidate subsequences $s_k^{\hat{w}}$ that have similar average distance with both $\tilde{n}_r^{\hat{w}}$ and $\tilde{c}_r^{\hat{w}}$ then repeat the steps above using this reduced set. Repeat this reduction for n_r times.

7. If the final set $s_k^{\hat{w}}$ is not empty output each of them as a motif seed M_s after concatenating any continuous subset of them.

Given the change point score $(\tilde{X}(t))$ calculated using RSST (see section 4.1) we can speed up Catalano algorithm by modifying the first three steps as follows:

1. Apply a Gaussian smoothing filter $(N(0, \sigma^2))$ to the original $\tilde{X}$ constraint which results on the smoothed constraint $\tilde{P}$.

2. Normalize $\tilde{P}$ so that $\sum_{t=1}^{n} \hat{p}(t) = 1$ and $0 \leq \hat{p}(t) \leq 1$.

3. Randomly select a subsequence s^w of length $w \geq l_{max}$ using $\hat{P}$ as the probability distribution.

4. Randomly select w values from X and concatenate them to form a noise sequence n^w using $1 - \hat{P}$ as the probability distribution.

5. Randomly select a set of n_c comparison subsequences of X $(\{c^w{}_i\})$ each of length w using $\hat{P}$ as the probability distribution.

For more details about this algorithm and extensive evaluation of its performance on synthetic and real world data please refer to [6].

4.3   Calculating Subdimensional Event Signals

The output of this phase is a set of m time series $E_{1:m}(t)$ where $0 \leq E_{1:m}(t) \leq 1$ called *event signals*. The way to calculate *event signals* depends on the value of *pattern-based* parameter.

If the causal relations are to be found using the instantaneous values of the input time series (*pattern-based* equals $FALSE$); then the *event signals* are equal to $\hat{X}_{1:n}$ and m equals to n.

If the causal relations are to be found using recurring patterns in the input time series (*pattern-based* equals $TRUE$); then the *event signals* are calculated by first discovering recurring patterns in every dimension using the algorithm described in section 4.2 which also finds the time steps at which each one of these recurring patterns begins and ends (also called motifs in this chapter) appear. The total number of patterns found is m_p and determines the number of *event signals* outputted from this phase. Every *event signal* is then calculated from the locations of recurring pattern occurrences using the following equation:

$$\hat{E}_i(t) = \begin{cases} t/\Delta_i + 1 - L_{ij}/\Delta_i & if \; max\,(L_{ij} - \Delta_i, B_{ij}) \leqslant t \leqslant L_{ij} \\ -t/\Delta_i + 1 + L_{ij}/\Delta_i & if \; L_{ij} < t \leqslant min\,(L_{ij} + \Delta_i, End_{ij}) \\ 0 & otherwise \end{cases} \quad (29)$$

where B_{ij}, L_{ij}, and End_{ij} are the beginning middle, and end time step of the occurrence number j of pattern i, Δ_i is a parameter specifying the confidence of the pattern detector. In all of the experiments done in this chapter we fixed Δ_i to ∞. This equation represents two lines intersecting at the value 1 in the center of the discovered occurrence. $E_i(t)$ then decreases monotonically with rate $1/\Delta_i$ until the two extremes of the discovered occurrence.

The final step is to combine event signals that are estimated to correspond to the same pattern but in multiple dimensions. First the correlation of every pair of *event signals* is calculated using Pearson correlation. If the correlation coefficient is over a predefined threshold and is statistically significant the two *event signals* are combined in a single *subdimensional event signal* using:

$$E_k(t) = \hat{E}_i(t) \times \hat{E}_j(t) \tag{30}$$

Finally if some *event signal* $(\hat{E}_l)$ was not combined with any other *event signal*; it is added to the final *subdimensional event signals* (E). This procedure is iterated until no *event signals* can be combined anymore. It should be noted that this procedure cannot distinguish *event signals* that correspond to occurrences of the same multidimensional pattern and immediate causal relations. If immediate causal relations are possible in the system then *event signals* should not be combined and in this case $E = \hat{E}$.

5   From Events to Causal Relations

The output of the previous phase is a set of m time series called the *subdimensional event signals* $(E_{1:m}(t))$. Each one of these signals represents one variable and the goal of this final phase of the algorithm is to discover the causal relations between these variables. In this section we present our proposed algorithm for this phase.

The first step in this phase is to search for pairwise dead-time for every pair of *subdimensional event signals*. The dead-time between any cause and its effect is defined as the time in time-steps required between the start of the cause and the start of the effect. To find the dead-time between a pair of signals E_i and E_j, we calculate the Pearson correlation coefficient between $E_i(t)$ and $E_j(t+\tau)$ for different values of τ and select the dead-time between variables i and j as:

$$d_{i \to j} = \arg\max_{\tau}\left(\rho\left(E_i(t), E_j(t+\tau)\right)\right)$$

$$d_{j \to i} = \arg\max_{\tau}\left(\rho\left(E_j(t), E_i(t+\tau)\right)\right)$$

After obtaining the dead-time $(d_{i \to j})$ between variables i and j we apply an extended version of Granger-Causality test as follows: First, we estimate the following equation also by OLS

$$\hat{X}_j(t) = \varepsilon_1 + u(t) + \sum_{l=1}^{m}\sum_{k=1}^{\rho}\alpha_{lk}\hat{X}_l(t - k - d_{l \to j}) \tag{31}$$

Second, we estimate the following restricted equation also by OLS:

$$\hat{X}_j(t) = \varepsilon_2 + e(t) + \sum_{l \neq i} \sum_{k=1}^{\rho} \beta_{lk} \hat{X}_l(t - k - d_{l \to j}) \tag{32}$$

We then calculate the sum of squared residuals (SSR) in both cases:

$$\begin{aligned} SSR_1 &= \sum_{i=1}^{T} u^2(t) \\ SSR_0 &= \sum_{i=1}^{T} e^2(t) \end{aligned} \tag{33}$$

We then calculate the test statistic S_ρ as:

$$S_\rho = \frac{(SSR_0 - SSR_1)/\rho}{SSR_1/(T - 2\rho - 1)} \tag{34}$$

If S_ρ is larger than the specified critical value then we estimate the following equation also by OLS:

$$\hat{X}_j(t) = \varepsilon_3 + f(t) + \sum_{l \neq i} \sum_{k=1}^{\rho} \lambda_{lk} \hat{X}_l(t - k - d_{l \to j}) + \sum_{k=1}^{\rho} \gamma_k \hat{X}_i(t + k + d_{j \to i}) \tag{35}$$

We then calculate the sum of squared residuals (SSR) in this case:

$$SSR_2 = \sum_{i=1}^{T} f^2(t) \tag{36}$$

We then calculate the test statistic S_ρ^f as:

$$S_\rho^f = \frac{(SSR_0 - SSR_2)/\rho}{SSR_2/(T - 2\rho - 1)} \tag{37}$$

Now if S_ρ^f is less than the specified critical value then we reject the null hypothesis that E_i does not cause E_j after $d_{i \to j}$ steps and in this case $X_i \xrightarrow{d_{i \to j}} X_j$. The basis of this decision is that having S_ρ greater than the specified critical value while S_ρ^f smaller than it indicates that the past of E_i gives information about E_j while the future of E_i does not give extra information. This means that the relation between E_i and E_j satisfies the necessary condition that a cause must precede its effect.

If S_ρ^f is greater than the specified critical value then we have a more complex situation as both past and future values of E_i are useful for predicting E_j. This can happen if there is a hidden variable that causes both E_i and E_j but with variable delays so that sometimes E_i precedes E_j and vice versa. In this case we add a hidden variable H and assume that $H \to E_j$ and $H \to E_i$.

After processing all pairs of *subdimensional event signals* ($2m(m-1)$ operations), the complete causal graph is built and outputted.

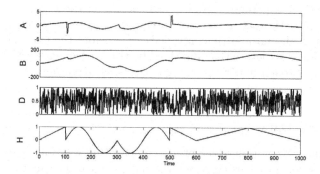

Fig. 4. Time series used to evaluate the effectiveness of the proposed extended Granger causality test. A seams to cause B if H is not taken into account using standard Granger Causality test. Using the proposed extended version the system infers the existence of H and correctly concludes that H causes both A and B.

To illustrate the usefulness of the extra steps of the proposed extended Granger-Causality check, consider the signals presented in Fig. 4. These signals were generated using the following linear systems:

$$A\,(t) = 4H\,(t-1) - 3H\,(t-6) + 0.5A(t-1)$$

$$B\,(t) = 8H\,(t-1-\kappa) - 6H\,(t-6-\kappa) + 0.99B\,(t-1)$$

where κ is a random number uniformly distributed between 0 and 9.

H is a sequence of linear and sinusoidal signals while D is a random signal that is not correlated with anything else.

Applying standard Granger Causality test (as explained in equations 9:12) to A and B to the data set containing (A, B, and D) results in falsely accepting that A causes B. On the other hand applying the proposed extended method to the same data detects that both A and B are caused by a common hidden variable that is not in the data set.

6   Evaluation

Evaluation of causal rule discovery algorithms is complicated by the difficulty to recover the causal structure of the domain at hand. Synthetic data evaluation is one way around this problem as the evaluator can design the underlying causal structure (even though some philosophical views may not accept the relation between computer variables to be causal in a strict sense [35]). The other way around this is to use a domain in which a tremendous amount of knowledge supports some known causal model (e.g. use a mechanical system known to be governed by Newtonian mechanics). The third possibility is to utilize a controlled experimental situation in which the causal structure is fixed by the designer.

In this paper we evaluate the proposed system using synthetic data in section 6.1 and using a controlled experiment in section 6.2. It should be noted that using a controlled experiment in this case does not entail any *manipulation* of the causal structure as we do not intend to discover the causal structure of the system in the normal (uncontrolled) case before the experiment but to discover the causal structure that was fixed by the experiment itself. This leaves the proposed approach under the category of *causality from data* rather than *causation from perturbation*.

6.1   Synthetic Data

First preliminary evaluation of the proposed algorithm was done using synthetic data that was generated using a deterministic model. The model consists of two signals X_1 and X_2 each of which is a concatenation of the outputs from five different patterns. The patterns constituting X_1 and X_2 are the same (generated from same processes). Fig. 5 shows the five patterns generated by these processes. Fig. 2 shows an example short sequence of these patterns in X_1 and X_2 along with the causal model used to generate them. In all cases the active process in X_1 *causes* the activation of one of these five processes in X_2 after some delay that is selected randomly using a uniform distribution.

To evaluate the algorithm we applied it using the two signals generated from 16 randomly generated causal models each was used to generate 100 different time series pairs of 1000 points each. In all cases the system was able to discover the five patterns correctly and to regenerate the same causality graph with a mean error of 6% in the discovered delays and standard deviation of 3%. It should be noted that the output of the algorithm in this example is not five patterns but twelve corresponding to the five patterns in the two input time series. To simplify the display we combined corresponding patterns (using simple Euclidean distance measure) together both in the generating and discovered graphs. This perfect discovery results just show that the algorithm can *in principle* work but gives no information about its accuracy as the patterns used are easily sparable and there was no noise, complex feedback, or unknown hidden factors in this experiment.

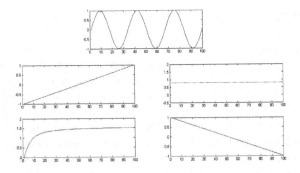

Fig. 5. The five patterns used in the generation of synthetic data

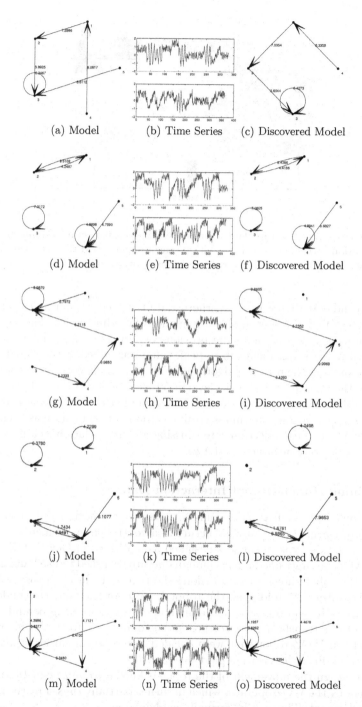

Fig. 6. Five example models used in the second experiment (after adding noise) with a part of the generated time series and the discovered model using the proposed algorithm

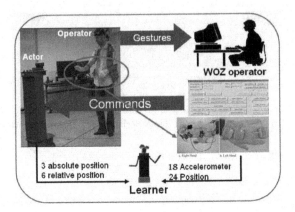

Fig. 7. The experiment setup. The actor is a WOZ operated robot. Command stream is 48 dimensions accelerometer and position sensor signals, and the action stream is 6 dimensional absolute and operator-relative location of the robot.

The second experiment using synthetic data targeted evaluating the accuracy of the system in the presence of noise in the input time series. The same generation process was used as in the previous experiment except that uniformly distributed noise of variable peak-to-peak value was added to the signal X_i and X_j before passing them to the algorithm. Fig. 6 shows five examples of the operation of the system at P-P noise level of 50% of the P-P values of the original signal. At this high noise level the maximum number of missed causal relations was a single one, the false alarm rate (taking a non-causal relation as causal) was still 0% while the false rejection rate (missing a causal relation as a non-causal) was 8.9% (detection accuracy is 91.1%).

6.2   Mining Human-Robot Interaction Data

This section presents a feasibility study to assess the applicability of the proposed approach in learning the causal structure in a controlled experiment.

The evaluation experiment was designed as a Wizard of Ooz (WOZ) experiment in which an untrained novice human operator is asked to use hand gestures to guide the robot shown in Fig. 7 along the two paths in two consecutive sessions. The subject is told that the robot is autonomous and can understand any gesture (s)he will do. A hidden human operator was sitting behind a magic mirror and was translating the gestures of the operator into the basic primitive actions of the WOZ robot that were decided based on an earlier study of the gestures used during navigation guidance [36].

In this design the movement of the robot is known to be *caused* by the commands sent by the WOZ operator which in turn is partially *caused* by the gestures of the participant. This can be formally explained as:

$$G_i \xrightarrow{T_1} W_j$$

$$W_j \xrightarrow{T_2} M_k$$

where G_i represent some gesture done by the participant, W_j represent some action done by the WOZ operator (e.g. pressing a button on the GUI of the control software), and M_k represents some pattern in the movement of the robot (e.g. moving toward the participant, stopping, etc).

Eight participants of ages 21 to 34 (all males) that have no prior knowledge of robots and do not study in the field of engineering were selected for this experiment. The total number of sessions conducted was 16 sessions with durations ranging from 5:34 minutes to 16:53 minutes.

The motion of the subject's hands (G) was measured by six B-Pack ([37]) sensors attached to both hands as shown in Fig. 7 generating 18 channels of data. The PhaseSpace motion capture system ([38]) was also used to capture the location and direction of the robot using eight infrared markers. The location and direction of the subject was also captured by the motion capture system using six markers attached to the head of the subject (three on the forehead and three on the back). 8 more motion capture markers were attached to the thumb and index of the right hand of the operator.

The following four feature channels were used as representation of robot's action (M):

- The directional speed of the robot in the XZ (horizontal) plane in the direction the robot is facing (by its cameras) (2 dimensions).
- The direction of the robot in the XZ plane as measured by the angle it makes with the X axis (1 dimension).
- The relative angle between the robot and the actor (1 dimension).
- The distance between the operator and the actor (1 dimension).

The time of button presses in the GUI used by the WOZ operator was collected in synchrony with both participant gestures and robot actions. The interface had seven buttons (related to robot motion) each of which can be toggled on and off and a single button can be on at any point of time. The WOZ operator's actions were represented by a single input dimension giving the ID of the currently active button (from 1 to 7).

This leads to a total of 67 input dimensions. The generating causal model (ground truth) consists of seven gestures, seven button press configuration and corresponding seven robot actions (21 total patterns).

Fig. 8 shows two example gestures discovered by the algorithm. The first gesture – corresponding to the command come-here needs – uses only three out of the 48 input gesture dimensions (corresponding to the sensor attached to the tip of the right hand). The second gesture – corresponding to the stop command – needs even less dimensions (only 2) because the acceleration in the Y direction does not matter for this command. This ability of the algorithm to discover patterns in a subset of the dimensions reduces its sensitivity to noise in the other distracting dimensions.

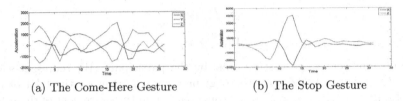

(a) The Come-Here Gesture   (b) The Stop Gesture

Fig. 8. Examples of discovered recurring patterns in the gestures of the operator (G). The first gesture was found in 3 dimensions of the 48 gesture input dimensions while the second one was found in only two dimensions.

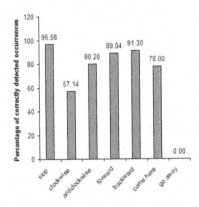

Fig. 9. The detection accuracy for the gestures used in the experiment

Fig. 9 shows the accuracy of the algorithm in discovering gestures and gesture occurrences. As shown in the Figure, the algorithm discovered all the types of gestures used by the subjects except the *go away* gesture which was mostly confused with the *Backward* gesture. The algorithm also made four false positive errors. The algorithm can then detect 85.7% (6 out of 7) of the gesture types and in average it finds 82% of the occurrences of discovered gestures.

The algorithm correctly discovered all the recurring patterns (corresponding to the seven actions) in the WOZ operator's time series data.

For robot's actions, the total number of basic actions discovered by the algorithm is eight actions. 87.5% of these actions correspond to authentic actions that were implemented in the interface of the WOZ software (7 out of 8). The accuracy of detecting the discovered actions was 88.3%.

Fig. 10 shows the generating model and the final discovered model. To simplify the display the nodes for patterns were numbered so that gestures correspond to nodes with numbers dividable by seven, WOZ operator actions corresponded to nodes with numbers that have a remainder of one when divided by seven, and Robot movement patterns corresponded to nodes with numbers that have a remainder of two when divided by seven. This way, the generating model is a seven cliques graph as shown in Fig. 10(a). In the discovered model, nodes that correspond to false patterns where removed from the graph as none of them had

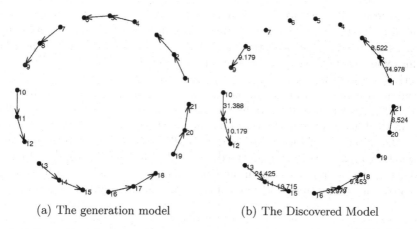

(a) The generation model   (b) The Discovered Model

Fig. 10. Comparison between the generation model (based on experimental setup) and discovered causal graph using the proposed algorithm

any causal connection to any other node in the system. As Fig. 10(b) shows, the discovered model contains no false alarms (non-causal relationships assumed causal) but it has four false rejections (causal relationships assumed non-causal) with a false-rejection rate of 28.57%.

7   Conclusion

In this chapter we reviewed the available techniques for discovering causal relations from data. The chapter also described in details a novel causal relation discovery algorithm that does not work directly on the instantaneous values of the input time series but discovers interesting events in these time series either as points of change in the underlying dynamics or recurring pattern occurrences.

The algorithm works in two phases. In the first phase it discovers interesting *events* in the input multidimensional data and then in the second phase it discovers the causal structure based on the relative locations of these *events* in time. During event discovery the algorithm makes use of a robust change point discovery algorithm developed by the author called the Robust Singular Spectrum Transform (RSST) as well as a recurrent pattern discovery algorithm that utilizes RSST output to work in linear time and space. During causal relation discovery the algorithm uses an extended Granger-Causality test to assess causal relations between the occurrences of events. The main difference between the proposed extended test over traditional Granger-Causality tests is the utilization of an extra statistic to confirm that the past of the cause contains information that helps in predicting its effect while the future of the cause does not contain equivalent information. Using this kind of confirmation it is possible for the proposed algorithm to discover the existence of hidden common causes.

The proposed system was evaluated using both synthetic and real world data. Synthetic evaluation showed that even with 50% noise level the system can still recover the causal structure with an accuracy of 91.1%. For real world evaluation we used controlled experiment that involved a human participant controlling a robot using free hand gestures through a hidden operator. Analysis of the result of this experiment shows that the system can recover the causal structure of the underlying system with accuracy 71.4% (10 out of 14 relations).

Extended evaluation of the proposed system using data from domains with known causal structure (e.g. mechanical systems, electrical systems etc) will be done in the future to confirm the applicability of the proposed system to general data. One direction of future research is to allow the system to utilize domain knowledge to reduce the errors in the discovered causal graph. Another direction is to extend the system to work in multiple time scales to produce a hierarchical model of causal structure at different time scales of the input time series.

References

1. Wang, X., Ye, L., Keogh, E., Shelton, C.: Annotating historical archives of images. In: JCDL 2008: Proceedings of the 8th ACM/IEEE-CS joint conference on Digital libraries, pp. 341–350. ACM, New York (2008)
2. Inokuchi, A., Washio, T.: Feasibility of graph sequence mining based on admissibility constraints. In: Thid International Workshop on Data Mining and Statistical Science, pp. 1–4 (2008)
3. Wang, X., Smith, K.A., Hyndman, R.J.: Dimension reduction for clustering time series using global characteristics. In: Sunderam, V.S., van Albada, G.D., Sloot, P.M.A., Dongarra, J. (eds.) ICCS 2005. LNCS, vol. 3516, pp. 792–795. Springer, Heidelberg (2005)
4. Kulic, D., Takano, W., Nakamura, Y.: Incremental on-line hierarchical clustering of whole body motion patterns. In: RO-MAN 2007 (2007)
5. Keogh, E., Lin, J., Fu, A.: Hot sax: efficiently finding the most unusual time series subsequence. In: Fifth IEEE International Conference on Data Mining, November 2005, pp. 226–233 (2005)
6. Mohammad, Y., Nishida, T.: Constrained motif discovery. In: International Workshop on Data Mining and Statistical Science (DMSS 2008), September 2008, pp. 16–19 (2008)
7. Aristotle: Metaphysics Book V Part 1
8. Glymour, C.: Learning, prediction and causalbayesnets. TRENDS in Cognitive Sciences 7(1), 43–48 (2003)
9. Menzies, P., Price, H.: Causation as a secondary quality. British Journal for the Philosophy of Science 4, 187–203 (1993)
10. Pearl, J.: Causality: Models, Reasoning, and Inferenc. Cambridge University Press, Cambridge (2000)
11. Han, J., Fu, Y., Wang, W., Chiang, J., Gong, W., Koperski, K., Li, D., Lu, Y., Rajan, A., Stefanovic, N., Xia, B., Zaiane, O.R.: Dbminer: A system for mining knowledge in large relational databases. In: Proc. 1996 Int'l Conf. on Data Mining and Knowledge Discovery (KDD 1996), pp. 250–255. AAAI Press, Menlo Park (1996)
12. Das, K., Lin, I., Mannila, H., Renganathan, G., Smyth, P.: Rule discovery from time series. In: The 4th International Conference of Knowledge Discovery and Data Mining, pp. 16–22. AAAI Press, Menlo Park (1998)

13. Hipp, J., Güntzer, U., Nakhaeizadeh, G.: Algorithms for association rule mining — a general survey and comparison. SIGKDD Explor. Newsl. 2(1), 58–64 (2000)
14. Lee, A.J., chuen Lin, W., sheng Wang, C.: Mining association rules with multidimensional constraints. Journal of Systems and Software 79, 79–92 (2006)
15. Sarker, B.K., Hirata, T., Uehara, K., Bhavsar, V.C.: Mining Association Rules from Multi-stream Time Series Data on Multiprocessor Systems. In: Pan, Y., Chen, D.-x., Guo, M., Cao, J., Dongarra, J. (eds.) ISPA 2005. LNCS, vol. 3758, pp. 662–667. Springer, Heidelberg (2005)
16. Hoover, K.: The logic of causal inference. Economics and Philosophy 6, 207–234 (1990)
17. Tian, J., Pearl, J.: Causal discovery from changes. In: Proceedings of UAI 2001, pp. 512–521. Morgan Kaufmann, San Francisco (2001)
18. Gelper, S., Croux, C.: Multivariate out-of-sample tests for granger causality. Comput. Stat. Data Anal. 51(7), 3319–3329 (2007)
19. Ding, M., Chen, Y., Bressler, S.: Granger causality: Basic theory and application to neuroscience. Wiley, Chichester (2006)
20. Karimi, K., Hamilton, H.J.: Distinguishing Causal and Acausal Temporal Relations. In: Advances in Knowledge Discovery and Data Mining., p. 569. Springer, Heidelberg (2003)
21. Spirtes, P., Glymour, C.N., Scheines, R.: Causation, prediction, and search. MIT Press, Cambridge (2001)
22. Basseville, M., Kikiforov, I.: Detection of Abrupt Changes. Printice Hall, Englewood Cliffs (1993)
23. Kadambe, S., Boudreaux-Bartels, G.: Application of the wavelet transform for pitch detection of speech signals. IEEE Transactions on Information Theory 38(2), 917–924 (1992)
24. Hirano, S., Tsumoto, S.: Mining similar temporal patterns in long time-series data and its application to medicine. In: ICDM 2002: Proceedings of the 2002 IEEE International Conference on Data Mining (ICDM 2002), p. 219. IEEE Computer Society, Washington (2002)
25. Gombay, E.: Change detection in autoregressive time series. J. Multivar. Anal. 99(3), 451–464 (2008)
26. Ide, T., Inoue, K.: Knowledge discovery from heterogeneous dynamic systems using change-point correlations. In: Proc. SIAM Intl. Conf. Data Mining (2005)
27. Zha, H., Simon, H.D.: On updating problems in latent semantic indexing. SIAM Journal on Scientific Computing 21(2), 782–791 (1999)
28. Ide, T., Tsuda, K.: Change-point detection using krylov subspace learning. In: Proceedings of the SIAM Internations Conference on Data Mining (2007)
29. Mohammad, Y., Nishida, T.: Robust singular spectrum transform. In: IEA/AIE, pp. 123–132 (2009)
30. Chiu, B., Keogh, E., Lonardi, S.: Probabilistic discovery of time series motifs. In: KDD 2003: Proceedings of the ninth ACM SIGKDD international conference on Knowledge discovery and data mining, pp. 493–498. ACM, New York (2003)
31. Oates, T.: Peruse: An unsupervised algorithm for finding recurring patterns in time series. In: International Conference on Data Mining, pp. 330–337 (2002)
32. Jensen, K.L., Styczynxki, M.P., Rigoutsos, I., Stephanopoulos, G.N.: A generic motif discovery algorithm for sequenctial data. BioInformatics 22(1), 21–28 (2006)
33. Minnen, D., Starner, T., Essa, I., Isbell, C.: Improving activity discovery with automatic neighborhood estimation. In: Int. Joint Conf. on Artificial Intelligence, pp. 6–12 (2007)

34. Catalano, J., Armstrong, T., Oates, T.: Discovering patterns in real-valued time series. In: Fürnkranz, J., Scheffer, T., Spiliopoulou, M. (eds.) PKDD 2006. LNCS (LNAI), vol. 4213, pp. 462–469. Springer, Heidelberg (2006)
35. Freedman, D., Humphreys, P.: Are there algorithms that discover causal structure. Synthese 121(1-2), 29–54 (2004)
36. Mohammad, Y., Nishida, T.: Human adaptation to a miniature robot: Precursors of mutual adaptation. In: The 17th IEEE International Symposium on Robot and Human Interactive Communication, RO-MAN 2008, pp. 124–129 (2008)
37. Ohmura, R., Naya, F., Noma, H., Kogure, K.: a bluetooth-based wearable sensing device for nursing activity recognition. In: 2006 1st International Symposium on Wireless Pervasive Computing, January 2006, pp. 1686–1693 (2006)
38. PhaseSpace Inc., http://www.phasespace.com

Author Index